Lecture Notes in Mathematics

Edited by A. Dold and B. Eckmann

775

Geometric Methods in Mathematical Physics

Proceedings of an NSF-CBMS Conference
Held at the University of Lowell,
Massachusetts, March 19–23, 1979

Edited by
G. Kaiser and J. E. Marsden

Springer-Verlag
Berlin Heidelberg New York 1980

Editors

Gerald Kaiser
Mathematics Department
University of Lowell
Lowell, MA 01854
USA

Jerrold E. Marsden
Department of Mathematics
University of California
Berkeley, CA 94720
USA

AMS Subject Classifications (1980): 53CXX, 58FXX, 73C50, 81-XX, 83CXX

ISBN 3-540-09742-2 Springer-Verlag Berlin Heidelberg New York
ISBN 0-387-09742-2 Springer-Verlag New York Heidelberg Berlin

Library of Congress Cataloging in Publication Data. Main entry under title:
Geometric methods in mathematical physics.
(Lecture notes in mathematics; 775)
Includes bibliographies and index.
1. Geometry, Differential--Congresses. 2. Mathematical physics--Congresses. I. Kaiser, Gerald. II. Marsden, Jerrold E. III. United States. National Science Foundation. IV. Conference Board of the Mathematical Sciences. V. Series: Lecture notes in mathematics (Berlin); 775.
QA3.L28 no. 775 [QC20.7.G44] 510s 80-332
ISBN 0-387-09742-2 [516.3'6]

Printing and binding: Beltz Offsetdruck, Hemsbach/Bergstr.
2141/3140-543210

Introduction

This volume represents invited papers presented at the CBMS regional conference held at the University of Lowell, March 19-23. The theme of the conference was geometric methods in mathematical physics and the papers were chosen with this in mind.

It is really only in the last couple of decades that the usefulness of geometric methods in mathematical physics has been brought to light. In other branches of mathematics their usefulness has been clearly demonstrated by Riemann, Poincaré and Cartan; a modern example is the use of symplectic geometry in group representations by Kirillov and Kostant. Save for general relativity, mathematical physics has been dominated primarily by analytical techniques. The excitement of the past few decades has been the complementing power of geometric methods.

The proper geometrization of classical mechanics started with Poincaré and continued with many workers, such as Synge (Phil. Trans. (1926)) and Reeb (C. R. Acad. Sci. (1948)). However, it wasn't until the analysis led to and became inextricably involved with geometry through the deep works of Kolmogorov, Arnold and Moser in celestical mechanics that a permanent bond became reality. The success of symplectic geometry in classical mechanics has motivated attempts to extend its use to the quantum domain. Some of these have borne rich fruit, such as the discovery of the geometry behind the WKB approximation (semiclassical mechanics) by Keller and Maslov and the quantization program of Souriau and Kostant. Symplectic geometry and classical mechanics have also revitalized linear partial differential equations through the work of Egorov, Hörmander, Nirenberg and Treves.

Much work is currently going on in gauge theory and supersymmetry that is, of necessity, geometric. Some believe that these geometric methods will finally close the circle with relativity as Einstein had dreamed.

G. Kaiser
J. Marsden
April, 1979

Acknowledgements

I wish to thank the following University of Lowell faculty members for their help in organizing this exciting conference: Alan Doerr, Lloyd Kannenberg, Eric Sheldon and Virginia Taylor. I am also grateful to the National Science Foundation and the Conference Board of the Mathematical Sciences for sponsoring the conference, and to Jerry Marsden for lighting the fire.

Gerald Kaiser

TABLE OF CONTENTS

Page

List of Participants

Dennis Aebersold, Physics & Chemistry Dept., Bennington College, Bennington, VT

Stuart Antman, Div. of Applied Math, Box F, Brown Univ., Providence, RI

Timothy Bock, 115 Broadmead, Princeton, NJ

R. Bolger, Fairfield Univ., Fairfield, CT

Bohumil Cenkl, Math Dept., Northeastern Univ., Boston, MA

William Crombie, P.O. Box 7025, Brown Univ., Providence, RI

Richard Cushman, Mathematics - Natural Science II, U of California, Santa Cruz, CA

Stanley Deser, Physics Dept., Brandeis U., Waltham, MA

Robert Devaney, Math Dept., Tufts Univ., Medford, MA

Alan Doerr, Math Dept., U. of Lowell, Lowell, MA

Alexander Doohovsky, 36 Brooks St., Concord, MA

Gerard Emch, Math Dept., U. of Rochester, Rochester, NY

Carlos Galvão, Physics Dept., Princeton Univ., Princeton, NJ

P. L. Garcia, Math Dept., Univ. of Salamanca, Salamanca, Spain

Maurice Gilmore, Math Dept., Northeastern Univ., Boston, MA

Daniel Goroff, Churchill College, Cambridge U., Cambridge, England

Mark Gotay, Physics Dept., U. of Maryland, College Park, MD

Morton Gurtin, Math Dept., Carnegie-Mellon U., Pittsburgh, PA

S. Hariharan, Math Dept., Carnegie-Mellon U., Pittsburgh, PA

H. Hattori, Math Dept., Rensselaer Polytechnic Institute, Troy, NY

Robert Hermann, Math-Sci Press, 53 Jordan Rd., Brookline, MA

Gerald Kaiser, Math Dept., U. of Lowell, Lowell, MA

Lloyd Kannenberg, Physics Dept., U. of Lowell, Lowell, MA

Boris Kupershmidt, Math Dept., M.I.T., Cambridge, MA

Henry Kurland, Math Dept., Boston Univ., Boston, MA

André Lichnerowicz, M.I.T. and College de France, Paris

Jerrold Marsden, Math Dept., U. of California - Berkeley, Berkeley, CA

Vincent Moncrief, Physics Dept., Yale Univ.,New Haven, CT

Harry Moses, U. of Lowell Research Foundation, Lowell, MA

Lea Murphy, Math Dept., Carnegie-Mellon U., Pittsburgh, PA

James Nester, Physics Dept., U. of Maryland, College Park, MD

A. Pérez-Rendón, Math Dept., Univ. of Salamanca, Salamanca, Spain

Kathleen Pericak, Math Dept., Carnegie-Mellon U., Pittsburgh, PA

Tudor Ratiu, Math Dept, U. C. - Berkeley, Berkeley, CA

David Reynolds, Math Dept., Carnegie-Mellon U., Pittsburgh, PA

David Rod, Math Dept., U. of Calgary, Calgary, Canada

Edwin Lee Rogers, Math Dept., Rensselaer Polytechnic Institute, Troy, NY

Richard Sacksteder, CUNY-Mathematics Graduate Center, 33 W. 42 St., New York,

Willy Sarlet, Harvard U., Science Center, Cambridge, MA

Garfield Schmidt, Math Dept., U. of Lowell, Lowell, MA

Patrick Shanahan, Math Dept., College of the Holy Cross, Worcester, MA

Henry Simpson, Math Dept., U. of Tennessee, Knoxville, TN

Marshall Slemrod, Math Dept., Rensselaer Polytechnic INstitute, Troy, NY

Scott Spector, Math Dept., U. of Tennessee, Knoxville, TN

Gilbert Strang, Math Dept., M.I.T., Cambridge, MA

Leonard Sulski, Math Dept., College of the Holy Cross, Worcester, MA

E.R. Suryanarayan, Math Dept., U. of Rhode Island, Kingston, RI

Virginia Taylor, Math Dept., U. of Lowell, Lowell, MA

W. Tulczyjew, Math Dept., U. of Calgary, Canada

Philip Yasskin, Physics Dept., Harvard U., Cambridge, MA

GEOMETRIC ASPECTS OF GLOBAL BIFURCATION IN NONLINEAR ELASTICITY

by

Stuart S. Antman

Lefschetz Center for Dynamical Systems
Division of Applied Mathematics
Brown University
Providence, Rhode Island 02912

and

Department of Mathematics
University of Maryland
College Park, Maryland 20742

CONTENTS

1. Introduction

The science of continuum mechanics treats the change of shape of material bodies under the action of forces. The basic problem of continuum mechanics is to find the position $\underset{\sim}{p}(\underset{\sim}{x},t)$ of each material point $\underset{\sim}{x}$ of a body B at each time t in some time interval, given the nature of the material of B, the force intensity per unit volume of B, and suitable conditions on the boundary ∂B of B. (We may identify the body B with the region of Euclidean 3-space $\mathbb{E}^3$ it occupies in some reference configuration and we may identify a material point with its position in this configuration. The reference configuration could be taken to be the configuration occupied by the body at some initial time, or could be taken to be a natural state, which is one in which the net force on each part of the body is zero.) Thus Euclidean geometry, which is not as simple as it sounds, enters continuum mechanics at the most primitive level. In this paper we examine some of the ways this underlying Euclidean geometry interacts with the geometrical or topological machinery used to analyze the governing equations.

Before embarking on the main theme of this paper, which is the study of this interaction in global bifurcation problems, it is worthwhile to pause to examine briefly a manifestation of this interaction in another and more fundamental setting. The requirement that the resultant force and moment on every (regular) subbody of B respectively equal the time derivative of linear and angular momentum yields the equations of motion for B. These involve the stress, which is the intensity of force over surfaces. The material properties of B are specified by constitutive equations, which prescribe the dependence of the stress on the deformation $\partial \tilde{p}/\partial \tilde{x}$. ($\partial \tilde{p}/\partial \tilde{x}$ is the tensor whose components are the partial derivatives of components of $\tilde{p}$ with respect to components of $\tilde{x}$.) The substitution of the constitutive equations into the equations of motion yields the governing equations for B. We represent these together with boundary, initial, and other subsidiary conditions in the abstract form

$$(1.1) \qquad \tilde{F}(\tilde{p}) = \tilde{0}.$$

Here $\tilde{F}$ is a quasilinear differential operator from one suitable collection of functions to another. (We shall derive concrete realizations of (1.1) for elementary, but important, mechanical problems in Section 2.) $\tilde{F}$ may profitably be interpreted as an infinite-dimensional vector field defined over its infinite-dimensional domain. $\tilde{p}$ is a solution of (1.1) if the vector field $\tilde{F}$ vanishes at $\tilde{p}$. In a formal way we have thus introduced geometric notions into our mechanical problem at a much deeper level.

Let us now examine how these two levels of geometric structure interact. To ensure that B is not torn apart by the forces acting on it, we require that $\tilde{p}(\cdot,t)$ be continuous for each t. To ensure that two material points of B do not simultaneously occupy the same point of space, we require that $\tilde{p}(\cdot,t)$ be invertible. But invertibility is a global restriction $\tilde{p}$, which is also required to satisfy local restrictions (1.1). Only now is work in analysis and geometry beginning to confront questions like that of finding physically reasonable restrictions on the data of (1.1) for its solutions to be invertible. We are accordingly motivated to replace the global requirement that $\tilde{p}(\cdot,t)$ be invertible by the local requirement that $\tilde{p}(\cdot,t)$ preserve orientation, i.e., that its Jacobian be positive:

$$(1.2) \qquad \det(\partial \tilde{p}/\partial \tilde{x}) > 0.$$

There are some serious questions relating to the definition of this Jacobian when $\underset{\sim}{p}$ is not continuously differentiable, which we pass over (cf. [16]). It is not difficult to show that the set of $\underset{\sim}{p}$'s satisfying (1.2), however it is defined, is neither convex nor closed (cf. [2]). Thus (1.2), an innocent restriction arising from Euclidean geometry, produces unpleasant geometric and topological consequences for the domain of $\underset{\sim}{F}$. This suggests that analytic treatments of problems involving (1.2) are likely to be difficult. This is indeed the case. The analysis of one-dimensional static problems involving (1.2), though delicate, is in a rather complete state (cf. [2,4,8]) but there are no corresponding results for one-dimensional dynamical problems. Significant progress for three-dimensional static problems has been made by Ball [16], yet much remains to be done on this very deep problem. We remark that the serious analysis of (1.2) has only recently occurred because only recently have scientists seriously studied large deformations.

Analogs of (1.2) will appear in our work where they will play but a subsidiary role. Our central concern is to describe recent work on the application of topological methods to equations like (1.1) in order to obtain detailed qualitative information about the deformed shapes of elastic structures. We limit our attention to static problems for nonlinearly elastic structures, especially bifurcation problems, that are described by systems of ordinary differential equations. The use of these equations not only enables us to avoid the severe technical difficulties presented by related partial differential equations, but also allows us to exploit the particularly rich analytic structure of ordinary differential equations that underlies our qualitative analyses.

In Section 2 we derive the equations for the equilibrium of rods because the derivation has two features that we wish to emphasize: simplicity and the absence of ad hoc geometric approximations. In the next four sections we discuss the buckling of straight rods under terminal thrust, of circular plates under edge thrust, of circular arches under hydrostatic pressure, and of spherical shells under hydrostatic pressure. We wish to examine the effects of the initial curvature of the arch and shell and the effects of singularities due to the use of polar coordinates for the plate and shell. Thereafter, we discuss a number of other problems.

2. The Equilibrium Equations for Nonlinearly Elastic Rods

We adopt a mathematical model for the deformation of nonlinearly elastic rods that is governed by a rich, yet tractable, quasilinear system of ordinary differential equations. To motivate our model, we suppose that the reference configuration of this rod is a prism of length $s_2 - s_1$ (see Fig. 2.1). We let s represent a coordinate measuring the length of the line of centroids C of the cross-sections of the prism. The cross-sections are identified by s. Under the action of forces this prism is deformed into the configuration shown in Fig. 2.2. In this configuration the material points of C lie along a space curve and the material points originally along a typical section σ now lie on a curved surface. To avoid confronting the full system of partial differential equations that a full description of this deformation would entail, we content ourselves with a less detailed picture of the deformed state. We imagine that the typical deformed section s can be approximated somehow by an oriented plane. We seek merely to determine the deformed image of C together with the orientation of this plane at each section. Now an oriented plane at s is determined by an orthonormal pair of vectors $\{\underset{\sim}{d}_1(s), \underset{\sim}{d}_2(s)\}$. (See Fig. 2.3.) Let $\underset{\sim}{r}(s)$ denote the deformed position of the material point originally lying on C at section s. We now define the geometrical structure of our mathematical model for a rod by defining a <u>configuration</u> of the rod to be a triple of functions

$$\underset{\sim}{r}, \underset{\sim}{d}_1, \underset{\sim}{d}_2 \tag{2.1}$$

from $[s_1, s_2]$ to $\mathbb{E}^3$ with $\underset{\sim}{d}_1$ and $\underset{\sim}{d}_2$ orthonormal. We denote derivatives with respect to s by primes. We set

$$\underset{\sim}{d}_3 = \underset{\sim}{d}_1 \times \underset{\sim}{d}_2 \tag{2.2}$$

and note that, in general, the tangent $\underset{\sim}{r}'(s)$ to the image $\underset{\sim}{r}(C)$ of C at s need not lie in the $\underset{\sim}{d}_3$ direction. We ensure that the local ratio of deformed to reference length of C (which is $|\underset{\sim}{r}'|$) is positive and that the body is not so severely sheared so as to cause the deformed section to be tangent to $\underset{\sim}{r}$ by requiring

$$\underset{\sim}{r}' \cdot \underset{\sim}{d}_3 > 0. \tag{2.3}$$

This is the one-dimensional analog of (1.2).

Note that the reference configuration of the body, shown in Fig. 2.1, need not be prismatic and that C need not be the curve of centroids in this configuration. C may merely be any prescribed curve of material points in the reference configuration. This versatility

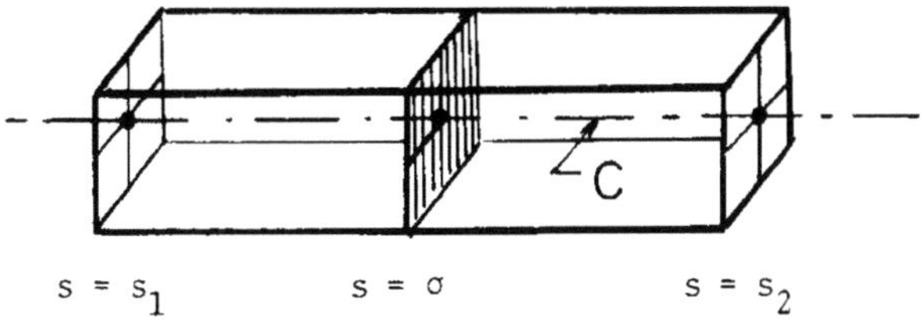

Figure 2.1

Reference Configuration of a Rod

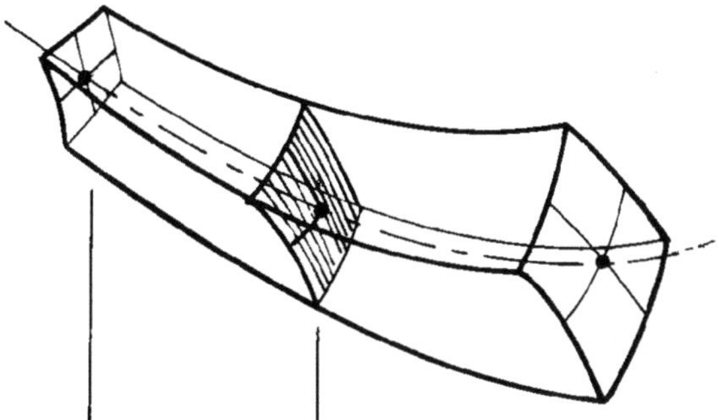

Figure 2.2

Deformed Configuration of Rod of Figure 2.1

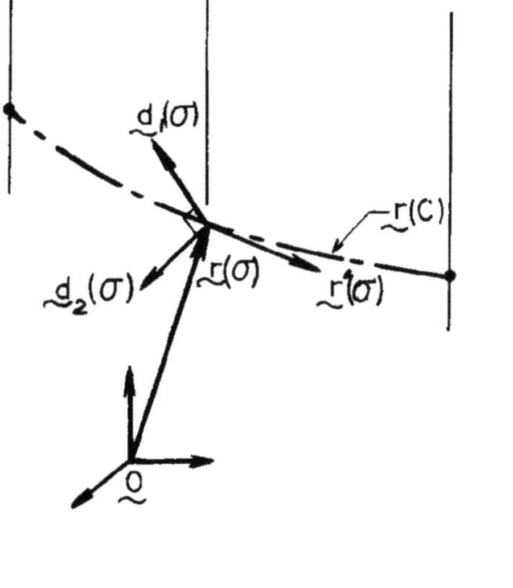

Figure 2.3

Deformed Configuration of the Model Characterized by (2.1)

will be useful in our analysis of the buckling of arches, in which it is convenient to take the reference configuration to be the natural state in which C is an arc of a circle.

We now obtain the equations of equilibrium of the rod. We consider the forces and moments acting on the material between the typical sections $s = \sigma_1 (> s_1)$ and $s = \sigma_2 (< s_2)$ in a deformed configuration. Let $\underset{\sim}{n}^+(\sigma_2)$ be the contact force exerted on the material of (σ_1, σ_2) by the material of $[\sigma_2, s_2]$ and let $-\underset{\sim}{n}^-(\sigma_1)$ be the contact force exerted on the material of (σ_1, σ_2) by the material of $[s_1, \sigma_1]$. (By definition of a contact force, $\underset{\sim}{n}^+(\sigma_2)$ is also the contact force exerted on the material of (τ_1, σ_2) by the material of $[\sigma_2, \tau_2]$ for all $\tau_1 \in [s_1, \sigma_2)$ and for all $\tau_2 \in (\sigma_2, s_2]$, i.e., the contact force depends only on the section of contact and not upon any other feature of the bodies in contact.) Let $\int_{\sigma_1}^{\sigma_2} \underset{\sim}{f}(s)ds$ be the force exerted on the material of (σ_1, σ_2) by any other agency. ($\underset{\sim}{f}$ is the intensity of this force per unit length of C.) Then the requirement that the resultant force on the material of (σ_1, σ_2) vanish yields the equation for the equilibrium of forces:

$$(2.4) \qquad \underset{\sim}{n}^+(\sigma_2) - \underset{\sim}{n}^-(\sigma_1) + \int_{\sigma_1}^{\sigma_2} \underset{\sim}{f}(s)ds = \underset{\sim}{0}.$$

This must hold for all σ_1 and σ_2 satisfying $s_1 < \sigma_1 < \sigma_2 < s_2$. Let us assume that $\underset{\sim}{f}$ is integrable. Then we can let $\sigma_2 \to \sigma_1$ to conclude that $\underset{\sim}{n}^+(\sigma_1) = \underset{\sim}{n}^-(\sigma_1)$ (for all $\sigma_1 \in (s_1, s_2)$). We may accordingly drop the plus and minus signs from $\underset{\sim}{n}$. If $\underset{\sim}{f}$ is continuous, then (2.4) can be differentiated with respect to σ_2 to yield the differential equation for the equilibrium of forces:

$$(2.5) \qquad \underset{\sim}{n}' + \underset{\sim}{f} = \underset{\sim}{0}.$$

The equilibrium of moments is treated the same way. Let $\underset{\sim}{m}^+(\sigma_2)$ and $-\underset{\sim}{m}^-(\sigma_1)$ represent the contact couples exerted on the material of (σ_1, σ_2) by that of $[\sigma_2, s_2]$ and $[s_1, \sigma_1]$, respectively, and let $\int_{\sigma_1}^{\sigma_2} \underset{\sim}{g}(s)ds$ represent the couple exerted on (σ_1, σ_2) by any other agency. The vanishing of the resultant moment (consisting of these couples and the moments of forces) on (σ_1, σ_2) yields the equation of equilibrium of moments:

$$(2.6)\quad \underset{\sim}{m}^{+}(\sigma_2) - \underset{\sim}{m}^{-}(\sigma_1) + \underset{\sim}{r}(\sigma_2) \times \underset{\sim}{n}(\sigma_2) - \underset{\sim}{r}(\sigma_1) \times \underset{\sim}{n}(\sigma_1) + \int_{\sigma_1}^{\sigma_2} [\underset{\sim}{g}(s) + \underset{\sim}{r}(s) \times \underset{\sim}{f}(s)]ds = \underset{\sim}{0}.$$

If $\underset{\sim}{f}$ and $\underset{\sim}{g}$ are integrable and if $\underset{\sim}{r}$ is continuous, then (2.6) implies that $\underset{\sim}{m}^{+}(s) = \underset{\sim}{m}^{-}(s) \equiv \underset{\sim}{m}(s)$. If $\underset{\sim}{f}, \underset{\sim}{g}$, and $\underset{\sim}{r}'$ are continuous, then we can use (2.5) to reduce (2.6) to the differential equation for the equilibrium of couples:

$$(2.7)\qquad \underset{\sim}{m}' + \underset{\sim}{r}' \times \underset{\sim}{n} + \underset{\sim}{g} = \underset{\sim}{0}.$$

We get a determinate system of equations for (2.1) by giving constitutive equations specifying how $\underset{\sim}{n}$ and $\underset{\sim}{m}$ depend on $\underset{\sim}{r}', \underset{\sim}{d}_1, \underset{\sim}{d}_2, \underset{\sim}{d}_1', \underset{\sim}{d}_2'$. These equations must ensure that the material properties of the rod are unaffected by rigid motions. (We indicate the nature of such laws in the special case treated below.)

Let $\underset{\sim}{i}, \underset{\sim}{j}, \underset{\sim}{k}$ be a fixed orthonormal basis for $\mathbb{E}^3$. For the sake of simplicity we restrict our attention to planar problems in which

$$(2.8)\qquad \underset{\sim}{f}\cdot\underset{\sim}{k} = 0,\ \underset{\sim}{g} = \underset{\sim}{0},\ \underset{\sim}{r}\cdot\underset{\sim}{k} = 0,\ \underset{\sim}{n}\cdot\underset{\sim}{k} = 0,\ \underset{\sim}{k} \times \underset{\sim}{m} = \underset{\sim}{0},\ \underset{\sim}{d}_2 = \underset{\sim}{k}.$$

We accordingly set

$$(2.9a)\qquad \underset{\sim}{a} \equiv \underset{\sim}{d}_3 \equiv \cos\theta\,\underset{\sim}{i} + \sin\theta\,\underset{\sim}{j},$$

$$(2.9b)\qquad \underset{\sim}{b} \equiv \underset{\sim}{d}_1 \equiv -\sin\theta\,\underset{\sim}{i} + \cos\theta\,\underset{\sim}{j},$$

$$(2.10)\qquad \underset{\sim}{r}' = (1+\nu)\underset{\sim}{a} + \eta\underset{\sim}{b},$$

$$(2.11)\qquad \underset{\sim}{n} = N\underset{\sim}{a} + H\underset{\sim}{b},$$

$$(2.12)\qquad \underset{\sim}{m} = M\underset{\sim}{k}.$$

Thus $\underset{\sim}{b}(s)$ gives the orientation of the deformed section s and $\theta(s)$ is the angle that the normal to this section makes with $\underset{\sim}{i}$. A <u>configuration</u> of a rod undergoing planar deformation is the pair of functions $\{\underset{\sim}{r}, \underset{\sim}{b}\}$ or equivalently the pair $\{\underset{\sim}{r}, \theta\}$. Inequality (2.3) reduces to

$$(2.13)\qquad 1 + \nu > 0.$$

Let $\theta_0(s)$ be the value of $\theta(s)$ in the reference configuration. Thus $\theta_0'(s)$ is the curvature of C at s in this configuration. We set

(2.14) $$\mu = \theta' - \theta_0'.$$

(θ' is not the curvature of $\underset{\sim}{r}$ because s is not the arc length parameter of $\underset{\sim}{r}$.) The variables

(2.15) $$\nu, \eta, \mu$$

are the <u>strains</u> for the planar deformation of the rod: By (2.9b) and (2.10) they determine the deformed configuration $(\underset{\sim}{r},\underset{\sim}{b})$ to within a rigid displacement. Moreover, they vanish in the reference configuration.

The material of this rod is homogeneous and <u>nonlinearly elastic</u> if there are functions

(2.16) $$(\nu,\eta,\mu) \mapsto \hat{N}(\nu,\eta,\mu), \hat{H}(\nu,\eta,\mu), \hat{M}(\nu,\eta,\mu)$$

such that

(2.17) $$N(s) = \hat{N}(\nu(s), \eta(s), \hat{\mu}(s)), \text{ etc.}$$

The form of (2.17) ensures that material properties are unaffected by rigid motions. Of all possible constitutive laws (2.17), we wish to single out those having the following physically desirable properties: An increase in tensile force produces an increase in length, an increase in shear force H produces an increase in the shear deformation η, and an increase in the bending couple M produces an increase in the bending deformation μ. These requirements as well as others arising from dynamical considerations are ensured by the strict monotonicity of (2.16). If these functions are continuously differentiable, as we shall assume, then this strict monotonicity is ensured by:

(2.18) $$\begin{pmatrix} \hat{N}_\nu & \hat{N}_\eta & \hat{N}_\mu \\ \hat{H}_\nu & \hat{H}_\eta & \hat{H}_\mu \\ \hat{M}_\nu & \hat{M}_\eta & \hat{M}_\mu \end{pmatrix} \text{ is positive-definite.}$$

Here the subscripts denote partial derivatives. We also require that an infinite compressive force is needed to violate (2.13) and an infinite tensile force is needed to produce an infinite extension. These and like restrictions are embodied in the growth conditions

$$(2.19) \qquad \hat{N}(\nu,\eta,\mu) \to \{{}^{\infty}_{-\infty}\} \quad \text{as} \quad \nu \to \{{}^{\infty}_{-1}\},$$

$$(2.20) \qquad \hat{H}(\nu,\eta,\mu) \to \pm\infty \quad \text{as} \quad \eta \to \pm\infty,$$

$$(2.21) \qquad \hat{M}(\nu,\eta,\mu) \to \pm\infty \quad \text{as} \quad \mu \to \pm\infty.$$

The symmetries of the problems we treat lead us to assume that

$$(2.22\text{a,b,c}) \qquad \hat{N}(0,0,0) = 0, \ \hat{H}(\nu,-\eta,\mu) = -\hat{H}(\nu,\eta,\mu), \ \hat{M}(\nu,\eta,0) = 0.$$

3. The Buckling of a Straight Rod

We assume that the reference state is natural and straight. We take C to lie along the horizontal $\underset{\sim}{i}$-axis so that $\theta_0 = 0$. We take $s_1 = 0$, $s_2 = 1$ without loss of generality. We assume that $\underset{\sim}{f} = \underset{\sim}{0}$ and $\underset{\sim}{g} = \underset{\sim}{0}$. We assume that the end section $s = 0$ is fixed at the origin and is there welded to a rigid vertical wall and that the end section $s = 1$ is constrained to be vertical and subjected to a compressive horizontal force of magnitude λ. (The end $s = 1$ is free to move within the $(\underset{\sim}{i},\underset{\sim}{j})$-plane.) Under these conditions, equations (2.5) and (2.7) reduce to

$$(3.1) \qquad \underset{\sim}{n}' = \underset{\sim}{0},$$

$$(3.2) \qquad M' + \underset{\sim}{k}\cdot(\underset{\sim}{r}'\times\underset{\sim}{n}) = 0,$$

which are subject to the boundary conditions

$$(3.3\text{a,b}) \qquad \theta(0) = 0, \quad \underset{\sim}{r}(0) = \underset{\sim}{0},$$

$$(3.4\text{a,b}) \qquad \theta(1) = 0, \quad \underset{\sim}{n}(1) = -\lambda\underset{\sim}{i}.$$

We now convert the boundary value problem (3.1)-(3.4), (2.17) to a mathematically convenient form. Equations (3.1) and (3.4b) imply that

$$(3.5) \qquad \underset{\sim}{n}(s) \equiv N(s)\underset{\sim}{a}(s) + H(s)\underset{\sim}{b}(s) = -\lambda\underset{\sim}{i}$$

for all s in $[0,1]$. The substitution of (3.5) into (3.2) and the use of (2.10) yields

$$M' + \lambda[(1+\nu)\sin\theta + \eta\cos\theta] = 0. \tag{3.6}$$

The substitution of (2.17) and (2.14) into (3.5) yields

$$\hat{N}(\nu(s),\eta(s),\theta'(s)) = -\lambda\cos\theta, \tag{3.7}$$

$$\hat{H}(\nu(s),\eta(s),\theta'(s)) = \lambda\sin\theta. \tag{3.8}$$

Now (2.18)-(2.20) support a global implicit function theorem to the effect that for given N,H,μ the algebraic equations

$$\hat{N}(\nu,\eta,\mu) = N, \quad \hat{H}(\nu,\eta,\mu) = H \tag{3.9}$$

have a unique solution for ν and η, which we denote by

$$\nu = \hat{\nu}(N,H,\mu), \quad \eta = \hat{\eta}(N,H,\mu). \tag{3.10}$$

(This implicit function theorem is a consequence of Brouwer degree theory, or its corollary, the Brouwer fixed point theorem. Conditions (2.19) and (2.20) ensure that there is a solution, while (2.18) ensures that it is unique.) The classical local implicit function theorem then ensures that $\hat{\nu}$ and $\hat{\eta}$ are continuously differentiable because $\hat{N}$ and $\hat{H}$ are. In virtue of these remarks we find that (3.7) and (3.8) are equivalent to

$$\nu(s) = \hat{\nu}(-\lambda\cos\theta(s), \lambda\sin\theta(s), \theta'(s)), \tag{3.11a}$$

$$\eta(s) = \hat{\eta}(-\lambda\cos\theta(s), \lambda\sin\theta(s), \theta'(s)). \tag{3.11b}$$

If we now replace M of (3.6) by its constitutive representation from (2.17) and if we replace ν and η wherever they appear in (3.6) by the representations of (3.11), we find that θ is governed by the second-order quasilinear ordinary differential equation

$$[\hat{M}(\hat{\nu},\hat{\eta},\theta')]' + \lambda[(1+\hat{\nu})\sin\theta + \hat{\eta}\cos\theta] = 0, \tag{3.12a}$$

where the arguments of $\hat{\nu}$ and $\hat{\eta}$ are

(3.12b) $$-\lambda \cos\theta,\ \lambda \sin\theta, \theta'.$$

Equation (3.12) is subject to the boundary conditions (3.3a) and (3.4a). From any solution θ of this boundary value problem we can find all the other geometrical and mechanical variables by using the various formulas developed above.

If we assume that the material of the rod is inextensible and unshearable so that $\hat{\nu} = 0$ and $\hat{\eta} = 0$ and if we assume that $\hat{M}$ is linear in μ, i.e., if $\hat{M}(\mu) = EI\mu$, where EI is a constant, then (3.12) reduces to the equation of the elastica

(3.13) $$EI\theta'' + \lambda \sin\theta = 0,$$

which Euler [21] so beautifully analyzed in 1744. In contrast to (3.13), our equation (3.12) has the parameter λ appearing throughout and has a far richer nonlinear structure.

Now (2.22b) and the argument leading to (3.10) shows that $\hat{\eta}(N,0,\mu) = 0$. This fact and (2.22c) imply that the boundary value problem (3.12), (3.3a), (3.4) has the trivial solution $\theta = 0$ for all real λ. We wish to determine the nature of nontrivial solutions of this boundary value problem and how they depend on λ. For this purpose it is useful to compare this problem with its linearization about $\theta = 0$, which is

(3.14) $$\psi'' + q(\lambda)\psi = 0, \qquad \psi(0) = 0 = \psi(1)$$

where

(3.15) $$q(\lambda) \equiv \frac{\lambda[1+\hat{\nu}(-\hat{\lambda},0,0) + \lambda\hat{\eta}_H(-\lambda,0,0)]}{\hat{M}_\mu(\hat{\nu}(-\lambda,0,0),0,0)}.$$

This problem has nontrivial solutions

(3.16) $$\psi_k(s) = \sin k\pi s$$

where k is a positive integer whenever λ is a real solution of

(3.17) $$q(\lambda) = k^2\pi^2.$$

Note that the linearization of (3.13) is (3.14) with $q(\lambda) = \lambda/EI$. Thus the linearization of (3.13) has a countable infinity of eigenvalues $\lambda_k = k^2\pi^2 EI$, $k = 1,2,\ldots$, which are positive and simple and which correspond to eigenfunctions $\sin k\pi s$, which have exactly $k-1$ zeros in $(0,1)$, each of which is simple. No such definitive conclusion holds for (3.17) when (3.15) is used. The number and location of roots of (3.17) depend on q, i.e., on the constitutive functions. Thus, for a given k, there may be no, one, several, or infinitely many solutions of (3.17). Each such eigenvalue, however, has the same eigenfunction (3.16). The only general result that can be stated about the ordering of solutions of (3.17) is that for a fixed $k = K$ the solutions of (3.18) lie in closed intervals (which generically are just points); each such interval is bordered by non-empty open intervals, which in turn touch closed intervals of solutions of (3.17) with $k = K - 1$, K, or $K + 1$. This remark is just a consequence of the continuity of q. We also note that (2.17) ensures that $\lambda^2\hat{n}_H/\hat{M}_\mu$ is positive. Thus it is very likely that (3.17) has negative solutions. These solutions correspond to shear instabilities produced by tension and will not concern us here (cf. [9]).

We say that a solution $\bar{\lambda}$ of (3.17) is <u>simple</u> if $q'(\bar{\lambda}) \neq 0$. We shall largely restrict our attention to such eigenvalues. (All solutions of (3.17) when $q(\lambda) = \lambda/EI$ are simple.)

Let us first analyze (3.13) in order to contrast the behavior of its solutions with those of (3.12). If $(\lambda,\theta) \in \mathbb{R} \times C^1([0,1])$ satisfies (a suitably generalized version of) (3.13) subject to boundary conditions $\theta(0) = 0 = \theta(1)$, then (λ,θ) is called a <u>solution pair</u> of this problem. A curve of solution pairs is called a <u>branch</u> of solutions. The set of points $\{(\lambda,0) : \lambda \in \mathbb{R}\}$ is a branch called the <u>trivial branch</u>. A point (solution pair) on a branch is called a <u>bifurcation point</u> of that branch if in every neighborhood of that point there are solution pairs not on that branch. (The topology used is that of $\mathbb{R} \times C^1$.) Let $\theta \mapsto h(\theta)$ be some convenient norm-like functional. It is customary to represent solutions of such problems on a bifurcation diagram, which is a plot of all points $(\lambda, h(\theta))$ in the plane for which (λ,θ) is a solution pair. Since it can easily be shown that $|\theta'(0)| = \max\{|\theta'(s)|, s \in [0,1]\}$ for solutions of this boundary value problem, we may take $h(\theta) = \theta'(0)$. Now (3.13) can be solved explicitly for θ in terms of elliptic functions (cf. [27],e.g.) from which one can obtain the bifurcation diagram shown in Figure 3.1. This diagram indicates that nontrivial solution branches bifurcate from the trivial branch at the points $(\lambda_k, 0)$ where $\{\lambda_k\}$ are eigenvalues of the linearized problem. Moreover, the representation of the solutions in terms of elliptic functions shows

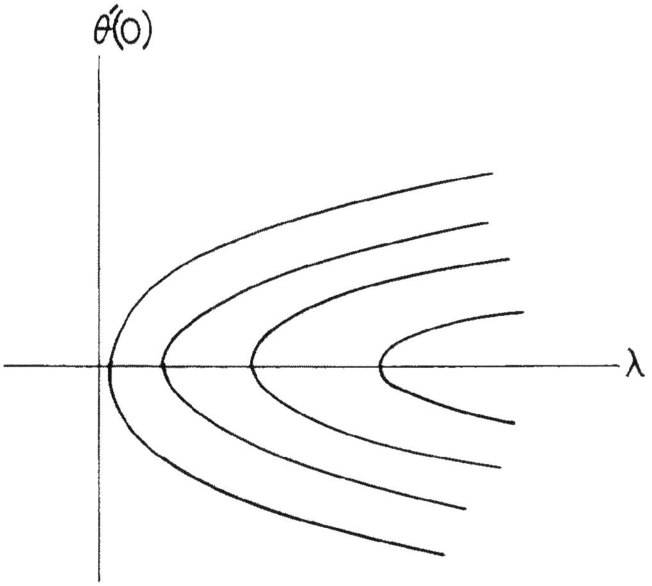

Figure 3.1. Bifurcation Diagram of (3.13), (3.3a), (3.4a).

that on the (nontrivial part of the) branch emanating from $(\lambda_k, 0)$, θ has exactly $k - 1$ zeros in $(0,1)$, i.e., it has exactly the same nodal structure as the eigenfunction ψ_k of the linearized problem.

We now show how to get these results without using the representation of solutions in terms of elliptic functions. This alternative approach is both useful and necessary because the elliptic function method fails for (3.13) when EI depends on s (i.e., when the rod is not of uniform thickness) and fails for (3.12), the equation of our central interest. We first observe that (3.13), (3.3a), (3.4a) is equivalent to the integral equation

$$(3.18) \qquad \begin{aligned} \theta(s) &= \lambda \int_0^1 G(s,t)\sin\theta(t)dt \\ &= \lambda \int_0^1 G(s,t)\theta(t)dt + \lambda \int_0^1 G(s,t)[\sin\theta(t) - \theta(t)]dt \\ &\equiv [L(\lambda)\theta](s) + [F(\lambda,\theta)](s) \end{aligned}$$

where the Green's function G is given by

$$(3.19) \qquad G(s,t) = \begin{cases} t(1-s) & \text{if } t \le s, \\ s(1-t) & \text{if } t \ge s. \end{cases}$$

It is well-known (cf. [18]) that $L(\lambda)$ is a completely continuous linear operator taking $C^1 (\equiv C^1([0,1]))$ into itself and F is a completely continuous operator taking $\mathbb{R} \times C^1$ into C^1 with $F(\lambda,\theta) = o(\|\theta\|_{C^1})$ as $\theta \to 0$, uniformly for λ in bounded sets of $\mathbb{R}$. The linearization of (3.18), corresponding to the linearization of (3.13), is $\psi = L(\lambda)\psi$. With this information we can invoke the following:

3.20. Global Bifurcation Theorem. Let $L(\lambda)$ be a completely continuous and linear mapping of a Banach space B into itself and let $L(\cdot)$ be continuously differentiable. Let F be a completely continuous mapping of $\mathbb{R} \times B$ into B with $F(\lambda,\theta) = o(\|\theta\|_B)$ as $\theta \to 0$, uniformly for λ in bounded sets of $\mathbb{R}$. Let $\bar{\lambda}$ be a simple eigenvalue of

$$\psi = L(\lambda)\psi. \tag{3.21}$$

Then the closure in $\mathbb{R} \times C^1$ of the set of all nontrivial solution pairs of

$$\theta = L(\lambda)\theta + F(\lambda,\theta) \tag{3.22}$$

contains a maximal connected subset $\Sigma(\bar{\lambda})$ that contains $(\bar{\lambda},0)$ and has at least one of the following two properties: (i) $\Sigma(\bar{\lambda})$ is not contained in any closed and bounded subset of $\mathbb{R} \times B$, (ii) $\Sigma(\bar{\lambda})$ contains a point of the form $(\kappa,0)$ where κ is another eigenvalue of (3.21). If, moreover, (λ,θ) belongs to $\Sigma(\bar{\lambda})$ and if (λ,θ) is near $(\bar{\lambda},0)$, then θ is approximated there in the norm of B by a constant multiple of the eigenfunction of (3.21) corresponding to $\bar{\lambda}$. $\Sigma(\bar{\lambda})$ is a curve near $(\bar{\lambda},0)$.

This theorem is a special case of that of Rabinowitz [25]. To handle (3.18) we simply identify B with C^1. This theorem then implies that connected sets of solutions bifurcate from the trivial branch at the eigenvalues of the linearized problem and that in a neighborhood of each bifurcation point the solution θ on the branch $\Sigma(\lambda_k)$ has exactly $k - 1$ zeros in $(0,1)$. This last result follows from the facts that ψ_k has exactly $k - 1$ zeros on $(0,1)$ with $\psi_k'(\zeta) \neq 0$ for each such zero ζ and that θ is approximated by a constant multiple of ψ_k in C^1.

Now suppose that the number of zeros of θ were to change as

(λ,θ) is varied on $\Sigma(\lambda_k)$. If $\Sigma(\lambda_k)$ were a branch, then it is easy to see that there is a solution pair (λ,θ) in $\Sigma(\lambda_k)$ with θ having a double zero, i.e., there would be a $\sigma \in [0,1]$ such that

$$\theta(\sigma) = \theta'(\sigma) = 0. \tag{3.23}$$

The same conclusion holds when $\Sigma(\lambda_k)$ is merely a connected set. But (3.13) subject to the initial condition (3.23) has the unique solution $\theta = 0$. Thus we conclude that the number of zeros of solutions θ on each $\Sigma(\theta_k)$ can change only at the trivial branch. We also observe that $\Sigma(\lambda_k)$ cannot contain $(\lambda_\ell,0)$, $\ell \neq k$, because if $(\lambda,\theta) \in \Sigma(\lambda_k)$ then θ has exactly $k - 1$ simple zeros on $(0,1)$ and cannot be approximated in C^1 by a constant multiple of a function ψ_ℓ with exactly $\ell - 1$ simple zeros on $(0,1)$. Thus we conclude that each $\Sigma(\lambda_k)$ is unbounded in $\mathbb{R} \times C^1$ and that on $\Sigma(\lambda_k)$ θ has exactly $(k-1)$ zeros in $(0,1)$. A more careful analysis using extremal characterizations of eigenvalues, Sturmian theory, and careful estimates would allow us to furnish a proof, not based on the use of elliptic functions, that the bifurcation diagram has the character indicated in Figure 3.1. This approach to bifurcation problems for second order ordinary differential equations was developed by Crandall and Rabinowitz [19]. We note that it relies upon (i) the conversion of the problem to an operator equation of the form (3.22) involving completely continuous operators from a Banach space into itself, (ii) the availability of a useful Sturmian theory for the linearized problem to locate eigenvalues and describe eigenfunctions, and (iii) a uniqueness theory for the differential equations subject to both boundary and initial conditions. (In the example just treated we obtained uniqueness from the initial conditions (3.23) alone.)

Let us now turn to (3.12). Condition (2.18) would enable us to carry out the differentiations in (3.12) and convert it into standard semilinear form: $-\theta'' = f(\lambda,\theta,\theta'')$, which is readily converted to an integral equation like (3.18). A far more elegant approach is to observe that (2.18)-(2.21) ensure that if N,H,M are given, then the algebraic equations

$$\hat{N}(\nu,\eta,\mu) = N, \quad \hat{H}(\nu,\eta,\mu) = H, \quad \hat{M}(\nu,\eta,\mu) = M \tag{3.24}$$

can be uniquely solved for ν,η,μ; we denote this solution by

$$\nu = \tilde{\nu}(N,H,M), \quad \eta = \tilde{\eta}(N,H,M), \quad \mu = \tilde{\mu}(N,H,M). \tag{3.25}$$

Then (3.12) is equivalent to the semilinear system

$$\theta' = \tilde{\mu}(-\lambda \cos\theta,\ \lambda \sin\theta, M), \tag{3.26a}$$

$$M' = -\lambda[(1+\tilde{\nu}(-\lambda \cos\theta,\ \lambda \sin\theta, M)\sin\theta + \tilde{\eta}(-\lambda \cos\theta,\ \lambda \sin\theta, M)\cos\theta]. \tag{3.26b}$$

This equation subject to the boundary conditions $\theta(0) = 0 = \theta(1)$ is readily converted to an operator equation for $(\lambda,\theta,M) \in \mathbb{R} \times C^0 \times C^0$, which has the same abstract form as (3.22). (Cf. [15, Sec. 7]). Thus the Global Bifurcation Theorem applies to (3.12). Noting that (2.18) and (2.22c) imply that $M(\sigma) = 0$ if and only if $\theta'(\sigma) = 0$, we see that if θ should change its number of interior zeros, then there is a $\sigma \in [0,1]$ such that $\theta(\sigma) = 0 = M(\sigma)$. The initial value problem consisting of (3.26) and these conditions has the unique solution $\theta = 0 = M$. Thus the number of simple zeros of solutions θ on $\Sigma(\bar{\lambda})$, where $\bar{\lambda}$ is a simple eigenvalue of (3.14), can change only at the trivial branch. In contrast to the situation for (3.13), the connected set $\Sigma(\bar{\lambda})$ could contain a point $(\kappa,0)$ where κ is another eigenvalue of (3.14) if either $\bar{\lambda}$ or κ is multiple or if $\bar{\lambda}$ and κ are each solutions of (3.17) for the same k. Thus the bifurcation diagram corresponding to (3.12) can differ considerably from Figure 3.1. Nevertheless, there remains the basic principle that the solutions preserve the nodal properties they inherit from the eigenfunctions of the linearized problem.

The simplicity of this development does not necessarily persist if the boundary conditions (3.3) and (3.4) are changed. E.g., if (3.4b) is replaced with

$$\underset{\sim}{r}(1)\cdot\underset{\sim}{j} = 0, \quad \underset{\sim}{n}(1)\cdot\underset{\sim}{i} = -\lambda, \tag{3.27}$$

then the governing equation (3.12) and its linearization (3.14) are replaced by equations that are much more complicated and the requisite uniqueness theorems used to ensure that nodal properties can change only at the trivial branch are quite delicate, rather than routine. The analyses of such problems are given in [15]. It is also possible to handle the spatial bifurcation of rods under terminal thrust and torsion. Cf. [12].

4. Buckling of a Circular Plate

We now study the axisymmetric buckling of a uniform, circular plate of reference radius 1 under the action of a uniform normal force applied to its circular edge. Since the axisymmetric deformation of a circular plate is determined by the deformation of any of its radial sections, an axisymmetric configuration of the plate is determined by the pair of functions $\{\underset{\sim}{r},\underset{\sim}{b}\}$ with $|\underset{\sim}{b}| = 1$ or equivalently by the pair of functions $\{\underset{\sim}{r},\theta\}$, where $\underset{\sim}{b}$ is related to θ by (2.9b). In short, the geometry is exactly analogous to that for the planar deformation of rods described in Section 2. There are, however, more strain variables besides $\{\nu,\eta,\mu\}$, which were introduced there. Let $\underset{\sim}{i}$ now denote unit vector in the radial direction of a typical section. Then

$$(4.1)\quad 1 + \tau(s) \equiv s^{-1}\underset{\sim}{r}(s)\cdot\underset{\sim}{i} = s^{-1}\int_0^s\{[1+\nu(t)]\cos\theta(t) + \eta(t)\sin\theta(t)\}dt$$

is the local ratio of deformed to reference length of a material circle with center at the center of the plate. The plate-theoretic analog of (2.1) consists of (2.13) and

$$(4.2)\qquad 1 + \tau > 0.$$

The strain μ measures the amount of bending about k. The strain σ given by

$$(4.3)\qquad \sigma(s) = \sin\theta(s)/s$$

likewise measures the amount of bending about $\underset{\sim}{i}$. Thus the full set of strains for the axisymmetric deformation of a plate is $\{\tau,\nu,\eta,\sigma,\mu\}$. The strains τ and σ can be determined from the other strains. (This is a consequence of the axisymmetry; they would otherwise be independent.)

The equilibrium equations are obtained from the requirement that the resultant force and moment on an arbitrary annular sector vanish (cf. [3].) A convenient version of these equations is

$$(4.4)\qquad sN(s) = -[\lambda g(\underset{\sim}{r}(1)\cdot\underset{\sim}{i}) + \int_s^1 T(t)dt]\cos\theta(s),$$

$$(4.5)\qquad sH(s) = [\lambda g(\underset{\sim}{r}(1)\cdot\underset{\sim}{i}) + \int_s^1 T(t)dt]\sin\theta(s),$$

$$(4.6)\qquad (sM)' - \Sigma\cos\theta + [\lambda g(\underset{\sim}{r}(1)\cdot\underset{\sim}{i}) + \int_s^1 T(t)dt]\cdot$$
$$[(1+\nu)\sin\theta + \eta\cos\theta] = 0.$$

Here T is the circumferential tension per unit radial distance s and Σ is the bending couple per unit of s about $\underset{\sim}{i}$. (Thus (T,N,H,Σ,M) are generalized forces corresponding to the generalized displacements $(\tau,\nu,\eta,\sigma,\mu)$ according to a suitable principle of virtual work.) $\lambda g(\underset{\sim}{r}(1)\cdot\underset{\sim}{i})$ is the prescribed value of the radial component of force applied to the edge. The vertical component is assumed to be zero. The presence of g allows this intensity to vary with the deformed radius and thus allows us to distinguish, e.g., between a prescribed intensity per unit original edge length (or area) and a prescribed intensity per unit deformed edge length (or area). In comparing (4.4)-(4.6) with (3.5)-(3.8) we see that the former are singular owing to the presence of s as a coefficient of the highest order terms and that they also contain integrals absent from the latter. In other respects the forms of these two systems are similar.

We restrict our attention to deformations satisfying the geometric boundary condition

$$\theta(1) = 0, \tag{4.7}$$

which ensures that the edges remain vertical. We do not spell out conditions at 0 that ensure that solutions be regular.

Our constitutive equations are analogous to (2.16) and (2.17) with $(T,N,H,\Sigma,M)(s)$ depending on $(\tau,\nu,\eta,\sigma,\mu)(s)$. In addition to (2.18) we require:

$$\begin{pmatrix} \hat{T}_\tau & \hat{T}_\sigma \\ \hat{\Sigma}_\tau & \hat{\Sigma}_\sigma \end{pmatrix} \text{ and } \begin{pmatrix} \hat{T}_\tau & \hat{T}_\nu & \hat{T}_\eta \\ \hat{N}_\tau & \hat{N}_\nu & \hat{N}_\eta \\ \hat{H}_\tau & \hat{H}_\nu & \hat{H}_\eta \end{pmatrix} \text{ are positive-definite.} \tag{4.8}$$

We supplement (2.19)-(2.21) with

$$\hat{T} \to \left\{ {\infty \atop -\infty} \right\} \text{ as } \tau \to \left\{ {\infty \atop -1} \right\}, \quad \hat{\Sigma} \to \pm\infty \text{ as } \sigma \to \pm\infty. \tag{4.9}$$

In place of (2.22) we require:

$$\hat{T},\hat{N},\hat{\Sigma},\hat{M} \text{ are even in } \eta, \ \hat{H} \text{ is odd in } \eta, \tag{4.10}$$

$$\overline{T} = \hat{T},\ \overline{N} = \hat{N},\ \overline{H} = \hat{H},\ \overline{\Sigma} = -\hat{\Sigma},\ \overline{M} = -\hat{M} \tag{4.11a}$$

where

$$\overline{T}(\tau,\nu,\eta,\sigma,\mu) \equiv \hat{T}(\tau,\nu,\eta,-\sigma,-\mu), \text{ etc.} \tag{4.11b}$$

Our boundary value problem consists of (4.1), (4.4)-(4.6) and the constitutive equations. We note that (4.10) and (4.11) ensure that this problem has a trivial solution of the form $\theta = 0$, $\eta = 0$ if τ

and ν can be shown to satisfy appropriately degenerate versions of (4.1) and of (4.4) with N and T given their constitutive representations. Let us assume that τ and ν can be found; we shall discuss this existence problem below. We also note that (2.18), (4.5), and (4.10) imply that $\eta = 0$ where $\theta = 0$.

We now confront the problem of converting our boundary value problem to the abstract form (3.22). The presence of integrals in (4.4) and (4.5) elevates these two equations beyond the level of mere algebraic equations, at which their counterparts (3.7) and (3.8) sit. Standard fixed point theorems do not seem to be effective in solving the system consisting of (4.1) and of (4.4) and (4.5) (with T,N,H replaced by their constitutive representations) for τ,ν, and η in terms of θ in a manner compatible with (2.13), (2.19), (4.2), and (4.9). We accordingly do not attempt to reduce our system to a single second-order equation for θ.

Now experience with axisymmetric solutions of such linear equations as Laplace's equation or the biharmonic equation (which governs the simplest linear theory of plates) might suggest that our governing equations are just nonlinear versions of Bessel's equation for J_1. It is well-known that the judicious placement of the factor $\sqrt{s}$ in a nonhomogeneous version of Bessel's equation for J_1 enables one to represent the solution as an integral operator acting on the nonhomogeneous term with the integral operator completely continuous from C^1 into itself and from L_2 into itself. Thus the conversion of our problem to the form (3.22) would seem routine. It is, however, anything but routine, and without further assumptions may be impossible for the common Banach spaces. The analogy between our problem and the biharmonic equation fails because we have not assumed the isotropy underlying the biharmonic equation. Without isotropy we cannot eliminate the coefficients of negative powers of s that destroy complete continuity and that arise from the singular character of our equations at the origin. In physical terms, the isotropy would prohibit the reinforcement of the plate by radially disposed fibers. The concentration of such fibers at the center of the plate is likely to cause a response markedly different from that which would occur in its absence. Since such singular behavior would be due to the lack of isotropy at the center and since θ, and hence η, must vanish there for regularity, we can prohibit this singular behavior by merely specifying isotropy conditions for our material when $\eta = 0$:

$$\hat{N}(\tau,\nu,0,\sigma,\mu) = \hat{T}(\nu,\tau,0,\mu,\sigma), \tag{4.12a}$$

(4.12b) $$\hat{M}(\tau,\nu,0,\sigma,\mu) = \hat{\Sigma}(\nu,\tau,0,\mu,\sigma).$$

The following observations indicate how (4.12) is to be used. The definitions (4.1) and (4.3) suggest that regular solutions of our problem satisfy

(4.13) $$\tau(0) = \nu(0), \quad \sigma(0) = \mu(0)$$

(because $\theta(0) = 0$ for regularity). The first two terms of (4.6) (with M and Σ replaced by their constitutive representations) have the form

(4.14) $$s\hat{M}' + \hat{M} - \hat{\Sigma} + \hat{\Sigma}(1-\cos\theta).$$

Condition (4.12) and Taylor's Theorem enable us to represent $\hat{M} - \hat{\Sigma}$ as a (nonhomogeneous) quadratic expression in the differences $\tau - \nu$ and $\sigma - \mu$ while condition (4.13) indicates that considerable "cancellation" will occur at the origin. The same process applies to the differential version of (4.4). This cancellation in fact removes the terms that obstruct the complete continuity. This development can be carried out in a rigorous fashion when the equations have been appropriately recast as integral equations. This is accomplished in the following way. We put (4.6) with the constitutive substitutions into the form

(4.15) $$(s\theta')' - s^{-1}\theta = f(s,\theta,\theta'\ldots.), \ \theta(1) = 0, \ (\theta(0) = 0).$$

(Similar conversions are applied to (4.4) and (4.5).) The operator on the left side of (4.15) corresponds to the Bessel function J_1 and has the Green's function

(4.16) $$k(s,t) = \begin{cases} (s-s^{-1})(t/2) & \text{for} \quad t \le s, \\ (t-t^{-1})(s/2) & \text{for} \quad s \le t. \end{cases}$$

Rather than converting (4.15) to the integral equation $\theta(s) = \int_0^1 k(s,t)f(t,\theta(t),\theta'(t),\ldots)dt$, which does not promote the analysis, we first introduce the auxiliary function ω by

(4.17) $$s^{-1/2}\omega = (s\theta')' - s^{-1}\theta.$$

Equation (4.15) ensures that ω satisfies the integral equation

$$\omega(s) = s^{1/2}f(s,(G\omega)(s),(G\omega)'(s),\ldots) \tag{4.18}$$

where

$$(G\omega)(s) = \int_0^1 k(s,t)t^{-1/2}\omega(t)dt. \tag{4.19}$$

In this setting (4.12) has the desired effect of ensuring that (4.18) is one component of an operator equation of the form (3.22) (for $\omega \in C^0$).

The isotropy conditions (4.12) also ensure that there is a trivial solution with $1 + \tau(s) = 1 + \nu(s) = cs$, where c is a constant depending upon λ. Moreover, this solution is unique.

The linearized equations are uncoupled. That corresponding to (4.6) has the form

$$(s\psi')' - s^{-1}\psi + q(\lambda)s\psi = 0,\ \psi(1) = 0,\ (\psi(0) = 0), \tag{4.20}$$

where q depends upon the constitutive functions (cf. (3.14) and (3.15)). The eigenvalues of (4.20) are solutions $\bar{\lambda}$ of $J_1(q(\lambda)) = 0$ and the corresponding eigenfunctions are $s \mapsto J_1(q(\bar{\lambda})s)$. Thus the requisite Sturmian theory for the linear problem is completely analogous to that for (3.14). The Global Bifurcation Theorem thus ensures that connected sets of solution pairs bifurcate from the trivial branch at simple eigenvalues $\bar{\lambda}$ of (4.20).

θ could change its nodal properties only where it has a double zero. This double zero could occur at 0 where the equations are singular. To show that the possession of a double zero causes θ to vanish, we resort to the theory of differential inequalities. Careful estimates show that $v = |s^{-1}\theta| + |\theta'|$ satisfies a differential inequality of the form

$$v' \leq Cs^{-1}v + \ldots . \tag{4.21}$$

The dots represent some complicated, but relatively unimportant terms. If θ has a double zero at the most difficult place, the origin, then $s^{-1}v(s) \to 0$ as $s \to 0$. If $C \leq 1$, then the only function v satisfying (4.21) and this initial condition is $v = 0$. (This is basically Nagumo's uniqueness theorem. Cf. [22, Chap. III].) The isotropy condition (4.12) ensures that $C \leq 1$.

Thus we conclude that the global behavior of buckled isotropic plates mirrors that for rods. The buckled states preserve the nodal behavior they inherit from simple eigenfunctions of the linearized problems. Connected sets of solution pairs can return to the trivial branch only under the special conditions described in the comments following (3.26). The details of the analysis described in this section are supplied in [3].

5. The Buckling of a Circular Arch

We return to the planar theory of rods developed in Section 2. We now assume that the reference configuration of the rod is a natural state in which C is a circular arc of radius 1 and length 2α. (See Figure 5.1.) Then $\theta_0' = 1$ and (2.14) reduces to

$$\mu = \theta' - 1. \tag{5.1}$$

We set $s_1 = -\alpha$ and $s_2 = \alpha$. We assume that the arch is subjected to a compressive hydrostatic pressure p in its deformed configuration. This is a loading in which the force acts normal to the deformed curve $\underset{\sim}{r}$ at each point with a constant intensity p per unit deformed length of $\underset{\sim}{r}$. Since the differential arc length of $\underset{\sim}{r}$ is $|\underset{\sim}{r}'(s)|ds$, the total hydrostatic force on the material of (σ_1, σ_2) is

$$p\int_{\sigma_1}^{\sigma_2} \frac{\underset{\sim}{k}\times\underset{\sim}{r}'(s)}{|\underset{\sim}{k}\times\underset{\sim}{r}'(s)|}|\underset{\sim}{r}'(s)|ds = p\int_{\sigma_1}^{\sigma_2} \underset{\sim}{k}\times\underset{\sim}{r}'(s)ds. \tag{5.2}$$

We therefore take $\underset{\sim}{f}$, introduced in the paragraph preceding (2.4), to

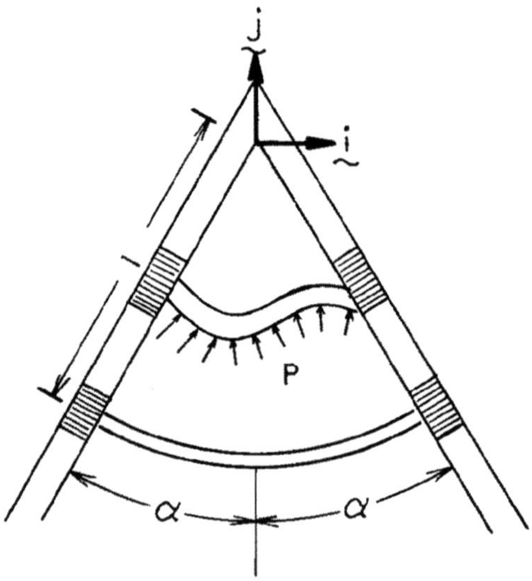

Figure 5.1

Reference and Buckled States of the Circular Arch

be given by $\underset{\sim}{f} = \underset{\sim}{k} \times \underset{\sim}{r}'$. Under the assumptions of (2.8), the governing equations (2.5), (2.7), (2.17) have the componential form

$$(5.3) \qquad \hat{N}(\nu,\eta,\mu)' = (1+\mu)\hat{H}(\nu,\eta,\mu) + p\eta,$$

$$(5.4) \qquad \hat{H}(\nu,\eta,\mu)' = -(1+\mu)\hat{N}(\nu,\eta,\mu) - p(1+\nu),$$

$$(5.5) \qquad \hat{M}(\nu,\eta,\mu)' = \eta\hat{N}(\nu,\eta,\mu) - (1+\nu)\hat{H}(\nu,\eta,\mu).$$

(Even though (2.5) can be integrated, it proves convenient to retain it in its differential and componential form as given in (5.3) and (5.4).) The full set of equations for this problem are obtained by supplementing (5.3)-(5.5) with the geometric relations (2.10) and (6.1). We can exploit the fact that (5.3)-(5.5) are uncoupled from the geometric relations, even though the boundary conditions may not be.

We assume that the ends of the arch are welded to blocks that are free to slide in radial grooves (see Figure 5.1). Thus our boundary conditions are that

$$(5.6) \qquad \theta(\pm\alpha) = \pm\alpha,$$

$$(5.7) \qquad \underset{\sim}{r}(\pm\alpha)\cdot[\cos(\pm\alpha)\underset{\sim}{i} + \sin(\pm\alpha)\underset{\sim}{j}] = 0,$$

$$(5.8) \qquad \underset{\sim}{n}(\pm\alpha)\cdot[-\sin(\pm\alpha)\underset{\sim}{i} + \cos(\pm\alpha)\underset{\sim}{j}] = 0.$$

Equations (2.11), (5.6), and (5.8) imply that

$$(5.9a) \qquad H(\pm\alpha) = 0,$$

or equivalently (by (2.18) and (2.22b)),

$$(5.9b) \qquad \eta(\pm\alpha) = 0.$$

Let us first study the critical points of the system (5.3)-(5.5). Condition (2.18) implies that such points occur where the right sides vanish. Note that the vanishing of the right sides of (5.3) and (5.4) imply that of the right side of (5.5). This means that the singular points lie along curves in the space of (ν,η,μ). This suggests in a rough way that our third order system may have solutions behaving like those of a second order system.

Now our experience with the straight rod of Section 2 and the plate of Section 3 correctly leads us to the belief that our present problem can be readily posed in the abstract form of (3.22).

The linearization of our problem about a uniformly compressed trivial state is straightforward and leads to equations not much worse than those of the straight rod. Thus we can apply the Global Bifurcation Theorem to our problem. To show that the connected sets of solution pairs that bifurcate from the trivial branch preserve some qualitative features of the eigenfunctions of the linearized problems, we must first find a dependent variable with the property that it can change the number of its zeros only at the trivial branch. We assert that this function is either η or H. To see this, we first note that (5.9) implies that η or H can change the number of its zeros only by having at double zero, say at σ: $\eta(\sigma) = 0$, $\eta'(\sigma) = 0$ or $H(\sigma) = 0$, $H'(\sigma) = 0$. The vanishing of $\eta(\sigma)$ or $H(\sigma)$ forces the right sides of (5.3) and (5.5) to vanish, while the vanishing of $H'(\sigma)$ causes the left, and therefore the right, side of (5.4) to vanish. Thus, if η has a double zero, then (ν,η,μ) is a critical point of (5.3)-(5.5). The requirement that (5.6) hold causes the constant value of μ, at this critical point to be zero. Thus this critical point corresponds to a trivial solution of our boundary value problem.

The treatment of other boundary conditions, e.g., those that would result from the replacement of (5.6) with $M(\pm\alpha) = 0$, is far more delicate. One indication of this difficulty is that (5.9) does not hold. The treatment of such conditions devolves on a careful analysis of the vector field defined by (5.3)-(5.5) in (ν,η,μ)-space. The full analysis of this problem is carried out in [10]. The problem of the buckling of a full ring affords certain simplifications. For these problems it can be shown that until the deformation becomes very large (in a way that can be made precise) every buckled state has at least two axes of symmetry and preserves the nodal properties of η,μ and the curvature inherited from the linear problem. Cf. [1].

6. The Buckling of a Spherical Shell

The global buckling problems for the plate and the arch each involve difficulties not present for the straight rod: The equations for the plate are singular at the origin and the equations for an arch are essentially of higher order than those of the straight rod. A problem that has both of these difficulties is the axisymmetric buckling of a spherical shell under hydrostatic pressure (whose equations we do not bother to exhibit). The tools that have been forged to treat the problems of the plate and arch can, with some adjustments, collectively handle this problem for a spherical shell (which need not be complete). The conclusion is of the same character as those for the other structures. From each simple eigenvalue of the problem linearized about the trivial spherical state there bifurcates a connected set of solution pairs. Within each such set η preserves the nodal character it inherits from the eigenfunctions of the linear problem, provided η remains within a large neighborhood of 0 that depends upon the constitutive functions. The analysis leading to this conclusion will appear in [11].

At this stage, it is worthwhile to emphasize the critical role of a geometrically exact formulation of the equations of equilibrium of the shell (along the lines of Section 2). Such a formulation exposes the basic mathematical structure of the physical problem. This mathematical structure is manifested in (2.18)-(2.22), (4.8)-(4.11) and in the quasilinear (divergence) form of the governing equations. It exposes the shear strain η or the shear force H as the variable that cannot change its nodal behavior away from the trivial branch. It would be very difficult (if it is indeed possible) to identify a variable that plays a comparable role in any of the numerous engineering theories of shells. (These theories are based upon the replacement of exact geometric relations, such as (2.10), with some ad hoc approximations and upon linear constitutive relations.)

7. Other Problems

In 1955, Kolodner [23] analyzed the steady states of a heavy, inextensible string attached to a spinning support at its upper end and free at its lower end. Using purely classical methods, especially the shooting method and the Sturmian theory, he was able to determine a complete qualitative description of all bifurcating solutions. This beautiful work had great influence on the subsequent development of bifurcation theory. In 1975, Stuart [27] studied the related problem of the whirling of a heavy, inextensible string whose ends are attached to a vertical axis at two points a fixed distance apart. The tension at the lower end is specified but the length of the string is left free. To handle this more difficult problem, Stuart employed the Global Bifurcation Theorem of Rabinowitz (giving this theorem its first application outside of [19]). The generalization of these problems to nonlinearly elastic strings was carried out in [6] by combining Sturmian theory with the Global Bifurcation Theorem (and in one instance with Krasnosel'skii's theory of genus). Now the bifurcation diagram for Kolodner's problem for inextensible strings is like Figure 3.1. The nature of the corresponding diagram for elastic strings is especially sensitive to whether the string is strong or weak, i.e., to whether the string can or cannot resist large centrifugal forces. On the other hand, the bifurcation diagram for Stuart's problem for elastic strings (which has a character different from Figure 3.1) is insensitive to the strength of the string. The reason for this dichotomy is suggested by the study of a complementary problem in which the upper end of a string of fixed length is attached to a fixed point on a spinning vertical axis, while the lower end, to which is attached a fixed weight, is free to slide up and down the axis. This problem is formally equivalent to Stuart's, but its bifurcation diagram, like that corresponding to Kolodner's problem for elastic strings, is very sensitive to the strength of the string. To discover what is happening, let u define a deformed configuration of the string. Let $S(u)$ denote the natural norm of u for Stuart's problem and let $C(u)$ denote the natural norm of u for the complementary problem. Then it turns out that

$$C(u)^2 = \frac{S(u)^2}{1+S(u)^2} . \tag{7.1}$$

Thus $C(u) \to 1$ as $S(u) \to \infty$. To each solution u of Stuart's problem there corresponds a solution u of the complementary problem, but not vice versa. The pronounced changes in the bifurcation diagram for the complementary problem due to the weakness of the string are only manifested after $C(u)$ becomes large. But there are no solutions of Stuart's problem here; the solutions of Stuart's problem have already attained infinite norms when $C(u)$ reached 1.

An obvious class of generalizations of these problems is obtained by replacing the strings by rods, which resist bending. The local existence of bifurcating branches for the elastica was shown in [24] by a variety of methods and the global existence for the general rod theory of Section 2 was shown in [14] by the Global Bifurcation Theorem. Neither of these papers was able to obtain a global qualitative theory for these problems. There is a problem for a whirling rod that readily admits such a global analysis (albeit with its characteristic peculiarities). In this problem a rod of unit length is welded to a rigid ring of radius R so that in the reference configuration the rod lies along the radius with the material points near the welded end lying inside the ring. The ring is then spun about its axis. The resulting centrifugal force may cause the rod to buckle either in the plane of the ring or out of this plane. If the buckled state lies in the plane determined by the reference state of the rod and the axis of rotation, then a full qualitative picture of the deformation can be found. One novel aspect of this problem is that the determination of the trivial state is not trivial: If $R > 1/2$ or, a fortiori, if $R > 1$, then we expect the centrifugal force to be compressive and to cause the rod to shorten. But we would also expect that one could stretch the rod so that its length exceeds the diameter of the ring and that certain such stretched states could also be maintained by centrifugal force. Related effects occur if $R < 1/2$. Thus there are serious questions of existence, multiplicity, and qualitative behavior concerning the trivial state, which is governed by the boundary value problem

$$(7.2) \qquad N' + \rho\omega^2[1+\hat{\nu}(N,0,0)] = 0, \; N(0) = 0, \; N'(1) = -\rho\omega^2 R.$$

Here ρ is the mass density per unit reference length and ω is the constant angular velocity of the ring. For each $(\rho\omega^2, R) \in (0,\infty) \times (0,\infty)$ we wish to determine the number and nature of solutions of (7.2). What makes (7.2) remarkable is that it is not difficult to apply virtually all the standard tools of nonlinear differential equations and nonlinear analysis and that each different tool gives distinctive and useful information about solutions of (8.2). This analysis is carried out in [14].

We conclude this section with a brief account of some related work. (i) A study [13] of the dynamical equations for the deformation of elastic rods in space shows that these problems have a very rich collection of travelling wave solutions and that travelling shocks must be of a very restricted form. The qualitative nature of solutions is determined by studying the projection of trajectories in a twelve-dimensional phase-space onto a certain two-dimensional plane. (ii) If a rod theory has enough geometric structure to characterize a change of thickness, then the application of terminal tensile forces can produce bifurcations in which there is a nonuniform change of thickness (necking) and/or shear. Cf. [9]. (iii) Elementary analytical means show that an elastic string subjected to a distributed vertical load of one sign admits a (stable, downward) solution for all values of parameters and admits (unstable, upward) solutions for certain values of parameters. The latter come in pairs. Similar results hold for the suspension bridge problem and a much richer collection of results holds for strings under hydrostatic pressure. Cf. [5, 20]. These multiplicity results are related to branching of solutions, a concept associated with bifurcation. (iv) The generalization of these results to geometrically exact theories of membranes involves technical problems more severe than those encountered in Sections 4 and 6 because the absence of bending stiffness magnifies the difficulties associated with the singularity at the origin. Leray-Schauder degree theory is used to obtain results comparable to those for strings (cf. [7]). (A beautifully detailed analysis of the corresponding problem for an engineering model of a membrane was carried out in [17].)

Acknowledgement

The preparation of this paper was supported in part by the National Science Foundation Grant MCS 77-03760 and by a J.S. Guggenheim Memorial Foundation Fellowship.

8. References

[1] S.S. Antman, Monotonicity and Invertibility Conditions in One-dimensional Nonlinear Elasticity, in Nonlinear Elasticity (ed. R.W. Dickey), Academic Press, New York, 1973, 57-92.

[2] S.S. Antman, Ordinary Differential Equations of Nonlinear Elasticity I: Foundations of the Theories of Nonlinearly Elastic Rods and Shells, Arch. Rational Mech. Anal., 61(1976), 307-351.

[3] S.S. Antman, Buckled States of Nonlinearly Elastic Plates, Arch. Rational Mech. Anal. 67(1978), 111-149.

[4] S.S. Antman, The Eversion of Thick Spherical Shells, Arch. Rational Mech. Analysis, 1979, to appear.

[5] S.S. Antman, Multiple Equilibrium States of Nonlinearly Elastic Strings, SIAM J. Appl. Math., to appear.

[6] S.S. Antman, Nonlinear Eigenvalue Problems for the Whirling of Heavy Elastic Strings, Proc. Roy. Soc. Edin., Scr. A, to appear.

[7] S.S. Antman, Multiple Equilibrium States of Nonlinearly Elastic Membranes, in preparation.

[8] S.S. Antman and H. Brezis, The Existence of Orientation-Preserving Deformations in Nonlinear Elasticity, in Nonlinear Analysis and Mechanics, Vol. II, ed. R.J. Knops, Pitman Research Notes in Mathematics, London, 1978, 1-29.

[9] S.S. Antman and E.R. Carbone, Shear and Necking Instabilities in Nonlinear Elasticity, J. Elasticity 7(1977), 125-151.

[10] S.S. Antman and J.E. Dunn, Qualitative Behavior of Buckled Nonlinearly Elastic Arches, J. Elasticity, to appear.

[11] S.S. Antman and J.E. Dunn, Large Buckled States of Spherical Shells, in preparation.

[12] S.S. Antman and C. Kenney, Greenhill's Problem for Nonlinearly Elastic Rods, in preparation.

[13] S.S. Antman and T.-P. Liu, Travelling Waves in Hyperelastic Rods, Quart. Appl. Math. 36(1979), 377-399.

[14] S.S. Antman and A. Nachman, Large Buckled States of Rotating Rods, J. Nonlin. Anal., to appear.

[15] S.S. Antman and G. Rosenfeld, Global Behavior of Buckled States of Nonlinearly Elastic Rods, SIAM Rev. 20(1978), 513-566. Corrections, to appear.

[16] J.M. Ball, Convexity Conditions and Existence Theorems in Nonlinear Elasticity, Arch. Rational Mech. Anal. 63(1977), 337-403.

[17] A.J. Callegari, E.L. Reiss and H.B. Keller, Membrane Buckling: A Study of Solution Multiplicity, Comm. Pure Appl. Math. 24(1971), 499-527.

[18] R. Courant and D. Hilbert, Methods of Mathematical Physics, vol. I, Interscience, New York, 1953.

[19] M.G. Crandall and P.H. Rabinowitz, Nonlinear Sturm-Liouville Eigenvalue Problems and Topological Degree, J. Math. Mech. 19 (1970), 1083-1102.

[20] R.W. Dickey, The Nonlinear String under a Vertical Force, SIAM J. Appl. Math. 17(1969), 172-178.

[21] L. Euler, Additamentum I de Curvis Elasticis, Methodus Inveniendi Lineas Curvas Maximi Minimivi Proprietate Gaudentes, Lausanne, 1744. Opera Omnia I, vol. 24, Füssli, 1960, 231-297.

[22] P. Hartman, Ordinary Differential Equations, Wiley, New York, 1964.

[23] I. Kolodner, Heavy Rotating String--A Nonlinear Eigenvalue Problem, Comm. Pure Appl. Math. 8(1955), 395-408.

[24] F. Odeh and I. Tadjbakhsh, A Nonlinear Eigenvalue Problem for Rotating Rods, Arch. Rational Mech. Anal. 20(1965), 81-94.

[25] P.H. Rabinowitz, Some Aspects of Nonlinear Eigenvalue Problems, Rocky Mountain J. Math. 3(1973), 161-202.

[26] E.L. Reiss, Column Buckling-An Elementary Example of Bifurcation in Bifurcation Theory and Nonlinear Eigenvalue Problems, ed. by J.B. Keller and S.S. Antman, Benjamin, New York, 1969.

[27] C.A. Stuart, Spectral Theory of Rotating Chains, Proc. Roy. Soc. Edinburgh, 73A(1975), 199-214.

THE BRANCHING OF SOLUTIONS OF EINSTEIN'S EQUATIONS*

Vincent Moncrief
Department of Physics
Yale University
New Haven, Connecticut 06520

INTRODUCTION

In this paper we shall discuss some recent results on the structure of the space of solutions to the vacuum Einstein equations in the context of spacetimes with compact Cauchy hypersurfaces. These results (obtained jointly with J. E. Marsden and A. E. Fischer[1]) characterize the solution space on a neighborhood of any solution admitting a Killing vector field as (homeomorphic to) a manifold × cone. Roughly speaking the manifold directions correspond to those nearby spacetimes which have the "same symmetry" as the original solution while the cone directions represent deformations towards solutions of lower symmetry. This result can be extended to the case of several Killing vector fields and leads to a structure of intersecting manifolds × cones. Conical singularities arise only at the symmetrical solutions and are closely related to the linearization instabilities of the Einstein equations which also arise in conjunction with Killing symmetries (see Fischer and Marsden[2,3], Moncrief[4,5] and Arms and Marsden[6]). The details of the conical structure for the several Killing field case have not yet been fully worked out so we shall discuss primarily the single Killing field case.

Our main results may be stated informally as follows. The space of solutions may be regarded as a subset of the contangent bundle $T^*\mathcal{M}$ of the space $\mathcal{M}$ of Riemannian metrics over a compact 3-manifold M ($\mathcal{M}$ is an open cone in the space of symmetric 2-tensors over M). This solution set is defined by the four (elliptic) initial value

*Research supported in part by NSF Grant PHY76-82353.

constraint equations of the Einstein theory. The solutions of these initial value constraints determine (though quite redundantly) the globally hyperbolic solutions of Einstein's equations on the manifold $M \times R$ (one can freeze out the redundancy in description by imposing suitable coordinate conditions). Within this context one finds that:

(1) The solution set is a manifold near any Cauchy data set for a vacuum spacetime with no Killing symmetries; the Einstein equations are linearization <u>stable</u> with respect to perturbations of any such non-symmetrical solution.

(2) Near any solution admitting a 1-dimensional isometry group and having a hypersurface of constant mean extrinsic curvature, the solution set is homeomorphic to a manifold $\times$ cone.

(3) Near any solution with a k-dimensional isometry group and having a hyper-surface of constant mean extrinsic curvature the solution space may be characterized as the intersection of k distinct (manifolds $\times$ cones). (An invariant characterization of the resulting structure has not yet been worked out.)

(4) The vacuum Einstein equations are linearization <u>unstable</u> with respect to perturbations about a symmetrical spacetime. The linearized equations must be supplemented by a system of k second order conditions whenever the background admits a k-dimensional isometry group.

The linearization instability problem for the vacuum Einstein equations was discovered by Brill[7] (see also Brill and Deser[8]). It was elaborated by Fischer, Marsden, Arms and Moncrief (see Refs. 1-6) who derived the main results cited above. The Einstein equations are said to be linearization stable with respect to a particular (background) solution provided every solution to the linearized equations is tangent to a curve of exact solutions. The linearization stability of non-symmetric solutions (1) has a simple geometric interpretation - the solution set is a manifold near any such particular solution and the manifold's tangent space is precisely defined by the solutions of the linearized equations.

Through any solution admitting a 1-dimensional isometry group there is a manifold of distinct solutions of the same symmetry type (i.e., a manifold of solutions with conjugate isometry groups). Off each point of this manifold branches a cone of solutions of lower (i.e., no) symmetry. The solution set is homeomorphic to a manifold × cone and the cone directions (the directions of symmetry breaking) are defined by the zeros of a certain explicitly given quadratic form.

For solutions with higher symmetry the situation is complicated by the different possibilities of (i) perturbing a solution so as to preserve its full symmetry, or (ii) breaking the solution's symmetry to any allowed subgroup. The case of breaking a k-dimensional symmetry down to a (k-1)-dimensional symmetry is very similar to the case of breaking a 1-dimensional symmetry down to the non-symmetric case. The solutions admitting a (k-1)-dimensional symmetry are homeomorphic to a (manifold × cone) where the manifold represents the solutions having full k-dimensional symmetry and the cones represent the branching to solutions of lower ((k-1)-dimensional) symmetry.

In (2) and (3) above we have included the assumption that the background spacetime have a hypersurface of constant mean extrinsic curvature. It can be shown that every sufficiently nearby spacetime will also admit such a hypersurface (see Choquet-Bruhat[9]). The existence of such surfaces in <u>arbitrary</u> (maximally extended, globally hyperbolic) vacuum spacetimes is still however an open question. The hypothesis of a constant mean curvature surface is not needed in (1) and (4).

Our discussion is limited to the case of pure gravity as described by the vacuum Einstein equations. It is clear however that the same methods may be applied to the study of other gauge fields (i.e., Yang-Mills fields) and to gauge fields coupled to gravity. Some aspects of the linearization stability problem for pure gauge fields on a flat spacetime were discussed by Moncrief[10]. An extensive treatment of the fully coupled Yang-Mills-Einstein system (including the extensions of (1) and (4) to this case) has been given by Arms[11].

The case of perfect fluids coupled to gravity has heretofore seemed quite different from that of pure gravity or of gauge fields coupled to gravity. D'Eath[12] showed that one could always solve the constraint equations algebraicly for certain of the fluid variables. Thus one does not encounter linearization instabilities even for highly symmetrical spacetimes such as the Robertson-Walker models considered in detail by D'Eath. A similar conclusion is implicit in the Hamiltonian treatment of perfect fluid dynamics given by the author[13]. However the Hamiltonian formalism suggests a significant further reduction which, it seems, would lead one back to linearization instabilities of precisely the Killing type (i.e., instabilities associated to Killing symmetries of the background solution).

The idea is that self-gravitating perfect fluids admit a large, explicitly-known symmetry group. The conserved quantities associated to this symmetry group (i.e., the Hamiltonian generators of the group) were found by Taub[14]. They are reexpressed in terms of the canonical variables in Ref. (13). Within this setting one can attempt a further <u>reduction</u> of the Hamiltonian system following the method of Marsden and Weinstein[15]. It is straightforward to show in specific examples that this reduction process (dropping down to a subset on which the conserved quantities are constant and moding out by the associated symmetry group) suffers an "instability" at particular solutions which admit Killing symmetries. Such examples lead us to conjecture that the reduced phase spaces defined by the Marsden-Weinstein reduction procedure have conical singularities precisely at those points which represent symmetrical solutions of the Einstein-Euler equations

II. SPACE OF SOLUTIONS

We first define some notation and recall some standard results from the literature. We then explain the reductions necessary to consider the solution set near a particular solution with a k-dimensional isometry group. We then show how the Liapunov-Schmidt procedure from bifurcation theory combined with a generalized Morse lemma due to Bott[16] (see also Tromba[17] and Buchner, Marsden and Schecter[18]) suffice to characterize the solution set completely in the case of 1-dimensional isometries. We conclude with a discussion of the problem of k-dimensional isometries.

A. Background and Standard Results

Let M be a fixed compact 3-manifold and let $\mathcal{M}$ denote the space of C^∞ Riemannian metrics on M (technical arguments require the use of Sobolev manifolds of metrics from with the C^∞ results may be recovered by a regularity argument). We may regard $T^*\mathcal{M}$ as the space of (unconstrained) Cauchy data on M. Each point of $T^*\mathcal{M}$ is a pair (g,π) consisting of a Riemannian metric g and a contravariant symmetric tensor density π.

The constraint subset $\mathcal{C}$ of $T^*\mathcal{M}$ is defined by

$$\mathcal{C} = \Phi^{-1}(0) \tag{1}$$

where

$$\Phi : T^*\mathcal{M} \to (\Lambda^0_d \times \Lambda^1_d) \tag{2}$$

$$= (\text{space of scalar densities on M})$$

$$\times\ (\text{space of one-form densities on M})$$

is given by

$$\Phi(g,\pi) = (\mathcal{H}(g,\pi),\ \mathcal{J}(g,\pi)) \tag{3}$$

with

$$\mathcal{H}(g,\pi) = \frac{1}{\mu_g}(\pi^{ij}\pi_{ij} - \tfrac{1}{2}(\mathrm{tr}\pi)^2) - \mu_g\ R$$

$$J(g,\pi) = -2\,\delta\pi. \quad (4)$$

In these formulas $\mathrm{tr}\pi = g_{ij}\pi^{ij}$, μ_g is the volume element, R is the scalar of curvature and δ is the divergence operator of g.

Any solution of $\Phi(g,\pi) = 0$ determines a vacuum spacetime $({}^{(4)}V, {}^{(4)}g)$, the maximal Cauchy development of the initial data (g,π).

A main result connecting linearization stability with the absence of Killing symmetries may be stated as:

Theorem (1): If $\Phi(g_o, \pi_o) = 0$ and $({}^{(4)}V, {}^{(4)}g_o)$ is a Cauchy development of (g_o, π_o) then the following are all equivalent

(i) ${}^{(4)}g_o$ has no global Killing vector fields,

(ii) the Einstein equations are linearization stable at ${}^{(4)}g_o$,

(iii) the solution set $\Phi^{-1}_{(0)}$ is a submanifold of T^*M near (g_o,π_o) with tangent space ker $D\Phi(g_o, \pi_o)$,

(iv) the L^2 adjoint $D\Phi(g_o, \pi_o)^*$ of $D\Phi(g_o, \pi_o)$ has trivial kernel.

Sketch of Proof: That (iv) implies (iii) follows from an application of the implicit function theorem and uses the fact that $D\Phi(g_o,\pi_o)^*$ is elliptic (see Fischer and Marsden[2,3]). A corrolary of this argument shows that the tangent space to the constraint submanifold at (g_o, π_o) coincides with the solutions of the linearized constraints (i.e., with ker $D\Phi(g_o, \pi_o)$). Standard results on the hyperbolic evolution equations (see, e.g., Ref. 3) show that linearization stability of the initial value equations extends to linearization stability of the evolution equations (at least on any compact set containing the initial surface). Thus (iii) implies (ii). That (ii) implies (i) follows from the fact that the existence of a Killing vector field of ${}^{(4)}g_o$ necessitates the imposition of nontrivial second order conditions upon the first order perturbations.

These second order conditions (see below) were derived in special cases by Brill and Deser[8] and in general by Fischer and Marsden[3]. They were related to Taub's conserved quantities (see Taub[19]) and thus shown to be hypersurface and gauge invariant by Moncrief[5]. In Brill and Deser's explicit examples the conditions were manifestly non-trivial. The general proof of non-triviality was given by Arms and Marsden[6].

That (i) is equivalent to (iv) was shown in Ref. (4). The argument entailed showing, by a direct computation, that any Killing field ${}^{(4)}X$ of a vacuum spacetime $({}^{(4)}V, {}^{(4)}g_0)$ induces on every Cauchy surface an element (N, X) in the kernel of $D\Phi(g_0, \pi_0)^*$, the adjoint operator associated with that surface. This element consists of the normal (N) and tangential (X) projections of ${}^{(4)}X$ at the hypersurface,

$$D\Phi(g_0, \pi_0)^* \cdot (N, X) = 0 \qquad (5)$$

Reference (4) also showed by employing certain projections of Killing's equations as evolution equations that every element in $\ker D\Phi(g_0, \pi_0)^*$ could be evolved to yield a Killing field ${}^{(4)}X$ on a Cauchy development of (g_0, π_0). This argument has subsequently been improved (through removing special coordinate conditions used in Ref. (4)) by Coll[20] and by Fischer, Marsden and Moncrief[1]. Thus the set of Killing fields of a vacuum spacetime is isomorphic to the kernel of the adjoint operator associated to any Cauchy surface of that spacetime. This last result does not depend on the compactness of M.

To show why Killing fields are associated with the breakdown of linearization stability (in the compact case) we shall sketch the derivation of the second order conditions. Let $\Phi(g_0, \pi_0) = 0$ and suppose that $(g(\lambda), \pi(\lambda))$, with $(g(0), \pi(0)) = (g_0, \pi_0)$, is a curve of exact solutions of the constraints, $\Phi(g(\lambda), \pi(\lambda)) = 0$. Differentiate this equation twice with respect to λ and set $\lambda = 0$ to obtain

$$D\Phi(g_o, \pi_o)\cdot(h', \omega') + D^2\Phi(g_o, \pi_o)((h, \omega), (h,\omega)) \tag{6}$$
$$= 0$$

where

$$(h,\omega) = \left(\frac{dg(\lambda)}{d\lambda}, \frac{d\pi(\lambda)}{d\lambda}\right)\Bigg|_{\lambda = 0} \tag{7}$$

$$(h', \omega') = \left(\frac{d^2g(\lambda)}{d\lambda^2} \quad \frac{d^2\pi(\lambda)}{d\lambda^2}\right)\Bigg|_{\lambda = 0}$$

Now contract equation (6) with an element (N, X) in ker $D\Phi(g_o, \pi_o)^*$ and integrate over the hypersurface. After an integration by parts the term in (h', ω') drops out since ker $D\Phi(g_o, \pi_o)^*\cdot(N, X) = 0$ by assumption. This leaves

$$\int_M (N, X)\cdot D^2\Phi(g_o, \pi_o)\cdot((h,\omega), (h,\omega)) = 0 \tag{8}$$

as a second order condition on the first order perturbation (h, ω). Since the elements of ker $D\Phi(g_o, \pi_o)^*$ are isomorphic to the Killing fields of $({}^{(4)}V, {}^{(4)}g_o)$, there are k such conditions when ${}^{(4)}g_o$ admits a k-dimensional isometry group.

Compactness of M was crucial in deriving Eq. (8). Had M been non-compact a surface integral involving the second order perturbations (h', ω') would have survived the integration by parts. In the asymptotically flat case such surface integrals measure the second order change in (asymptotically defined) conserved quantities induced by the first order perturbations (h, ω). The surface integral measures the second order correction to the energy, momentum or angular momentum according to whether ${}^{(4)}X$ (the Killing field inducing (N, X)) is timelike, spacelike translational or rotational respectively.

The second order conditions (8) may be reexpressed in terms of the four dimensional perturbation ${}^{(4)}h$ of ${}^{(4)}g_o$ induced by its Cauchy data (h,ω) as follows. Let Σ be any Cauchy surface in the spacetime and let ${}^{(4)}Z$ be the (future pointing) unit normal field to Σ and $d^3\Sigma$ be the Riemannian volume element induced in Σ. Then

$$\int_{\Sigma} {}^{(4)}X \cdot D^2\text{Ein}({}^{(4)}g_o)\cdot({}^{(4)}h,\ {}^{(4)}h)\cdot{}^{(4)}Z\ d^3\ \Sigma \qquad (9)$$

$$= 0$$

(where $\text{Ein}({}^{(4)}g)$ is the Einstein tensor of ${}^{(4)}g$ and $D^2\text{Ein}$ its second (Frechet) derivative) is equivalent to (8). In this notation ${}^{(4)}h$ satisfies the linearized equations

$$D\ \text{Ein}({}^{(4)}g_o)\cdot\ {}^{(4)}h = 0 \qquad (10)$$

and induces (h,ω) as its Cauchy data on the initial hypersurface.

As discussed in Ref. (5) the integrals (9) are hypersurface and gauge invariant and thus independent of the choice of Cauchy surface Σ.

B. Bifurcation Analysis

Let us suppose that $\Phi(g_o,\ \pi_o) = 0$, $(\text{tr}\ \pi_o/\mu_{g_o}) = k = \text{const.}$ and that $\ker D\Phi(g_o,\ \pi_o)^*$ is one-dimensional (i.e. that $(g_o,\ \pi_o)$ is (constant mean extrinsic curvature) Cauchy data for a vacuum spacetime with a one-dimensional isometry group). One can show (see Ref. (1)) that $\ker D\Phi(g_o,\ \pi_o)^*$ must be spanned by either

(a) $(N,X) = (1,0)$ if $\pi_o = 0$ and g_o is flat

or

(b) $(N,X) = (0,X)$ for some X such that $L_X g_o = L_X\pi_o = 0$

where L_X is the Lie derivative with respect to X. For simplicity we shall concentrate on the spacelike case (b) though the timelike case can be handled in a similar way.

The group $\mathcal{D}^3$ of (smooth) diffeomorphisms of M acts on $T^*\mathcal{M}$ by "coordinate transformation". Thus if $\phi\in\mathcal{D}^3$ the action gives $(\phi,\ (g,\pi)) \to (\phi^*g,\ \phi^*\pi)$. This action preserves the constraint equations since the Φ map is three-dimensionally covariant,

$$\Phi(\phi^*g,\ \phi^*\pi) = \phi^*\Phi(g,\pi) \tag{11}$$

and preserves the condition $(\mathrm{tr}\ \pi_o/\mu_{g_o}) = k = \mathrm{const}$. Cauchy data on the same orbit as $(g_o,\ \pi_o)$ generates a spacetime diffeomorphic to that determined by $(g_o,\ \pi_o)$.

We propose to "freeze out" this three-dimensional gauge freedom of the constraint equations by passing to a slice for the $\mathcal{D}^3$ action on T^*M. Within the slice we shall simultaneously impose the timelike coordinate condition

$$(\mathrm{tr}\pi/\mu_g) = (\mathrm{tr}\pi_o/\mu_{g_o}) = k = \mathrm{const} \tag{12}$$

and the projected constraints

$$(I - P)\Phi(g,\pi) = 0 \tag{13}$$

where $P\Phi(g,\pi) = 0$ (the complementary projection) is equivalent to

$$\int_M (0,X)\Phi(g,\pi) = 0. \tag{14}$$

The point of passing to the slice and of imposing the timelike coordinate condition is that it removes certain directions of degeneracy which would otherwise interfere with an application of the (generalized) Morse lemma. The point of splitting the constraints is that we can prove that the solutions of the projected constraints

$$(I - P)\Phi(g,\pi) = 0$$

define a manifold (within the slice). Thus the problem is reduced to solving

$$f(g,\pi) \equiv \int_M (0,X)\cdot\Phi(g,\pi) = 0 \tag{15}$$

within the submanifold defined by the coordinate conditions and the complementary projection of the constraints. This technique of splitting the constraints is known as the Liapunov-Schmidt procedure. The final step, showing that the solutions of $f(g,\pi) = 0$ define a manifold $\times$ cone, is handled by the Morse lemma. We shall briefly discuss these steps in more detail.

1. Construction of the Slice

One can construct a slice for the $\mathcal{D}^3$ action on T^*M by following the methods of Ebin[21] and Palais[22] (who constructed slices for the $\mathcal{D}^3$ action on M). Fix (g_o, π_o) and define a weak Riemannian metric on T^*M by

$$\left\langle (h,\omega), (h,\omega) \right\rangle_{(g,\pi)} = \int_M \mu_{g_o} \; [(h\cdot h) + (\omega'\cdot\omega')]$$

where (h,ω) is contained in $T_{(g,\pi)}T^*M$, ω' is the "tensor part" of ω ($\omega = \mu_{g_o}\omega'$), and where $(\cdot)$ signifies constraction using g_o. Note that (g,π) is variable but that the contraction and integration are made using g_o. By construction this metric is invariant under the isotropy group of g_o.

The orbit $\mathcal{O}_{(g_o,\pi_o)}$ through (g_o, π_o) has, at (g_o, π_o) the tangent space

$$T_{(g_o,\pi_o)}\mathcal{O}_{(g_o,\pi_o)} = \left\{ (h,\omega) = (L_Y g_o, L_Y \pi_o) \;\middle|\; Y \text{ is a vector field on } M \right\} \tag{17}$$

One can use elliptic theory to split $T_{(g_o,\pi_o)}T^*M$ into $T_{(g_o,\pi_o)}\mathcal{O}_{(g_o,\pi_o)}$ and a $\langle\;,\;\rangle$-orthogonal complement.

One can exponentiate this complementary subspace using the (flat) metric $\langle\,,\rangle$ to obtain an "affine" submanifold of T^*M passing through (g_o, π_o) orthogonally to $T_{(g_o,\pi_o)}$. It is possible to show that a sufficiently small neighborhood of (g_o, π_o) within this affine manifold is a slice for the $\mathcal{D}^3$ action. One uses a Sobolev norm invariant under the isotropy group of g_o to define such a neighborhood. The details of this argument are given in Ref. (1).

2. Solving the Projected Constraints

Let $S_{(g_o,\pi_o)}$ be the slice through (g_o, π_o) discussed above. Within this slice define the function

$$\Gamma : S_{(g_o,\pi_o)} \rightarrow \left((\Lambda^o_d \times \Lambda^1_d)_X \times \Lambda^o \right) ,$$

$$\Gamma(g,\pi) = \left((I-P)\,\Phi(g,\pi),\; (tr\pi/\mu_g) - k \right) \tag{18}$$

where Λ^k (Λ^k_d) represents the k-forms (k-form densities) over M and where $(\Lambda^o_d \times \Lambda^1_d)_X$ is the L^2 orthogonal projection of $(\Lambda^o_d \times \Lambda^1_d)$ perpendicular to $(0,X) \in \ker D\Phi(g_o, \pi_o)^*$.

One can now show that Γ has surjective derivative at (g_o, π_o) and thus, from the implicit function theorem, that the solutions of $\Gamma(g,\pi) = 0$ define a submanifold of $S_{(g_o, \pi_o)}$ near (g_o, π_o). The Liapunov-Schmidt procedure has thus reduced the problem of solving $\Phi(g,\pi) = 0$ to that of solving

$$f(g,\pi) \equiv \int_M (0,X)\,\Phi(g,\pi) = 0 \tag{19}$$

We note that f does <u>not</u> have surjective derivative at (g_o, π_o) since

$$\begin{aligned} Df(g_o, \pi_o)\cdot(h,\omega) &= \int_M \left[(0,X)\cdot D\Phi(g_o, \pi_o)\cdot(h,\omega)\right] \\ &= \int_M \left\langle (h,\omega), D\Phi(g_o, \pi_o)^*\cdot(0,X)\right\rangle \equiv 0 . \end{aligned} \tag{20}$$

Indeed there is a whole manifold of critical points of f through (g_o,π_o). This manifold is given by

$$B_X \equiv A_X \cap S(g_o, \pi_o) \tag{21}$$

where A_X is the "affine" submanifold of T^*M defined by

$$A_X = \{(g,\pi) \in T^*M \mid L_X g = L_X \pi = 0\} \tag{22}$$

Our "affine" construction of the slice $S(g_o, \pi_o)$ ensures that the intersection of A_X with $S_{(g_o, \pi_o)}$ is in fact a manifold.

One can show either by another application of the implicit function theorem or by a transversality argument that $\Gamma^{-1}(0)$ intersects B_X in a manifold near (g_o, π_o).

From the definitions of Φ, f and B_X it follows that

$$f\Big|_{B_X} = \int_M \pi\cdot L_X g\Big|_{B_X} \equiv 0. \tag{23}$$

Thus $B_X \cap \Gamma^{-1}(0)$ is a manifold of solutions of $\Phi(g,\pi) = 0$ (within the slice) which satisfy $L_X g = L_X \pi = 0$ and $(\mathrm{tr}\pi/\mu_g) = k$. The elliptic character of $D\Phi(g,\pi)^*$ ensures the locally decreasing character of ker $D\Phi(g,\pi)^*$ and thus that $(0,X)$ spans the kernel of $D\Phi(g,\pi)^*$ for all (g,π) contained in $\mathcal{B}_X \cap \Gamma^{-1}(0)$ sufficiently near (g_o, π_o).

To summarize the above we have found a manifold

$$\mathcal{N} \equiv \mathcal{B}_X \cap \Gamma^{-1}(0) \tag{24}$$

of solutions of the constraints within the slice which each have $(0,X)$ spanning the kernel of their associated adjoint maps. One may regard $\mathcal{N}$ as the manifold of nearby solutions with the "same symmetry" as (g_o, π_o). $\mathcal{N}$ is a manifold of critical points (and zeros) of the remaining constraint function $f(g,\pi)$.

3. Applying the Morse Lemma

We must finally impose $f(g,\pi) = 0$ within the manifold $\Gamma^{-1}(0)$ and determine the resulting structure. We cannot use the standard Morse lemma since (g_o, π_o) is a degenerate critical point of f. Indeed we have already found an entire manifold $\mathcal{N}$ of critical points of f passing through (g_o, π_o). However we can apply the generalized Morse lemma (Bott[16], Tromba[17], Buchner, Marsden and Schecter[18]) provided that the Hessian $d^2f(g_o, \pi_o)$ has its degeneracy space coinciding with $T_{(g_o, \pi_o)}\mathcal{N}$, the tangent space of the degenerate critical manifold. (see Ref. (1) for a more precise statement including the technical conditions needed in the infinite dimensional case). In that case $d^2f(g_o, \pi_o)$ will be non-degenerate on a complement to $T_{(g_o, \pi_o)}\mathcal{N}$ and the (generalized) Morse lemma asserts that the set of zeros of f near (g_o, π_o) will be homeomorphic to a (manifold) $\times$ (cone). The manifold here is $\mathcal{N}$ and the cone directions are defined, at each point of $\mathcal{N}$, by the zeros of $d^2f(g,\pi)\cdot((h,\omega), (h,\omega))$ (with (h,ω) restricted to a complement of $T_{(g,\pi)}\mathcal{N}$).

In the above remarks f really means $f|_{\Gamma^{-1}(0)}$ and $d^2f(g,\pi)$ refers to the Hessian of this restriction. However, since each point of $\mathcal{N}$ is a critical point of f we can always compute the Hessian in the ambient space $T^*\mathcal{M}$ and then restrict its application to vectors (h,ω) contained in $T_{(g_o, \pi_o)}\Gamma^{-1}(0)$.

The proof that the degeneracy space of the quadratic form $d^2f(g_o, \pi_o)$ coincides with the tangent space to N at (g_o, π_o) is rather lengthy and is given in full in Ref. (1). The idea is to show that the degeneracy space of $d^2f(g_o, \pi_o)$ consists of those sets of Cauchy data (h,ω) for the linearized equations which preserve (to first order) the condition that $(0,X) \in \ker D\Phi(g,\pi)^*$. The full argument is essentially a linearized version of that which, in the non-linear case, characterizes the data with $\ker D\Phi(g,\pi)^* \neq \phi$ as the symmetrical solutions of the Einstein equations.

The picture that emerges from this argument is that, within the slice $S_{(g_o, \pi_o)}$, the solution set of $\Phi(g,\pi) = 0$, $(tr\pi/\mu_g) = k$ is homeomorphic to the product of a manifold N of solutions with the "same symmetry" (i.e., conjugate isometry group) as (g_o, π_o) and a cone defined by the zeros of the quadratic form $d^2f(g_o, \pi_o)$. There is a close relationship between the Hessian $d^2f(g_o, \pi_o)((h,\omega), (h,\omega))$ and Taub's conserved quantity which we discussed earlier. The former is simply a restriction of the latter to tangent vectors (h,ω) which satisfy the linearized form of the coordinate conditions implicit in $S_{(g_o, \pi_o)}$ and $tr\pi/\mu_g = k$.

Finally, one can remove the coordinate conditions and allow the conical structure to flow out of $S_{(g_o, \pi_o)}$ into a full neighborhood of T^*M. This is accomplished by letting the group $\mathcal{D}^3$ act on $S_{(g_o, \pi_o)}$ and by evolving the solutions of $\Phi(g,\pi) = 0$ with the Einstein evolution equations. The latter step removes the condition $(tr\pi/\mu_g) = k = \text{const}$. Both steps are discussed in detail in Ref. (1).

C. Extensions and Generalizations

One can easily extend the above argument to the (timelike) case which has $\pi_o = 0$ and g_o flat (where $\ker D\Phi(g_o, 0)^*$ is spanned by $(N,X) = (1,0)$). In this case one must allow the constant k in the time coordinate condition $(tr\pi/\mu_g) = k$ to "float" with the perturbation. The background solution has $k = 0$ but nearby solutions (in particular those of lower symmetry) may have only surfaces with $k = \text{const.} \neq 0$. (The existence of at least the latter is assured by the argument of Choquet-Bruhat and Marsden[9].) In the timelike case the symmetrical solutions are stationary and hence flat. The manifold $\mathcal{N}$ of nearby stationary solutions is always finite dimensional. By contrast $\mathcal{N}$ may be infinite dimensional in the spacelike case even if more than one Killing field occurs.

The case of several Killing fields may be treated in much the same way as that discussed above. For definiteness consider the case of n spacelike Killing fields. On a hyper-surface with $(\mathrm{tr}\pi/\mu_g) = k = \text{const.}$ we find that $\ker D\Phi(g_o, \pi_o)^*$ is spanned by $\{(0,X_a)\}$, $a = 1 \ldots n$, where $\{X_a\}$ are vector fields on M. The conditions

$$D\Phi(g_o, \pi_o)^* \cdot (0,X_a) = 0 \tag{25}$$

are equivalent to

$$L_{X_a} g_o = L_{X_a} \pi_o = 0 \tag{26}$$

One can construct a slice $S_{(g_o, \pi_o)}$ for the $\mathcal{D}^3$ action on $T^*\mathcal{M}$ as before and solve, within the slice the simultaneous equations

$$\begin{aligned} \Gamma(g,\pi) &= \left((I-P)\Phi(g,\pi),\ (\mathrm{tr}\pi/\mu_g) - k \right) \\ &= (0,0) \end{aligned} \tag{27}$$

where $P\Phi(g,\pi) = 0$ (the complementary projection) is equivalent to the n constraints

$$f_a(g,\pi) = \int_M (0,X_a) \cdot \Phi(g,\pi) = 0 \tag{28}$$

Again the implicit function theorem may be used to show that $\Gamma^{-1}(0)$ is a submanifold of the slice. One can now define the spaces

$$A_{(X_a)} = \{ (g,\pi) \in T^*\mathcal{M} \mid L_{X_a} g = L_{X_a} \pi = 0) \} \tag{29}$$

and

$$\mathcal{B}_{\{X_a\}} = A_{\{X_a\}} \cap S_{(g_o, \pi_o)} \tag{30}$$

and show that

$$N \equiv \mathcal{B}_{\{X_a\}} \cap \Gamma^{-1}(0) \tag{31}$$

is a manifold. It is straightforward to show that

$$f_a\Big|_N = 0 \tag{32}$$

so that N is in fact a manifold of solutions of $\Phi = 0$.

A direct (but awkward) way to study the structure of the set of solutions of $f_a = 0$ (within $\Gamma^{-1}(0)$) is to impose these conditions one at a time and then to intersect the resulting family of (manifolds x cones). The shortcoming of this approach is that one needs to do a considerable amount of additional work (using, e.g., transversality arguments) to characterize the nature of the resulting intersection. In addition this method would not be obviously invariant under a change of basis of the $\{X_a\}$. One seems to need a further generalization of the Morse lemma to treat the case of $\mathbb{R}^k$ - valued functions (on infinite dimensional manifolds) which have degenerate critical manifolds. One effort towards such a generalization has been made by Buchner, Marsden and Schecter[18].

There are a variety of other problems in mechanics and classical field theory wherein bifurcations of the type considered here might arise. If one reduces a Hamiltonian system with symmetries, following the methods of Marsden and Weinstein[15], then singularities may be expected to occur whenever some subgroup of the symmetry group has fixed points. Such fixed points are the analogues of the Cauchy data for spacetimes with non-trivial isometry groups. The singularities would appear in the level sets of the "moment map" for the symmetry group action (i.e., in the level sets of the Hamiltonian generating function for the symmetry group). In general such singularities need not be "conical" but could instead be of higher order. The conical singularities however seem to be common in practice for equations arising in physics.

As a simple example from mechanics we consider N point particles moving in $\mathbb{R}^3$. The phase space for this system is $T^*\mathbb{R}^{3N} \approx \mathbb{R}^{6N}$ and the moment map for (say) the usual SO(3) action on this space is simply the total angular momentum vector $\vec{J}$. In reducing this system (for any Hamiltonian invariant under SO(3)) one considers the level sets of $\vec{J}$. It is straightforward to show that the level set $\vec{J}^{-1}(0)$ has conical singularities precisely at those points which are fixed under a subgroup of the SO(3) action. Such fixed points correspond to particle configurations in which all the position and momentum vectors are co-alligned.

Another, more complicated, example of reduction is that of perfect fluids coupled to gravity which we discussed in the introduction. The moment map for the symmetry group of this system consists of the particle number density and the vorticity field of the fluid (see Ref. (13) for details). The (product) symmetry group corresponding to these generators consists of deformations of the initial hypersurvace along the flow lines of the fluid and of diffeomorphisms within the initial hypersurface. It appears (from the study of numerous examples) that the fixed points of k-dimensional subgroups of this symmetry group correspond precisely to the solutions of the Einstein-Euler equations admitting k-dimensional isometry groups. The singularities in the level sets of this moment map would then represent the branching of solutions of a fixed particle number density and vorticity distribution which occurs at any particular, symmetrical solution.

REFERENCES

1. A. Fischer, J. Marsden and V. Moncrief, "The Structure of the Space of Solutions of Einstein's Equations. I. One Killing Field", unpublished.
2. A. Fischer and J. Marsden, Bull. Am. Math. Soc. 79, 997 (1973).
3. A. Fischer and J. Marsden, Proc. Symp. Pure Math. 27, 219 (1975).
4. V. Moncrief, J. Math. Phys. 16, 493 (1975).
5. V. Moncrief, J. Math. Phys. 17, 1893 (1976).
6. J. Arms and J. Marsden, Ind. Math. J. 28, 119 (1979).
7. D. Brill, "Isolated Solutions in General Relativity", University of Maryland Technical Report No. 71-076 (1971).
8. D. Brill and S. Deser, Commun. Math. Phys. 32, 291 (1973).
9. Y. Choquet-Bruhat, C. R. Acac. Sci. Paris 280, 169 (1975).
10. V. Moncrief, Ann. Phys. 108, 387 (1977).
11. J. Arms, J. Math. Phys. 20, 443 (1979).
12. P. D'Eath, Ann. Phys. 98, 237 (1976).
13. V. Moncrief, Phys. Rev. D. 16, 1702 (1977).
14. A. Taub, Commun. Math. Phys. 15, 235 (1969).
15. J. Marsden and A. Weinstein, Rep. Math. Phys. 5, 121 (1974).
16. R. Bott, Ann. of Math. 60, 248 (1954).
17. A. Tromba, Canad. J. Math. 28, 640 (1976).
18. M. Buchner, J. Marsden and S. Schecter, "Differential topology and Singularity Theory in the Solution of Nonlinear Equations", (to appear).
19. A. Taub, contribution to "Relativistic Fluid Dynamics", edited by C. Cattaneo (Lectures at the Centro Internazionale Matematico Estivo, Bressanone, 1970), Edizioni Cremonese, Rome (1971). See also : A. Taub, J. Math. Phys. 2, 787 (1961).
20. B. Coll, J. Math. Phys. 18, 1918 (1977).
21. D. Ebin, Symm. Pure Math., Amer. Math. Soc. 15, 11 (1970).
22. R. Palais (unpublished) has constructed an affine slice for the

action of $\mathfrak{D}^3$ on $\mathcal{M}$. An analogous slice for the $\mathfrak{D}^3$ slice on T*M was constructed in Ref. (1).

WHAT DOES SUPERGRAVITY TEACH US ABOUT GRAVITY?

S. Deser
Brandeis University
Waltham, Massachusetts 02254

I must warn you that relative to those in the audience whom I know, I am an experimental physicist, and so when I speak of "supersymmetry" it may not look like what was discussed in the previous lecture. My main thesis in this talk will be the usefulness of supergravity[1] to ordinary classical general relativity rather than the theory for its own intrinsic interest. With this in mind I will discuss three topics: (1) The positive energy problem in GR (a problem of longstanding interest), (2) Properties of graviton-graviton scattering, or more properly the self-interactions of the gravitational field, and (3) perhaps a few remarks about quantum gravity.

It turns out that the first two problems are surprisingly soluble as a consequence of the fact that GR possesses a "square root" structure,

$$\text{Supergravity} = \sqrt{\text{GR}}$$

as first shown by Teitelboim, et al.[2] This is not a metaphor but a rigorous statement in a sense to be clarified; because GR possesses a square root in the Dirac sense we learn a great deal about it quite apart from whether we are interested in supergravity or not. However, I would like to incite those people who are interested in the mathematical aspects of GR and who have developed such powerful methods in the last years to look at the similar problems in supergravity. There is a close relationship; still, there is a great deal about supergravity we don't yet know and it seems to me an obvious field of endeavor either for the people concerned or else for their graduate students. Not all the theorems necessarily carry over but it will be interesting to see which ones do and which ones don't.

We start then by reviewing (from the experimental physicist's point of view) supersymmetry, i.e., the grading of the Poincare algebra which is the basis of ordinary GR. As you know, GR can be considered as the gauging of the Poincare algebra, i.e., turning the global properties of that algebra into local ones; in

+Supported in part by NSF Grant PHY-78-09644

other words building curved space out of tangent spaces which satisfy the Poincare algebra. So supergravity is in exactly the same sense the gauging of the graded Poincare algebra. Ordinary Poincare algebra is defined for systems based in flat space, that is it has realizations in terms of dynamical fields, local fields in ordinary Minkowski space. It is defined in terms of ten generators: The momenta P^μ, i.e., the translation generators, and the angular momenta $J_{\alpha\beta}$, the rotation generators in Minkowski four-space. These generators obey the commutation relations (CR)

$$[P_\mu, P_\nu] = 0$$

$$[P_\mu, J_{\alpha\beta}] = i(\eta_{\mu\alpha} P_\beta - \eta_{\mu\beta} P_\alpha)$$

$$[J_{\alpha\beta}, J_{\gamma\delta}] = i(\eta_{\alpha\gamma} J_{\beta\delta} - \eta_{\beta\gamma} J_{\alpha\delta} + \eta_{\beta\delta} J_{\alpha\gamma} - \eta_{\alpha\delta} J_{\beta\gamma})$$

(note: the indefinite Minkowski metric $\eta_{\mu\nu} = \delta_{\mu\nu} - 2\delta_{\mu o}\delta_{\nu o}$). In order to understand what a square root or grading of this algebra might be, we would like to "take the square root" of P^μ (the analogous operation on $J_{\alpha\beta}$ is less physical). Thus we invent a Fermion operator $Q^{(\alpha)}$, (α) being a spinorial index which must therefore have anticommutation rather than commutation relations with spinorial objects, in particular with itself; the square root of P^μ means then that some bilinear combination of Q's is proportional to P^μ; we choose

$$\{Q^{(\alpha)}, \bar{Q}_{(\beta)}\} = -2\gamma^{\mu(\alpha)}{}_{(\beta)} P_\mu$$

where the usual 4X4 Dirac matrices γ^μ in Majorana representation are employed and we use real spinors so that

$$\bar{Q} = \tilde{Q}\gamma^o$$

The above commutator is the fundamental one from which our major consequences flow, in particular the positive-energy theorem. To establish the grading however we must close the algebra with the CR

$$[Q^{(\alpha)}, P_\mu] = 0$$

$$[Q^{(\alpha)}, J_{\mu\nu}] = \frac{i}{2}(\sigma_{\mu\nu} Q)^{(\alpha)}$$

The first relation simply says that Q^α is conserved (or it wouldn't be interesting), the second says that it transforms as a spinor under Lorentz rotations generated by $J_{\mu\nu}$.

This last CR is important for obtaining graviton-graviton scattering results. This is a global algebra - we have not yet gauged the theory. Before seeking representations of this algebra we examine the consequences of the fundamental commutator. If you can take the square root of something it implies that that something is essentially positive. In the present case we seek the energy P^0 which comes by projecting out P^0 in the commutator. From the fundamental property of the Dirac matrices

$$\{\gamma^\mu, \gamma^\nu\} = -2\eta^{\mu\nu}$$

we have

$$\mathrm{Tr}\ \gamma^0\gamma^\mu = 4\eta^{0\mu}$$

whence

$$P^0 = \tfrac{1}{4}Q^2 \geqslant 0$$

where Q^2 is a real operator. Thus an immediate consequence is that any system which realizes the algebra automatically has positive energy, however complicated its interactions, however nontransparent the Hamiltonian of the system may be. The crux of the supergravity proof of the energy theorem will be that I will be able to deduce $P^0 > 0$ as a consequence of the existence of the Fermionic generators $Q^{(\alpha)}$. This "no hands" result is in marked contrast with earlier approaches to the problem on the positivity of the gravitational energy, which were by no means devoid of calculation. A second result (shortly to appear in Phys. Rev.[3]) is that there are no tachyonic solutions in the theory, i.e., any system satisfying the graded algebra is such that the Casimir operator $(P^\mu)^2$ satisfies

$$(P^\mu)^2 = -M^2 \leq 0$$

To see this, suppose we have a tachyon (i.e., any solution whose translation generators are spacelike); then there is a Lorentz frame for which $P^0=0$; but if this is true, then Q vanishes also, whence $\gamma^\mu P_\mu = 0$ or $\vec{\gamma}\cdot\vec{P}=0$, i.e., $P^\mu=0$, a form-invariant statement. This is interesting, because York[4] has pointed out that in the famous proof by Schoen and Yau[5] of the gravitational positive-energy conjecture, it was necessary to assume the energy-momentum of gravity is timelike (the Cauchy data were such that it was timelike or null) and the procedure as stated couldn't handle spacelike energy-momentum.

We now return to global supersymmetry. The simplest examples of a representation of this global algebra are a massless spin 1 or spin 0 field together with their fermionic partners: in the general case it has been shown[6,7] that the realizations of this system are provided by adjoining fields whose spins differ by a half integer. The situation is different according as the mass does or does not vanish. I will be interested here in the m=0 case because I am talking about gravity. For the massless case you take a system of spin s and add a system of spin s+1/2, and this supermultiplet provides a theory which realizes the graded Poincare algebra. In particular the most familiar example is spin 1/2 plus spin 1- a massless neutrino along with the Maxwell field. The relevant noninteracting Lagrangian is

$$\mathcal{L} = -\tfrac{i}{2}\bar{\lambda}\not{\partial}\lambda - \tfrac{1}{4}F_{\mu\nu}^{\,2}$$

The action corresponding to this Lagrangian is invariant under a certain set of supersymmetry transformations generated by supercharges $Q^{(\alpha)}$ defined as follows:

$$Q^{(\alpha)} = \int j^{(\alpha)0}\,d^3x$$

since every generator is built from a conserved current which in this case is given by

$$j^{(\alpha)\mu} = 2F_{\alpha\beta}(\sigma^{\alpha\beta}\gamma^{\mu}\lambda)^{(\alpha)}$$

Now this unprepossessing quantity is in fact conserved as a consequence of the field equations of the theory. This hardly seems obvious from the way it is written (although it is more obvious in two-component spinor notation). The basic relations needed are that in

$$\sigma^{\alpha\beta}\gamma^{\mu} = \tfrac{1}{2}\{\sigma^{\alpha\beta},\gamma^{\mu}\} + \tfrac{1}{2}[\sigma^{\alpha\beta},\gamma^{\mu}],$$

$$\{\sigma^{\alpha\beta},\gamma^{\mu}\} \sim \epsilon^{\alpha\beta\mu\rho}\gamma_5\gamma_\rho \qquad \text{and} \qquad [\sigma^{\alpha\beta},\gamma^{\mu}] \sim \eta^{\mu\alpha}\gamma^{\beta} - \eta^{\mu\beta}\gamma^{\alpha}$$

These result in the term $\partial_\mu F_{\alpha\beta}$ being multiplied by quantities which reduce it to a combination of the charge-free Maxwell's equations, to ensure that $\partial_\mu j^\mu = 0$. Furthermore, by directly calculating the anticommutator, one finds

$$\iint \{j^0(x), j^0(x')\}\,d^3x\,d^3x' = 2\gamma_\mu \int T^{0\mu}\,d^3x$$

upon using the canonical commutation relations for the fields, where $T^{\mu\nu}$ has the form

$$T^{\mu\nu} \sim \bar{\lambda}\gamma^{\mu}\partial^{\nu}\lambda + (F^{\mu\rho}F^{\nu}{}_{\rho} + {}^{*}F^{\mu\rho}\,{}^{*}F^{\nu}{}_{\rho})$$

This example is a toy theory. The important point is that the kinematics assures us that the addition of arbitrarily complicated interaction structures does not alter these results.

Q. If you have a spin s plus a spin s+1/2 field, do you say that you can write down a Lagrangian of the system?

A. If there are to be any realizations at all within representations of the Poincare group, then for m=0 these consist of adjoining spin pairs. The example just given is the existence theorem for the 1/2+1 pair and also for 1/2+1 pairs with appropriate interactions; there are also existence theorems for 0+1/2, 1+3/2, and 3/2+2. Beyond this, e.g., 2+5/2, things are not so clear, not because the kinematics don't allow it but because there is no satisfactory theory for coupling systems of spin greater than 2, in particular, their couplings to gravity seem to be restricted.[8] Thus, there is an upper limit at least for non-trivial, interacting systems. Of course it is the 3/2+2 case which is of interest to us and there supergravity is the consistent realization, (aside from quantum correction problems).

Now, in what context am I going to do supergravity? I cannot go through either the motivation or how one would derive the theory, or show that in fact it accomplishes the realization. I will simply be forced to write down SG and assert that it is to the graded Poincare algebra exactly what the ordinary Einstein theory is to the ordinary Poincare algebra, that is to say, this algebra is satisfied on tangent "flat spaces." SG is then <u>the</u> realization corresponding to 3/2+2 with zero mass. The action of supergravity looks very simple:

$$I=I_E+I_{RS}$$

where the Einstein action is

$$I_E=1/2\kappa\int {}^*R^*(\omega)^{\mu\nu ab}\, e_{\mu a}\, e_{\nu b}\, d^4x \ ,$$

${}^*R^{*\mu\nu ab}$ being the double dual

$${}^*R^{*\mu\nu ab} = \frac{1}{4}\, \epsilon^{\mu\nu\rho\sigma} R_{\rho\sigma cd}\, \epsilon^{abcd}$$

of the vierbein curvature

$$R_{\mu\nu ab} = \partial_\mu \omega_{\nu ab} - \partial_\nu \omega_{\mu ab} + \cdots$$

It is <u>not</u> assumed to be torsion-free; $e_{\mu a}$ are the vierbein components. The Rarita-

Schwinger action which governs the vector-spinor field $\psi_\mu^{(\alpha)}$ is

$$I_{RS} = -\frac{i}{2}\int \bar{\psi}_\mu \gamma_5 \gamma^a e_{\nu a} {}^*f^{\mu\nu}(\psi;\omega)\, d^4x$$

where

$$f_{\mu\nu} = D^{1/2}_\mu \psi_\nu - D^{1/2}_\nu \psi_\mu \ , \quad {}^*f^{\mu\nu} \equiv \frac{1}{2}\epsilon^{\mu\nu\alpha\beta} f_{\alpha\beta}$$

and

$$D^{1/2}_\mu \psi_\nu \equiv \partial_\mu \psi_\nu - \frac{1}{2}\omega_{\mu ab}\sigma^{ab}\psi_\nu$$

This is the minimal coupling to gravity of spin 3/2. The claim is that this is precisely the needed realization.

Having disposed of what supergravity is, let me now review for the nonexperts what the energy problem is in ordinary classical Riemannian Einstein theory; then we will come back and see how the apparently far more complicated problem of showing that the energy of the above system is positive turns out to be trivial as a consequence of the supersymmetry algebra. Finally, we will show what trick is required to shave off the "super" part of SG and thus reduce the proof to ordinary classical gravity.

First then we review the energy problem in ordinary gravity (in particular, Einstein theory without sources; the result holds also for theories with any well-behaved sources). Historically the road to positive energy in the theory is easy to see. The linearized version of the Einstein theory, in which one expands the theory around Minkowski space and retains only the leading non-trivial terms, has positive energy as first shown systematically by Pauli and Fierz.[9] Then Araki[10] showed that in fact the energy is positive for any weak excitation near flat space - not quite the same thing as linearized theory, for technical reasons. For particular examples positivity had been known over the years; for example there is a particularly beautiful paper by Brill[11] on axially symmetric solutions. All this has to do with the initial value problem only, because the energy is of course defined entirely on the initial Cauchy surface. Arnowitt, Deser, and Misner[12] investigated various cases and Brill and Deser[13] gave some general arguments which are certainly mathematically wrong but which also show that at least metaphorically the energy is positive. Y. Choquet and Marsden[14] then gave a series of results on positive energy quite recently and there is now a super-fancy proof (which I cannot claim to follow fully) by Schoen and Yau[5] which establishes it for the full classical theory modulo the tachyon absence which is taken care of here, anyway. Rather let me explain why we're concerned with energy, which is the same reason we are concerned with energy for any physical system; the stability of any system requires that the Hamiltonian be bounded from below. The reason you are worried about Einstein theory as you are not worried about other systems like Maxwell or Yang-Mills theory is twofold: for Maxwell and Yang-Mills you can show that the energy is manifestly nonnegative, which helps, i.e., it is $\int(E^2+B^2)$ so that however complicated E and B are it is their squares that generate (time) translations, and second, in GR not only do you not have an explicit expression for the energy, but in principle in order to establish the properties of the energy in a brute-force way you would have to solve the four initial value equations of the theory and use those solutions in a detailed way by plugging them into the "energy density" and so establish positivity. That way has never and probably will never be completed.

What then does the energy mean? It is obviously a property which is meaningful only within some sort of Minkowski tangent space context; that is to say, for GR it is a global property which can only apply if the system (i.e., a Riemannian space) which has whatever curvature you might like in the interior also has an exterior, i.e., it is asymptotically flat in the sense that you can get sufficiently far from a given region for the metric to approach the Minkowski metric at an appropriate speed. There is a lot of fundamental technical work on the appropriate speed which is absolutely crucial to the whole energy concept. I can't go into it here;[12] suffice it to say that it is necessary that there exist an asymptotic set of coordinates such that

$$g_{\mu\nu} \sim \eta_{\mu\nu} + O(1/r)$$

where r is whatever Cartesian distance you use. If you violate this, i.e., if the system is asymptotically flat but with behavior slower than this then the concept of energy is not well defined, nor should it be well defined. It is also meaningless, for example, in the case of deSitter space, since in deSitter space P^2 is not a Casimir operator and thus the notion of mass-energy is not very interesting. Energy is thus strictly for asymptotically flat systems; but those are of great importance, certainly for quantum gravity since it is only in the case of asymptotic flatness that we would understand its quantization; but it is also important in classical theory. In any case a dynamical system is satisfactory only if for this class of solutions the energy has the right properties. Once asymptotic flatness is established there are a number of more or less equivalent definitions of energy (the differences have to do with some fine print, of interest only to experts, about rates of decay). We examine two: we can define P^0 through

$$g^T \sim M/r$$

where g^T is a particular combination of metric components and the M is the traditional notation for what actually is P^0. One of the beauties of the theory is that you need only assume the $O(1/r)$ asymptotic behavior and the constraint equations immediately convert it into a strict $1/r$ dependence. In this expression M- or P^0 - is the energy and is defined in this highly implicit way, which is the precise analogue of the electromagnetic statement that the asymptotic behavior of the longitudinal part of the electric field,

$$E^L \sim Q/r^2$$

is necessarily $\sim 1/r^2$ and the coefficient of the $1/r^2$ term is the total charge. In both cases we have an asymptotic way of obtaining the <u>value</u> of the charge or mass; but these rules don't tell you how to calculate the sign of P^0, since it is the sum of all the interior contributions. An alternative way to define the energy is to say that it is the integral on a two-sphere at asymptotically flat spatial infinity of the gradient of g^T,

$$P^0 = -\oint d\vec{S} \cdot \vec{\nabla} g^T$$

just as in electromagnetism

$$Q = \oint d\vec{S} \cdot \vec{\nabla} E^L$$

The two definitions of energy are equivalent, as I have said, modulo the fine print about decay rates which need not concern us here. The energy is thus a flux integral.

The important feature of a gauge theory is that the total "charge" (here the energy) is on the one hand to be counted by taking a volume integral, on the other hand it can be obtained in any particular case if you have the right probe at infinity, e.g., a little test mass to tell you how much you are attracted to the system. As in electrodynamics P^o (or the total charge Q) could be either positive or negative as far as the definition is concerned. This sign must be obtained by knowing everything there is to know about the interior. The reason the problem of its sign cannot be attacked frontally is that g^T is in fact defined in terms of the four initial value constraints

$$G^{o\mu} = 0$$

of the Einstein theory. Loosely put, one has

$$G^{oo}(\text{linear}) = \nabla^2 g^T$$

and thus

$$G^{oo}(\text{linear}) = -G^{oo}(\text{nonlinear})$$

where G^{oo} (nonlinear) is some horrible function of all the g's including g^T itself. The solution of this differential equation is then an infinite series in the potential g^T. Thus one must seek ways around this difficulty.

So much for the energy problem in ordinary GR. How are we to approach it, and more particularly, what can SG possibly do to help us, since if anything it's going to make life worse; the constraint equations for SG read

$$G^{o\mu} = T^{o\mu}(\psi_\nu)$$

after all. The answer is that if you now consider a system which not only has asymptotic Killing vectors appropriate to Minkowski space but also has asymptotic "super"-Killing vectors, i.e., a set of solutions of the coupled SG equations which emerge from variation of the above SG action and for which

$$g_{\mu\nu} \sim \eta_{\mu\nu} + O(1/r)$$

and

$$\psi_\mu \sim O(1/r^2)$$

then of course the energy can be defined because it doesn't care what the sources are so long as the potentials (fields) fall off sufficiently rapidly. The energy is always the same thing, now to be calculated according to

$$\nabla^2 g^T = -G^{00}(\text{nonlinear}) + T^{00}(\psi_\mu)$$

For SG there now exists a further initial value constraint, which I will write as

$$R^0(\psi_\mu) = 0$$

for the spin 3/2 fields. This requires a moments' explanation. Every gauge theory (i.e., m=0) has too many components compared to the number of real fields present and thus there will necessarily be some constraints on them. In electrodynamics these are just

$$\vec{\nabla}\cdot\vec{E} = 4\pi\rho$$

and for our spin 3/2 field there will be constraints as well, which corresponds to the zeroth component of the Rarita-Schwinger equation, which is

$$R^\mu = \epsilon^{\mu\nu\alpha\beta}\gamma_\nu \partial_\alpha \psi_\beta = 0$$

in flat space. Because of the alternating symbol the $R^0=0$ component of this equation is not a time development equation. We write its full coupled form as

$$R^0(\text{linear}) \sim \sigma^{ij}\partial_i\psi_j = -R^0(\text{nonlinear})$$

which can be cast into a form appearing as the "square root" of the Gauss' law as follows

$$\sigma^{ij}\partial_i\psi_j = (\vec{\gamma}\cdot\vec{\nabla})(\vec{\gamma}\cdot\vec{\psi}^T) = -R^0(n.l.)$$

where $\vec{\psi}^T$ is the spatially transverse part of ψ_μ. Now just as for gravity the total energy can be written as the flux integral over certain components of the solutions of the constraint equations at spatial infinity, so the total supercharge $Q^{(\alpha)}$ of the system can also be written as the flux integral at spatial infinity of this quantity:

$$Q^{(\alpha)} = 2\oint dS_i\, \sigma^{ij}\psi_j = 2\oint d\vec{S}\cdot\vec{\gamma}\,(\vec{\gamma}\cdot\vec{\Psi}^T)$$

Thus, if $\sigma^{ij}\psi_j$ falls as $O(1/r^2)$ and I have asymptotic flatness this is of course finite. The beautiful thing about this system is that without any calculations, just from general arguments about the properties of gauge theories, you can show that the $Q^{(\alpha)}$ so defined and the P^o, similarly defined above, as well as the Lorentz generators $J_{\alpha\beta}$, satisfy the global graded Poincare algebra. Thus even though this is a local gauge system so that there is no real significance to the local densities, if I look only at systems tied down at infinity in the sense of particular solutions with asymptotically "super" Poincare Killing vectors, then for those systems there exist well defined, finite spinorial charges $Q^{(\alpha)}$, four-momentum P^μ, and angular momentum $J_{\alpha\beta}$ which from the dynamics of gravitation and the ψ_μ field satisfy the global graded algebra. Therefore I know that $P^o = \frac{1}{4}Q^2 > 0$ as an operator statement in SG, without any explicit calculations. That is our proof.[15]

Q. You obtained a realization by going to GR; is there a way of obtaining a realization "purely algebraically" a la Bargmann-Wigner, just closing your eyes to everything else?

A. What Bargmann and Wigner did for the Poincare group was achieved for this system in.[6] I gave you the Reader's Digest version of their conclusions for m=0 at the beginning. For m≠0 the realizations are obtained by adjoining to a system of spin s two systems of spin s+1/2 and one of spin s+1; but massive systems are not of interest here.

Q. Well, on one side you don't have to appeal to GR at all and on the other side you have a realization from relativity.

A. To explain the connection in detail requires another lecture. To summarize it, you can prove the following: suppose you start from[6]. They tell you to use a Poincare invariant spin 2 Pauli-Fierz field plus a spin 3/2 free field (the linearized versions of the SG action). The question of what are the allowed interactions within this framework, what are the allowed (and required) nonlinearities, has only been answered in recent months.[16] The answer is that the interacting theory we have here is the unique theory we can reach from the noninteracting theory (a) in a finite number of steps and (b) consistently. The procedure is very cute and hinges on the masslessness of the two fields; the one sacred condition on each of them is that they are only allowed helicity 2 and 3/2 components. Thus whatever the sources of these fields may be they must respect the transversality of the fields, i.e., the currents must be conserved. Now: what sort of current can I make which can be the source of ψ_μ and of $g_{\mu\nu}$ made up of these fields themselves? The answer is the Noether current of the global conservation laws of supersymmetry and the global Poincare group I started with. So I start coupling $h_{\mu\nu}=g_{\mu\nu}-\eta_{\mu\nu}$ to the global $T_{\mu\nu}$ of the whole system plus $\psi_\mu^{(\alpha)}$ to $J_\mu^{(\alpha)}$ global (the spin 3/2 analogue of $j^{(\alpha)\mu}$ displayed before for the 1/2+1 case), and it turns out that if you pick the right variables and are clever enough, you iterate this procedure just twice to produce the coupled SG. Furthermore the result is unique. The same holds a fortiriori for GR itself, i.e., you can reach[17] GR purely from Pauli-Fierz plus the statement that you want the theory to be nontrivial, i.e., allow interactions. There is then only one way it can be nonlinear and it is GR. For the massless case everything is in the kinematics by the requirement that at each step you must keep only helicity 2 and 3/2. So that's the connection.

I now know that $Q^{(\alpha)}$, P^μ, and $J^{\alpha\beta}$ satisfy the graded algebra and hence that the energy is a positive operator; but in a sense we have overshot, because SG has to be considered at least a first quantized theory and really strictly speaking a second quantized theory because half integral spins make no sense except in the Clifford algebra context

$$\{\psi^\dagger(\vec{r}),\psi(\vec{r}\,')\}\sim\delta^3(\vec{r}-\vec{r}\,')$$

Given all this, what then can I learn from this about plain old ordinary classical GR, without "super," without spin 3/2, and without quantum corrections?

Here our results[15] on positivity in SG were used to reduce to positivity of classical gravity by Grisaru.[1] To sketch the proof I will lapse into diagram language. We have shown that $P^0=Q^2/4$ as an operator relation so that whatever space of states you have, any matrix element of P^0 is positive,

$$\langle\alpha|P^0|\beta\rangle\geq 0$$

where the states $\langle\alpha|$ and $|\beta\rangle$ contain gravitons and some (even number of) spin 3/2 particles, $|\beta\rangle=|g_{\mu\nu};\psi_\alpha\rangle$. The appropriate diagram is perhaps

and the matrix element is evaluated on a background classical metric and semi-classical spin 3/2 field. In this case "graviton" simply means that the metric defining the state satisfies the Einstein equations; the two things together satisfy the SG equations. Now suppose that of all possible matrix elements (diagrams) of this type I consider the subclass in which there are no external spin 3/2's - the spin 3/2 vacuum. This is consistent since these just constitute the vacuum Einstein solutions and

$$\langle g_{\mu\nu}; 0 | P^0 | g_{\alpha\beta}; 0 \rangle \geqslant 0$$

still. This matrix element is still a quantum object, it just lacks external spin 3/2 particles. Because the theory is quantized, however, if you look with a magnifying glass at the blob it includes <u>all possible</u> closed loops (quantum corrections) e.g.

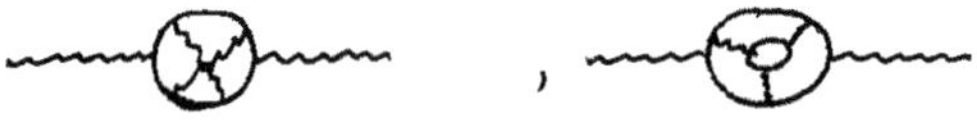

Thus we have eliminated as of no interest to us all external spin 3/2 particles but they are still there inside; they are virtual and cannot be excluded. Luckily however, the theory contains not only Einstein's constant but also Planck's constant $\hbar$, and it's well-known (to physicists) that from loop expansion these closed loops have powers of $\hbar$ as coefficients, the power being appropriate to the number of internal lines. Thus if I let $\hbar$ go to zero I throw out <u>all</u> internal loops, both spin 2 <u>and</u> spin 3/2, and still

$$\lim_{\hbar \to 0} \langle g_{\mu\nu}; 0 | P^0 | g_{\alpha\beta}; 0 \rangle \geqslant 0$$

Now however, these matrix elements are simply a fancy way of expressing the classical energy spectrum, and so P^0 is the Hamiltonian of the classical Einstein theory. This verifies therefore that the energy of classical gravity is positive. The beautiful part of this is that although a true field theorist would be worried about some of the formal manipulations in the SG proof because the theory is not too well-behaved, the limit of classical interest, i.e., $\hbar \to 0$, avoids all that.

Q. But you have introduced an apparatus only to remove it.

A. Exactly. I have used the potentiality of taking the square root to discover that a realization for which that is possible has a positive P^0; then I proceeded to show that the square root possibility existed for GR. The reason that I have to do a <u>little</u> more work is that having produced the square root I have a real quantum theory; so to remove the apparatus consists of getting rid of the external ψ_μ's and then taking $\hbar \to 0$ to kill off the internal quantum corrections.

Q. But how do you know that this quantum theory has a well defined scattering operator?

A. The scattering operator is defined as a perturbative expansion (the appropriate sum of diagrams), the terms of which are terrifying; but each is multiplied by an appropriate power of $\hbar$ and thus on taking $\hbar\to 0$ only the coefficient of $(\hbar)^0$ survives. You might still worry about the order of performing the operations of expansion and $\hbar\to 0$ but even that is probably all right.

Q. No one has ever proved that an appropriate Hilbert space exists and these are operators on some Hilbert space.

A. It is all scaffolding, because the Hilbert space I need is really that of the tree approximations, the truly nonquantum contributions. I only proceed this way because I'm too lazy or insufficiently clever to proceed directly. I conjecture that just the potentiality of taking the square root should suffice to settle the positivity question without the need of detouring through quantum field theory.

Q. It seemed that you indicated that SG required quantum theory; but the theory of graded manifolds allows you to make perfect sense of anti-commuting Fermion fields.

A. There are two levels of quantization. The theory must certainly be first quantized, i.e., the Fermion fields must anticommute (with no delta function) or else nothing makes sense, and no one should be concerned about this. What bothers people is the full second quantized version (with a δ-anticommutator), in which these probably not well defined expressions appear (diagrams) and therefore no Hilbert space strictly speaking. Suppose then I consider an intermediate first quantized theory; is it consistent to have first quantized Fermion fields but with classical $g_{\mu\nu}$ (i.e., they are not Hilbert space operators)? If so I have a well defined Hilbert space as there is nothing horrible happening. In that sense I think the scaffolding is removable.

Q. Your proof seems to exclude bound states.

A. Actually nothing has been excluded. There are no negative energy states here (remember the usual bound states also have positive energy = $2Mc^2$ - binding).

I want now to mention another application of the square root idea, perhaps of less direct mathematical interest but which is still very beautiful and serves as an introduction to my concluding remarks on quantum corrections, a topic of importance to physicists at the moment. Suppose one were interested in graviton-graviton scattering. If we carry out a perturbative expansion of the Einstein action its general character is

$$\int R \sim \int (\partial h)^2 + \int h(\partial h)^2 + \int h^2(\partial h)^2 + \cdots$$

where h is, symbolically, the deviation of the metric from its Minkowski background. Forgetting that this is a Riemann space and regarding it simply as an ordinary flat space theory with a nasty self interaction, the natural question to ask is, what do these nonlinear terms imply for the scattering of gravitons by gravitons? Diagrammatically the terms $\int h(\partial h)^2$ - the cubic vertex-laid end-to-end contribute a structure like

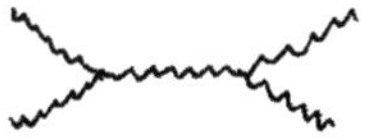

while the quartic piece, $\int h^2(\partial h)^2$ is described by a diagram like

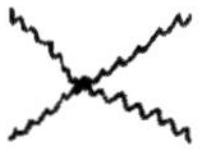

DeWitt pointed out that the quartic diagram represents about 2500 terms and since there are about 10^2 contributions to the cubic vertex, two laid end-to-end also add up to a healthy number. Thus if people were to attempt to calculate this scattering (as they did), first they got it wrong and second they spent far too long on it. The difficulty is that we haven't enough insight into the details to make general remarks about this scattering; but it turns out that SG again saves the day[19] this time using the innocuous CR

$$[Q^{(\alpha)}, J_{\mu\nu}] = \frac{i}{2}(\sigma_{\mu\nu}Q)^{(\alpha)}$$

which imposes restrictions on the helicities of the incoming and outgoing particles. Thus it allows scatterings like $\langle 2,2 | 3/2, 3/2\rangle$ and relates these amplitudes to $\langle 2,2 | 2,2\rangle$ but forbids negative helicity amplitudes such as $\langle 2,2 | -2,-2\rangle$, a result which is extraordinarily difficult to obtain directly. Using such simple relations obtained entirely from the kinematics you can evaluate the sum of all these thousands of terms of each diagram, again by virtue of SG which may not interest you but serves as an intermediate scaffolding. All this emerges directly and beautifully and if nothing else gives you a great deal of insight into the structure of the classical Einstein action.

Let me now conclude with something brand new which is rather off this topic but which I should mention because it's a central problem in SG; it has to do with the infinities of the theory. One of the motivations for SG was to make gravity less horrible in respect of those loop diagrams that diverge in a more and more complicated way. Unfortunately, SG has not quite lived up to its promise; it turns out that although it's better than gravity, it's not quite good enough; beyond a certain level the infinities seem to reappear, although in a rather peculiar form. I'll put the question as follows: Suppose you wanted to construct all possible local invariants corresponding to the local algebra. What does Cartan tell you in GR? He says to take any powers of the curvature and any even number of derivatives to match up the indices and integrate, e.g.,

$$\int R^n \quad , \int R^m D D R^n$$

and so you get the different local invariants. The same question can be raised (and has now been answered) for SG. Of course pieces such as $\int R^n$ are included, but since every spin s has a spin s+1/2 partner, only those Cartan terms are allowed for which there exist appropriate spin 3/2 partner terms such as $\int ffR$ which satisfy the supersymmetry gauging (something I haven't discussed here). The question of the super Cartan program is important because I cannot calculate those horrible infinities; all we know is that they must be proportional to higher and higher super Cartan invariants, which is at least one nice property. One can now seem to get a handle on such super Cartan invariants in a way that has some mathematical interest,[20] I think. To see it consider continuing onto a Euclidean metric ($\eta_{\mu\nu} \to \delta_{\mu\nu}$) and define the self-dual and antiself-dual Weyl tensors as

$$C_{\pm} \equiv C \pm {}^{*}C$$

and similarly for the spin 3/2 analogue,

$$f^{\pm}_{\mu\nu} \equiv f_{\mu\nu} \pm {}^{*}f_{\mu\nu}$$

The following turns out to be true (as can be seen most easily in two-component spinor notation): The super Cartan invariants can be expanded as sums of terms in which appear m powers of C_+, n powers of C_-, and p powers each of f_+ and f_-, i.e.,

$$\int (C_+)^m (C_-)^n (f_+ f_-)^p$$

but no integral invariant exists which contains terms with only self-dual or antiself-dual pieces. We derived it by using helicity properties, i.e., particular properties of the massless spin 3/2 fields. All such (not purely self-dual) terms vanish for all self-dual spaces, e.g., if $C_+ = 0$ and $f_+ = 0$ or $f_- = 0$, and indeed in Euclidean space it suffices that either $C_+ = 0$ or $C_- = 0$ to kill all such terms. Thus although the theory is bad, at least in the configuration of half-flat (i.e., self-dual) Riemannian and super-Riemannian spaces these quantum corrections all seem to vanish like magic (when the SG field equations are also satisfied). This is a fundamental difference from Einstein theory since we know perfectly well that not all Cartan invariants can be written in this form, e.g., R^3 certainly contains C_+ - only or C_- - only terms, so it doesn't necessarily vanish on a half flat background. The difference between the Cartan invariants of ordinary gravity and SG is thus that in SG they all vanish when half-flatness (self-duality or antiself-duality) holds, whereas this is not true of the generic Cartan invariants of differential geometry. Whether this is a hint that the theory is not all that bad (as a quantum thoery) as it might be is of course not clear.

Q. Is the point that in a path integral approach you wouldn't have to integrate over these spaces?

A. Hawking has conjectured that, in ordinary (Euclidean metric) Einstein theory, the only nonsingular locally asymptotically Euclidean solutions to the Einstein equations are in fact, either self-dual or antiself-dual. The statement is that after you have finished summing over the path integrals, which includes all possible spaces (not just solution spaces), then upon evaluating the result at a half flat space all the infinite contributions cancel out. Of course it is also true that you don't have to count many half flat spaces; remember "half flat" is not a property of the field equations. The set of half flat spaces is of measure zero but if Hawking is right in terms of interesting background solution spaces those are the only ones that survive. I have no reason either to believe or to disbelieve Hawking's conjecture (either in GR or as extended to SG), but if it were true it would add a great deal of interest to the result.

Setting aside for the moment my remarks on physics and quantum field theory, the fact is that just because Einstein's theory has this very peculiar (spin 3/2) source, which is a source and nothing else, then unlike coupling to any other form of matter you discover that there is an enormous restriction on the invariants you can write down, even in the purely gravitational sector. Never mind the terms that have f's in them (i.e., look at the p=0 terms). I can write down the no-f invariants and look at the ghosts of their pieces which don't vanish when there's no spin 3/2; these invariants are a small subclass of all possible normal Riemannian space invariants. That's a purely kinematical statement, never mind why I want to use this quantum mechanically.

There are still a great many lessons to be learned from this: even if you're not interested in the square root for its own sake, but only interested in what's under the square root, the simple fact that the relation exists must contain enormous further information.

Q. Let me ask a question about the tachyons and the nature of the sign of the Casimir operator. You said that spacelike momentum was one of the cases Yau couldn't handle. I was a little confused about that because I had the impression that that had to do with the existence of maximal hypersurfaces.

A. This is all very recent. In fact I had the no-tachyon result for some time and was not particularly surprised by it. Then York sent out a preprint pointing out that the existing proof by Yau and Co. implicitly assumed timelike momentum. I gather that it's not an earthshaking difficulty, they are probably going to be able to remove it. It's just that from the present point of view you have excluded tachyons. I also conjecture (but don't know how to prove) that there exist no null solutions either; that is if P^μ is well defined, i.e., the system has bounded energy, then the system's four-momentum is necessarily timelike and not null, never mind spacelike. That simply is based on the fact that null requires that you have a plane wave basically, and that has infinite energy. Thus either you have unbounded plane waves in which case the energy doesn't exist or you necessarily turn timelike due to the interactions. That's something which I think should be provable, it is sort of the obverse of the question about bound states.

Q. Is the self-duality property you spoke of at the end at all related to the difficulties Penrose's program ran into? He handles things something like that.

A. No. Penrose did find he was better equipped to handle half flat. That's a problem of the twistor program. He was able to define sort of twisted

gravitons on a half flat space but not in the generic case. "Half flat" is as the name implies a much easier situation, but it is not true, and I don't mean to imply that, if I look at just the field equations for coupled Einstein and spin 3/2, they somehow predict the Weyl tensor is half flat.
That had better not be true; the Weyl tensor had better be left more or less alone. You could imagine that regularity might do something, I don't know. That's another interesting question; it would take another lecture to discuss the question of whether half flatness "propagates." In Yang-Mills theory something of the sort seems to be happening as far as we can tell, but we really don't understand it. It would be very interesting, maybe fatal to the theory if SG predicted with the right regularity conditions that the only Weyl tensors it allows globally are self-dual or antiself-dual. I shouldn't even speculate on it, but certainly if you just look at the equations they tell you nothing about the Weyl tensor; but regularity might tie you down. In particular you might find that if you look at initial conditions in which asymptotically at infinity you came in with only gravitons of one helicity (corresponding to a linearized half flat Weyl tensor) the only thing that emerged would be a half flat Weyl tensor. That sort of thing might happen, but right now there's no real information. In any case, the vanishing of all higher invariants of half-flat is not the same as removal of all divergence difficulties by any means, even if such solutions had a privileged classical standing, and no such story conclusion should be drawn yet.

Acknowledgement

I am grateful to Lloyd Kannenberg for invaluable help with the preparation of this manuscript.

REFERENCES

1. D. Freedman, P. van Nieuwenhuizen & S. Ferrara, Phys. Rev. D13, 3214 (1976); S. Deser and B. Zumino, Phys. Lett. 62B, 335 (1976).

2. C. Teitelboim, Phys. Rev. Lett. 38, 1106 (1977); Phys. Lett. 69B, 240 (1977); R. Tabensky & C. Teitelboim, ibid 69B, 453 (1977).

3. S. Deser, Phys. Rev. D19, 3165 (1979).

4. J. W. York, Jr. (North Carolina preprint).

5. R. M. Schoen & S. T. Yau, Comm. Math. Phys. 65, 45 (1979).

6. A. Salam and J. Strathdee, Nucl. Phys. B76, 477 (1974).

7. R. Haag, J. Lopuszanski and M. Sohnius, ibid B88, 513 (1975).

8. C. Aragone and S. Deser, Phys. Lett. B (in press).

9. W. Pauli and M. Fierz, Proc. Roy. Soc. A173, 711 (1939).

REFERENCES (Cont.)

10. H. Araki, Ann. Phys. 7, 456 (1959).

11. D. Brill, Ann. Phys. 7, 466 (1959).

12. R. Arnowitt, S. Deser and C. W. Misner, Phys. Rev. 117, 1595 (1960); 118, 1100 (1960); 122, 997 (1961) and in Gravitation: An Introduction to Current Research, L. Witten, ed. (Wiley, New York, 1962).

13. D. Brill and S. Deser, Ann. Phys. 50, 548 (1968); with L. D. Fadeev, Phys. Lett. 26A, 538 (1968); S. Deser, Nuovo Cimento 55B, 393 (1968).

14. Y. Choquet and J. Marsden, C. R. Acad. Sci. 282, 609 (1976).

15. S. Deser and C. Teitelboim, Phys. Rev. Lett. 39, 249 (1977).

16. D. Boulware, S. Deser and J. H. Kay, Physica 96A, 141 (1979).

17. S. Deser, Gen. Rel. & Grav. 1, 9 (1970); D. G. Boulware and S. Deser, Ann. Phys. 89, 193 (1975).

18. M. T. Grisaru, Phys. Lett. 73B, 207 (1978).

19. M. T. Grisaru and H. N. Pendelton, Nucl. Phys. B124, 81 (1977); with P. van Nieuwenhuizen, Phys. Rev. D15, 996 (1977).

20. S. Christensen, S. Deser, M. Duff and M. T. Grisaru, Phys. Lett. B (in press).

CLASSICAL $\frac{1}{2}$ SPIN PARTICLES INTERACTING WITH GRAVITATIONAL FIELDS: A SUPERSYMMETRIC MODEL

Carlos A. P. Galvão (*), *Joseph Henry Laboratories, Princeton University, Princeton, New Jersey 08544*

INTRODUCTION

We present a classical description of the motion of massive $\frac{1}{2}$ - spin particles in a curved spacetime. To describe the spinning particle we start with the quantum Dirac equation and interpret it as a first class constraint on the physical states. We associate a set of anticommuting classical variables with the particle and write the Dirac bracket relations that they satisfy. Upon quantization these relations reproduce the quantum commutation relations satisfied by the Dirac γ-matrices. The algebra of constraints closes giving rise to the constraint associated with Klein-Gordon equation.

It is shown that the constraint associated with Dirac equation is the generator of supersymmetry transformations. With the help of the anticommuting variables we define a spin tensor for the classical particle which is supersymmetric invariant and obeys the Lie algebra of Lorentz group. Finally we obtain the equations of motion for the classical $\frac{1}{2}$ - spin particle, which coincide with the equations of motion for the pole-dipole particle interacting with an external gravitational field obtained by Papapetrou.

In section I we briefly discuss the Dirac equation in a Riemannian spacetime. In section II we obtain the classical constraint equations; we show that the contraint associated with Dirac equation generates supersymmetry transformations and define the spin tensor. In section III we obtain the equations of motion and comment on the results. The appendix contains a summary of notations and conventions used throughout the paper.

The results we are going to present here are part of a research work developed by the present author and Claudio Teitelboim (Institute for Advanced Study, Princeton). A detailed presentation of this work including the analysis for the case of a massless $\frac{1}{2}$ - spin particle will be published elsewhere.

(*) *On leave of absence from the High Energy Physics Laboratory, CBPF, Rio de Janeiro, and Dept. of Physics, UFRN, Natal. This work is supported by CNPq, Brazil.*

I - THE DIRAC EQUATION IN A RIEMANN SPACETIME

We will consider the spacetime $\mathcal{M}$ to be a Riemann manifold with metric tensor g of hyperbolic signature. According to the theory of general relativity spacetime will be locally Minkowskian, that is at each point in $\mathcal{M}$ there exists a local coordinate system in which the metric tensor assumes the form of the constant Minkowski metric $\eta_{AB} = \mathrm{diag}(-1,+1,+1,+1)$.

As it is known it is possible to define a basis $\{L^{\mu}{}_{(A)}(x)\}$, $\mu = 0,1,2,3$, $A = 0,1,2,3$ for the local tangent space such that

$$L^{\mu}{}_{(A)}\, L^{\nu}{}_{(B)}\, g_{\mu\nu} = \eta_{AB} \tag{I-1}$$

$$L_{\mu}{}^{(A)}\, L_{\nu}{}^{(B)}\, \eta_{AB} = g_{\mu\nu} \ . \tag{I-2}$$

It is clear that these vectors, usually called tetrad vectors, are determined up to a local Lorentz rotation.

We shall suppose that a local spinor structure can be defined on $\mathcal{M}$, the 4-components spinors $\psi(x)$ being elements of the vector space associated with the $(0,\frac{1}{2})\oplus(\frac{1}{2},0)$ representation of the local Lorentz group. In this space the constant Dirac matrices $\gamma^{A} = (\gamma^{Aa}{}_{b})$ constitute a representation of the Clifford algebra associated with the Minkowski metric

$$\gamma^{A}\gamma^{B} + \gamma^{B}\gamma^{A} = 2\eta^{AB} \tag{I-3}$$

or

$$\{\gamma^{A},\gamma^{B}\} = \eta^{AB} \ .$$

A representation for the Clifford algebra associated with the spacetime metric $g_{\mu\nu}(x)$ can be obtained from (I-3) with the help of the tetrad vector fields,

$$\{\gamma^{\mu}(x),\gamma^{\nu}(x)\} = g^{\mu\nu}(x) \tag{I-4}$$

with

$$\gamma^{\mu}(x) = L^{\mu}{}_{(A)}(x)\gamma^{A} \tag{I-5}$$

The generalization of the special relativistic Dirac equation

$$\gamma^{\mu}\partial_{\mu}\psi(x) + m\psi(x) = 0 \tag{I-6}$$

to General Relativity is obtained[1] by means of the minimal coupling to the gravitational field in the sense that we make the substitution $\partial_\mu \longrightarrow \nabla_\mu$ where ∇_μ is the operator of covariant derivative for the spinor field $\psi(x)$. It can be shown that

$$\nabla_\mu = \partial_\mu - \Gamma_\mu \tag{I-7}$$

where $\Gamma_\mu(x)$ are the Ricci rotation coefficients (or spin connections) given by

$$\begin{aligned}\Gamma_\mu &= -\frac{1}{8} L^\nu{}_{(A)}\, L_{(B)||\nu}(\gamma^A\gamma^B - \gamma^B\gamma^A) \\ &= -\frac{1}{4}\,\omega_{\mu AB}\,\gamma^A\gamma^B \quad .\end{aligned} \tag{I-8}$$

In the above expression the double bar denotes covariant differentiation with respect to the metric $g_{\mu\nu}(x)$. Thus, the general relativistic Dirac equations will be written as

$$\gamma^\mu(\partial_\mu + \frac{1}{4}\,\omega_{\mu AB}\gamma^A\gamma^B)\psi(x) + m\psi(x) = 0 \quad . \tag{I-9}$$

II - CLASSICAL DESCRIPTION OF $\frac{1}{2}$ - SPIN PARTICLES INTERACTING WITH GRAVITATIONAL FIELDS

Our starting point for obtaining the classical equations of motion of Dirac particles interacting with a given gravitational field is the quantum equation (I-9). In order to do this we introduce a set of anticommuting classical variables[2,3]

$$\theta^A = \theta^A(\tau) \quad , A = 0, 1, 2, 3 \quad ,$$

$$\theta_5 = \theta_5(\tau) \quad ,$$

τ is a parameter, satisfying the Dirac bracket relations

$$\{\theta^A,\theta^B\}^* = i\eta^{AB} \tag{II-1}$$

$$\{\theta_5,\theta_5\}^* = i \tag{II-2}$$

These variables satisfy the anticommutation property $\theta^A\theta^B = -\theta^B\theta^A$ and are usually called "odd variables". Within this context, classical commuting variables are called "even variables". We will require that upon quantization the following relations to hold:

$$\theta^A = i\sqrt{\frac{1}{2}}\gamma_5\gamma^A \tag{II-3}$$

$$\theta_5 = \sqrt{\frac{1}{2}} \tag{II-4}$$

It follows from (II-3,4) and (II-1,2) that the canonical quantization procedure

(DIRAC BRACKETS) ⟶ -i(QUANTUM COMMUTATOR)

leads to the correct commutation relations for the Dirac γ-matrices.

The key of our procedure is to reinterpret equation (I-9) as a first class constraint on the quantum states,

$$\$\psi(x) \approx 0 \tag{II-5}$$

with

$$\$ = \gamma^\mu(\partial_\mu + \frac{1}{4}\omega_{\mu AB}\gamma^A\gamma^B) + m \tag{II-6}$$

The classical analogue of the constraint equation (II-5) which follows from the definitions (II-3,4) is

$$\mathcal{S} = \theta^\mu(p_\mu - \frac{i}{2}\omega_{\mu AB}\theta^A\theta^B) + m\theta_5 \approx 0 \tag{II-7}$$

In order to obtain the algebra of the constraints we use the conventional Poisson bracket relations for the canonical variables x^μ and p_ν to obtain the following Dirac brackets:

$$\{\theta^\mu, \mathcal{P}_\alpha\}^* = -\{^{\mu}_{\alpha\beta}\}\,\theta^\beta \tag{II-8}$$

$$\{\mathcal{P}_\alpha, \mathcal{P}_\beta\}^* = \frac{i}{2} R_{\alpha\beta\mu\nu}\theta^\mu\theta^\nu \tag{II-9}$$

$$\{\theta^C, \mathcal{P}_\alpha\}^* = \omega_\alpha{}^C{}_A\theta^A \tag{II-10}$$

where $\theta^\mu = L^\mu{}_{(A)}\theta^A$, and we used the definition

$$\mathcal{P}_\alpha = p_\alpha - \frac{i}{2}\omega_{\alpha AB}\theta^A\theta^B \tag{II-11}$$

With the help of the above relations we obtain the following closed algebra for the constraints

$$\{\mathcal{S}, \mathcal{S}\}^* = i\mathcal{H} \tag{II-12}$$

$$\{\mathcal{S}, \mathcal{H}\}^* = 0 \tag{II-13}$$

$$\{\mathcal{H},\mathcal{H}\}^* = 0 \tag{II-14}$$

with

$$\mathcal{H} = g^{\alpha\beta}\mathcal{P}_\alpha\mathcal{P}_\beta + m^2 \approx 0 \quad . \tag{II-15}$$

One recognizes the above first class constraint as the classical analogue of Klein-Gordon equation which must be satisfied by any Dirac spinor.

The transformation generated by the constraint (II-7) led to the following changes on the dynamical variables:

$$\delta x^\mu = \{x^\mu, i\alpha(\tau)\mathcal{S}\} = i\alpha(\tau)\theta^\mu \tag{II-16a}$$

$$\delta\theta^\alpha = -\alpha(\tau)\mathcal{P}^\alpha \tag{II-16b}$$

$$\delta\mathcal{P}_\nu = i\alpha(\tau)\{{}^{\alpha}_{\nu\mu}\}\theta^\mu\mathcal{P}_\alpha \tag{II-16c}$$

$$\delta\theta_5 = -\alpha(\tau)m \tag{II-16d}$$

where $\alpha(\tau)$, an odd function of τ , is the parameter of the transformation. We see that the constraint $\mathcal{S}$ considered as the generator of a gauge transformation has the property of mixing odd and even variables, which is characteristic of the generators of local supersymmetries[4].

Finally, we shall define the spin of the classical Dirac particle as

$$S^{\mu\nu} = i\theta^\mu\theta^\nu \tag{II-17}$$

which is based on the flat space model[3,5] for the $\frac{1}{2}$ - spin particle with $m \neq 0$. In that case, it can be shown that the invariance of the action functional under Lorentz transformation leads to the conservation of the total angular momentum $J^{\mu\nu}$ defined by

$$J^{\mu\nu} = x^\mu p^\nu - x^\nu p^\mu + i\theta^\mu\theta^\nu \quad .$$

The spin tensor defined by (II-17) is invariant under supersymmetry transformations and spacetime translations, and satisfies the Lie algebra of Lorentz group.

III - THE EQUATIONS OF MOTION

The total Hamiltonian[6] for the classical Dirac particle is

$$H = N(\tau)\mathcal{H} + iM(\tau)\mathcal{S} \approx 0 \tag{III-1}$$

where $N(\tau)$ and $M(\tau)$ are arbitrary functions of the parameter τ . (Note that $M(\tau)$ is an odd function. The factor of i has been introduced in order to make the last term in (III-1) real.) We could also write an action functional corresponding to (III-1) but we shall omit it as it is not needed here.

The equation of motion for a dynamical variable A reads

$$\frac{dA}{d\tau} = \{A,H\}^{*} \quad .$$

It follows that

$$\frac{d\theta_5}{d\tau} = M(\tau)m \quad . \tag{III-2}$$

In order to fix the gauge of the odd variables we chose the gauge constraint

$$\theta_5 \approx 0 \tag{III-3}$$

which due to (III-2) implies that $M(\tau) = 0$.

The equation for x^μ results in

$$\frac{dx^\mu}{d\tau} = 2N\mathcal{P}^\mu \quad . \tag{III-4}$$

Combining this result with the constraint (II-15) we get

$$\frac{dx^\mu}{d\tau}\frac{dx_\mu}{d\tau} = -4N^2m^2 \quad . \tag{III-5}$$

Introducing the proper time t with

$$\dot{x}^\mu\dot{x}_\mu = -1$$

where $\dot{x}^\mu = dx^\mu/dt$, it follows from (III-4) that

$$\dot{x}^\mu = \frac{1}{m}\mathcal{P}^\mu \tag{III-6}$$

We observe that as a consequence of (III-6) the constraint (II-7) implies that the spin tensor $S^{\mu\nu}$ satisfies the conditions

$$S_{\mu\nu}\mathcal{P}^{\nu} = m\,S_{\mu\nu}\dot{x}^{\nu} = 0 \quad . \tag{III-7}$$

The equations of motion that result for the other dynamical variables are

$$\dot{\mathcal{P}}^{\alpha} + \{^{\alpha}_{\rho\nu}\}\dot{x}^{\rho}\,{}^{\nu} = \frac{1}{m} R^{\alpha}{}_{\beta\mu\nu} S^{\mu\nu}\mathcal{P}^{\beta} \tag{III-8}$$

$$\dot{\theta}^{\alpha} + \{^{\alpha}_{\mu\nu}\}\dot{x}^{\mu}\theta^{\nu} = 0 \tag{III-9}$$

$$\ddot{x}^{\mu} + \{^{\mu}_{\alpha\beta}\}\dot{x}^{\alpha}\dot{x}^{\beta} = \frac{1}{2m} R^{\mu}{}_{\lambda\rho\sigma} S^{\rho\sigma}\dot{x}^{\lambda} \tag{III-10}$$

From the definition of the spin tensor and equation (III-9) it follows that

$$\dot{S}^{\alpha\beta} + \{^{\alpha}_{\mu\nu}\}\dot{x}^{\nu}S^{\mu\beta} + \{^{\beta}_{\mu\nu}\}\dot{x}^{\nu}S^{\alpha\mu} = 0 \tag{III-11}$$

which tells us that the spin tensor is covariantly constant as it should be.

The result that a classical spinning particle does not follow a geodesic in spacetime due to the coupling of the spin tensor to the curvature, equation (III-10), is not a new one. In fact, this problem has been studied by Papapetrou[7] and equations (III-8,10,11) are just the equations obtained by him. However, his equations must be supplemented by subsidiary conditions on the spin tensor in order to have a non-redundant system of equations. As a consequence of our approach there is no need to impose ad hoc conditions as we already have conditions (III-7). Finally we mention that our procedure can be applied to the case of massless $\frac{1}{2}$ - spin particles, and is general enough to be applied to other physical systems (strings, etc.) as well. These results will be published elsewhere.

ACKNOWLEDGMENTS

It is a pleasure to thank Dr. Gerald Kaiser for the hospitality of the Department of Mathematics of the University of Lowell during the realization of this conference. We also wish to express our deep gratitude to Dr. Claudio Teitelboim, Institute for Advanced Study, Princeton, and Dr. Yavuz Nutku, Dept. of Physics, Princeton University, for their constant encouragement and constructive criticisms.

APPENDIX: Summary of notations and conventions

In a coordinate basis the covariant derivative of a vector field $V_\lambda(x)$ with respect to the spacetime metric $g_{\alpha\beta}(x)$ is given by

$$V_\lambda(x)_{||\mu} = \partial_\mu V_\lambda - \{^{\alpha}_{\mu\lambda}\} V_\alpha$$

where $\{^{\alpha}_{\mu\lambda}\}$ are the Christoffel symbols. The Riemann tensor $R^{\alpha}{}_{\rho\mu\nu}$ is defined by

$$V_{\lambda||\mu||\nu} - V_{\lambda||\nu||\mu} = R^{\alpha}{}_{\lambda\mu\nu} V_\alpha$$

We use Majorana representation for the Dirac γ-matrices which is, explicitly

$$\gamma_0 = \left(\begin{array}{cc|cc} 0 & 1 & & \\ -1 & 0 & \multicolumn{2}{c}{0} \\ \hline & & 0 & -1 \\ \multicolumn{2}{c|}{0} & 1 & 0 \end{array}\right) \qquad \gamma_1 = \begin{pmatrix} 1 & & & 0 \\ & -1 & & \\ & & -1 & \\ 0 & & & -1 \end{pmatrix}$$

$$\gamma_2 = \left(\begin{array}{cc|cc} & & -1 & 0 \\ \multicolumn{2}{c|}{0} & 0 & -1 \\ \hline -1 & 0 & & \\ 0 & -1 & \multicolumn{2}{c}{0} \end{array}\right) \qquad \gamma_3 = \left(\begin{array}{cc|cc} 0 & -1 & & \\ -1 & 0 & \multicolumn{2}{c}{0} \\ \hline & & 0 & 1 \\ \multicolumn{2}{c|}{0} & 1 & 0 \end{array}\right)$$

Thus, $(\gamma_0)^2 = -1$, $(\gamma_k)^2 = -1$, $k = 1, 2, 3$. The adjoint of any of these matrices is defined by $\gamma_A^\dagger = \gamma_0\gamma_A\gamma_0$, and is also equal to its transpose. The γ_5 matrix is defined by

$$\gamma_5 = \gamma_0\gamma_1\gamma_2\gamma_3$$

and it satisfies the relation $\gamma_5\gamma_A = -\gamma_A\gamma_5$.

We use a system of units in which $\hbar = c = 1$.

REFERENCES

1. A. Lichnerowicz, Bull. Soc. Math., France, 92, 11 (1969)
2. R. Casalbuoni, Il Nuovo Cimento 33A, 1, 115 (1976) and 3, 289 (1976)
3. C. Teitelboim and Carlos A. P. Galvão, to be published.
4. P. Fayet and S. Ferrara, Phys. Reports 32c, nº 5 (1977)
5. A. Barducci, R. Casalbuoni and L. Lusanna, "Supersymmetry and the Pseudoclassical Relativistic Electron", preprint, Firenze (1976)
6. P.A.M. Dirac, "Lectures in Quantum Mechanics", Belfer Graduate School of Science Monograph Series, Yeshiva University, New York (1964)
7. A. Papapetrou, Proc. Roy. Soc., London, A209, 248 (1951)

GENERALIZED CONSTRAINT ALGORITHM
AND
SPECIAL PRESYMPLECTIC MANIFOLDS

*Mark J. Gotay** *James M. Nester*

Center for Theoretical Physics
Department of Physics and Astronomy
University of Maryland
College Park, Maryland 20742

Abstract

A generalized constraint algorithm is developed which provides necessary and sufficient conditions for the solvability of the canonical equations of motion associated to presymplectic classical systems. This constraint algorithm is combined with a presymplectic extension of Tulczjew's description of constrained dynamical systems in terms of special symplectic manifolds. The resultant theory provides a unified geometric description as well as a complete solution of the problems of constrained and a priori presymplectic classical systems in both the finite and infinite dimensional cases.

I. *Introduction*

Recently, Tulczyjew has given a description of constrained classical systems in terms of special symplectic manifolds [1–5]. This elegant theory adequately describes the dynamics of first-class systems in which (in the sense of Dirac[6]) no secondary constraints appear.

In a different approach [7–9], we have developed a geometric constraint algorithm which completely solves the problem of defining, obtaining and solving "consistent" canonical equations of motion for presymplectic dynamical systems. This algorithm is phrased in the context of global infinite-dimensional presymplectic geometry, and generalizes as well as improves upon the local Dirac-Bergmann theory of constraints [6]. The algorithm is applicable to the degenerate Hamiltonian and Lagrangian formulations of constrained systems [10] as well as to a priori presymplectic systems.

In this paper, we consolidate Tulczyjew's theory and our presymplectic techniques obtaining a complete unified geometric treatment of constrained and a priori presymplectic dynamical systems in terms of special presymplectic manifolds. This combined approach has several advantages over either method taken individually. The notion of special symplectic manifold, as Tulczyjew has pointed out, allows a uniform treatment of classical physics including relativistic and nonrelativistic dynamics as well as provides a basis for generalization to field theories, encompassing in particular the Poincaré-Cartan (multisymplectic) formalism [5, 11, 12]. Besides yielding geometrical insight into the mechanics of the presymplectic constraint algorithm, special symplectic techniques are indispensible in the consideration of singular dynamical systems, where, for instance, they may be used to "unfold" singular constraint submanifolds (cf. §VIII).

On the other hand, our presymplectic methods are capable of treating completely general constrained and a priori presymplectic dynamical systems. Specifically, given a physical system described by a presymplectic phase space (M, ω) and a Hamiltonian H on M, the algorithm finds whether or not there exists a submanifold N of M along which the canonical equations of motion

$$i(X)\omega = - dH \tag{1.1}$$

hold; if such a submanifold exists, the algorithm provides a *constructive* method for finding it. Moreover, the "final constraint submanifold" N is *maximal* in the sense that it contains any other submanifold along which (1.1) is satisfied.

In contrast, Tulczyjew's program is not constructive, that is, Tulczyjew does not consider the "Dirac constraint problem" per se, but rather only describes the finished product. Except under very special conditions (viz., when no secondary constraints appear in the theory), one must be given the final constraint submanifold N before Tulczyjew's techniques can be applied. The presymplectic constraint algorithm therefore can be used to extend Tulczyjew's theory of constrained dynamical systems to those in which secondary constraints are present.

There is, however, one profound difference between the synthetic approach of this paper and that proposed by Menzio and Tulczyjew [4], centering on the role of the integrability conditions in the theory. In the formulation of Menzio and Tulczyjew, certain integrability conditions are imposed which effectively demand that the final constraint submanifold N be first class. The integrability conditions associated with the presymplectic constraint algorithm, however, place no restriction on the class of N.

It is our contention that the integrability conditions of Menzio and Tulczyjew are inappropriate for a description of the dynamics of constrained classical systems. In fact, it turns out that these conditions are sufficient but not necessary for solutions of (1.1) to exist. Consequently, the imposition of such integrability conditions will artificially eliminate from consideration a great many systems of genuine physical interest (e.g., the Proca field).

Menzio and Tulczyjew claim that discarding constrained classical systems which are not first class *a priori* is acceptable, since such systems can never be the classical limits of consistent quantum theories. While this latter remark is -- strictly speaking -- true, there seems to be no compelling reason to eliminate such systems from consideration on the classical level. Furthermore, a theorem of Śniatycki [13] shows that it is usually possible to reformulate the dynamics of constrained systems in a manner such that the resulting dynamics is first class. Failing this, one may of course quantize the reduced phasespace [9].

Therefore, we feel that Menzio and Tulczyjew's dictum that the dynamics of constrained classical systems be first class a priori is unnecessarily severe. It is our opinion that there is much to be gained, and little to be lost, by developing techniques which are capable of treating constrained systems of arbitrary class.

*****

The language used throughout this paper is that of infinite-dimensional presymplectic geometry. Notation and terminology are summarized in the Appendix.

II. Presymplectic Geometry and Classical Mechanics

Let M be a Banach manifold, and suppose that ω is a closed 2-form on M. Then (M, ω) is said to be a *strong symplectic* manifold if the map $\flat : TM \to T^*M$ defined by $\flat(X) := i(X)\omega$ is a toplinear isomorphism. However, it may happen that $\flat$ is injective but not surjective, in which case (M, ω) is called a *weak symplectic* manifold, ω being *weakly* nondegenerate. Generically, $\flat$ will be neither injective nor surjective and ω is then *degenerate*. For brevity, weakly nondegenerate and degenerate manifolds will often be referred to simply as *presymplectic*. When M is finite-dimensional, there is of course no distinction between weak and strong symplectic forms.

Physically, M represents the phasespace of a classical system, while ω is a generalization of the Poisson (or Lagrange) bracket [14].

The standard example of a symplectic manifold is the cotangent bundle $\pi_Q: T^*Q \to Q$ of any Banach manifold Q. Indeed, on T^*Q there exists a canonical 1-form Θ_Q (the *Liouville form*) defined by the universal property

$$\alpha^*(\Theta_Q) = \alpha, \tag{2.1}$$

where α is any 1-form on Q. Alternatively, since the diagram

$$\begin{array}{ccc} TT^*Q & \xrightarrow{\ \tau_{T^*Q}\ } & T^*Q \\ \Big\downarrow{\scriptstyle T\pi_Q} & & \Big\downarrow{\scriptstyle \pi_Q} \\ TQ & \xrightarrow[\ \tau_Q\ ]{} & Q \end{array}$$

commutes, Θ_Q may be characterized as follows:

$$\langle v | \Theta_Q \rangle = \langle T\pi_Q(v) | \tau_{T^*Q}(v) \rangle, \tag{2.2}$$

where $v \in T(T^*Q)$. The Liouville form determines the exact symplectic structure

$$\Omega_Q = d\Theta_Q. \tag{2.3}$$

It is not difficult to show that Ω_Q so defined is weakly nondegenerate, and moreover that (T^*Q, Ω_Q) is strongly symplectic iff Q is reflexive [15].

The mechanics of the cotangent bundle case can be better understood by examining the local representatives of the above formulas. Let $U \subset F$ be a chart, where F is the model space for Q. The local representative of $m \in T^*Q$ is $(x, \sigma) \in U \times F^*$, and for $v \in T_m(T^*Q)$, one has $v = (x, \sigma) \oplus (a, \pi)$ in $(U \times F^*) \oplus (F \times F^*)$. It follows that

$$\tau_{T^*Q}(v) = (x, \sigma) \in U \times F^*$$

and

$$T\pi_Q(v) = (x, a) \in U \times F.$$

Therefore, (2.2) becomes, employing the shorthand notation $a \oplus \pi := (x, \sigma) \oplus (a, \pi)$,

$$\Theta_Q(x, \sigma)\cdot(a \oplus \pi) = \langle a|\sigma\rangle. \tag{2.4a}$$

Similarly, one calculates that

$$\Omega_Q(x, \sigma)\cdot(a \oplus \pi, b \oplus \tau) = \langle b|\pi\rangle - \langle a|\tau\rangle . \tag{2.4b}$$

In the finite-dimensional case, these formulas are not nearly so mysterious. If $(T^*U; q^i, p_i)$ is a natural bundle chart for T^*Q, (2.4a) and (2.4b) become

$$\Theta_Q|T^*U = p_i dq^i \tag{2.5a}$$

and

$$\Omega_Q|T^*U = dp_i \wedge dq^i. \tag{2.5b}$$

Physically, the weak and strong symplectic manifolds one almost always encounters are cotangent bundles. Indeed, *physics in the Hamiltonian formulation is none other than mechanics on cotangent bundles*. The manifold Q is the configuration space of the physical system, its cotangent bundle T^*Q is momentum phasespace and the canonical 1-form Θ_Q is the integrand in the Principle of Least Action.

There do, however, exist physically interesting systems whose phasespaces are <u>not</u> cotangent bundles and whose symplectic forms are <u>not</u> exact. An example of such a system was given by Souriau [16], who investigated the dynamics of a freely spinning massive particle in Minkowski spacetime from a symplectic viewpoint (here,

$M = R^6 \times S^2$). Systems of this type do not possess configuration manifolds and consequently do not admit Hamiltonian or Lagrangian formulations (at least in the usual sense).

Furthermore, the geometry of classical systems need not be strongly symplectic. This phenomenon is characteristic of systems with an infinite number of degrees of freedom, where ω may be presymplectic even when there are no constraints (e.g., the Klein-Gordon field [9, 15]). An example of an a priori presymplectic dynamical system has been provided by Künzle [17], who obtained genuinely presymplectic phase spaces for spinning particles in curved spacetimes.

The most important application of presymplectic geometry is to the theory of constrained classical systems. Typically, (e.g., electromagnetism, gravity), the constraints take the form of internal consistency conditions on the dynamics of the system.

Such constraints appear when one transforms from the Lagrangian to the Hamiltonian formalism. A physical system, described by a configuration space Q and a Lagrangian L, is cast into canonical form by "changing variables" from $(q^i, \dot{q}^i)$ to (q^i, p_i) and replacing L by the Hamiltonian H through $H(q, p) = p_i\dot{q}^i - L(q, \dot{q})$. Mathematically, this transition is accomplished via the *Legendre transformation* $FL: TQ \to T^*Q$ defined by

$$\langle w|FL(z)\rangle := \frac{d}{ds} L(z + sw)\Big|_{s=0} , \qquad (2.6)$$

where $z, w \in TQ$.

Presymplectic manifolds arise when FL is not a diffeomorphism [14], in which case the Legendre transformation defines a submanifold $FL(TQ)$ of T^*Q. This is the starting point of the Dirac-Bergmann constraint theory [6], in which $FL(TQ)$ is called the *primary constraint* submanifold. $FL(TQ)$ will inherit a presymplectic structure from T^*Q by pulling Ω_Q back to M via the inclusion $j: FL(TQ) \to T^*Q$). The degree of degeneracy of $\omega = j^*\Omega_Q$ depends entirely upon the behavior of FL. On $FL(TQ)$ Hamilton's equations take the form (1.1).

Another example of an a priori presymplectic system is provided by Lagrangian dynamics, where the fundamental dynamical arena is not *momentum* phasephase T^*Q, but rather *velocity* phasespace TQ. Whereas T^*Q carries a canonical exact symplectic structure, TQ does not. Nonetheless, it is always possible to transfer the exact symplectic structure Ω_Q on T^*Q to TQ by pull back via FL. Generically, however, this induced structure will not be symplectic, but merely presymplectic, depending upon the regularity properties of FL.

III. *Canonical Systems and their Classification*

It is useful to have a classification scheme for generalized submanifolds of presymplectic manifolds which is both mathematically convenient and physically meaningful. Dirac first developed a local classification of submanifolds of strongly symplectic manifolds by describing them in terms of certain types of constraint functions (see refs. [4], [6], [9] and [18] for details concerning this approach). Tulczyjew and Śniatycki [13] have found an intrinsic generalization of Dirac's classification scheme, which is extended here to the presymplectic case. This classification is of the utmost significance insofar as the physical interpretation of the constraint algorithm is concerned, and has important applications to both the gauge theory and the quantization of presymplectic dynamical systems [8, 9].

Let N be a g-submanifold of the presymplectic manifold (M, ω) with inclusion j. The manifold N is called a *constraint* g-submanifold, and the triple (M, ω, N) a *canonical system*. Define the *symplectic complement* $TN^{\perp}$ of $\underline{TN}$ in TM to be

$$TN^{\perp} = \{Z \in T_N M \text{ such that } \omega(X, Z) = 0 \text{ for all } X \in \underline{TN}\}.$$

The *annihilator* $TN^{\vdash}$ of $\underline{TN}$ in T^*M is

$$TN^{\vdash} = \{\alpha \in T^*_N M \text{ such that } \langle X|\alpha\rangle = 0 \text{ for all } X \in \underline{TN}\}.$$

The constraint g-submanifold N is said to be

(*i*) *isotropic* if $\underline{TN} \subset TN^{\perp}$,

(*ii*) *coisotropic* or *first class* if $TN^{\perp} \subset \underline{TN}$,

(*iii*) *weakly symplectic* or *second class* if $\underline{TN} \cap TN^{\perp} = \{0\}$, and

(*iv*) *Lagrangian* if $TN = TN^{\perp}$ [19].

If N does not happen to fall into any of these categories, then N is said to be *mixed* constraint g-submanifold.

From the point of view of the g-submanifold N, this classification reduces to a characterization of the naturally induced presymplectic structure ω_N on N. Indeed, $TN^{\perp} \cap \underline{TN} = \underline{ker\ \omega_N}$, where $\omega_N := j^*\omega$. In particular, N is isotropic iff $j^*\omega = 0$.

As an illustration, let $C \subset Q$. Then T^*C is a second class submanifold of (T^*Q, Ω_Q). Furthermore, the constraint submanifold $\pi_Q^{-1}(C) \subset T^*Q$ is first class.

Let α: $Q \to T^*Q$ be a closed 1-form. By virtue of the definition (2.1) of Θ_Q, it follows that the image $\alpha(Q)$ of Q under α is an isotropic submanifold of (T^*Q, Ω_Q):

$$\alpha^*\Omega_Q = d\alpha^*\Theta_Q = d\alpha = 0.$$

In fact, $\alpha(Q)$ is maximally isotropic and hence Lagrangian. If α were only densely defined, however, then the image of α would be merely isotropic.

Thus the zero-section Q of T^*Q provides a natural example of a Lagrangian constraint submanifold. Also, for each $m \in Q$, the fiber $\pi_Q^{-1}(m)$ is a Lagrangian submanifold. A Lagrangian submanifold, as these examples indicate, generalizes the classical coordinate and momentum representations.

IV. *Canonical Dynamics of Presymplectic Systems*

The presymplectic form ω and the phasespace M have only kinematical significance -- the ***dynamics*** of the physical system (M, ω) is determined by specifying on M a closed 1-form α, the *Hamiltonian form*. One then solves the *generalized Hamilton equations*

$$i(X)\omega = \alpha \tag{4.1}$$

for the ***evolution vectorfield*** X. Once X has been determined, one appeals to the standard results of differential equation theory in order to integrate X, thereby obtaining the dynamical trajectories of the system in phasespace.

When (M, ω) is strongly symplectic, the induced map $\flat : TM \to T^*M$ is an isomorphism. Consequently, in this case (4.1) possesses a unique solution $X = \flat^{-1}(\alpha)$. Since X is every where defined and smooth, it gives rise to a unique local flow [20].

We now calculate the local representative of X in the strongly symplectic case. Let V be a (contractible) chart on M, and suppose for simplicity that M is Hilbertable. Then, Darboux's Theorem [21] asserts the existence of a reflexive Banach space F and a chart $U \subset F$ such that

$$(V, \omega|V) \approx (T^*U, \Omega_U).$$

Furthermore, since V is contractible, $\alpha|V = -dH$, where H is the ordinary Hamiltonian. If $m = (x, \sigma) \in T^*U$, and $Y = b \oplus \tau \in F \times F^*$ is a vector at (x, σ), then

$$\begin{aligned} i_Y\alpha(m) &= - DH(x, \sigma)\cdot(b \oplus \tau) \\ &= - \bar{D}H(x, \sigma)\cdot b - \dot{D}H(x, \sigma)\cdot\tau. \end{aligned}$$

Similarly, writing $X(x, \sigma) = a \oplus \pi \in F \times F^*$, (2.4b) yields

$$\begin{aligned} i_Y i_X\omega(m) &= \Omega_U(x, \sigma)\cdot(a \oplus \pi, b \oplus \tau) \\ &= \langle b|\pi\rangle - \langle a|\tau\rangle. \end{aligned}$$

Comparing this expression with the previous one, equation (4.1) implies that the local representative of X is

$$X(x, \sigma) = \dot{D}H(x, \sigma) \oplus -\bar{D}H(x, \sigma).$$

In the finite-dimensional case, this reduces to

$$X = \frac{\partial H}{\partial p_i}\frac{\partial}{\partial q^i} - \frac{\partial H}{\partial q^i}\frac{\partial}{\partial p_i},$$

the integral curves of which are found by solving Hamilton's equations:

$$\frac{dq^i}{dt} = \frac{\partial H}{\partial p_i}, \quad \frac{dp_i}{dt} = -\frac{\partial H}{\partial q^i}.$$

Turning now to the presymplectic case, there are four major difficulties one encounters when trying to solve the generalized Hamilton equations associated with a presymplectic dynamical system (M, ω, α):

(*i*) These equations are typically inconsistent and consequently will not possess globally defined solutions; if an evolution vectorfield X exists at all, then in general it will be defined only on some g-submanifold N of M [22];

(*ii*) X does not necessarily define a differential equation on N, that is, $X \notin \underline{T_N \overline{N}}$ in general;

(*iii*) The solution X, if it exists, need not be unique; and

(*iv*) X will usually be discontinuous so that it may not possess even a locally defined flow.

Difficulty (*i*), the *existence problem*, is encountered even in well-behaved systems, e.g., the Klein-Gordon field, for which $M = H^1 \oplus L^{2*}$ and $N = H^2 \oplus H^{1*}$. Physically, N is to be regarded as a *constraint g-submanifold*, that is, $\overline{N} \subseteq M$ consists of those states of the system which are *physically realizable*. The implication is that states in M not contained in $\overline{N}$ are *dynamically inaccessible* to the system, since the equations of motion cannot be integrated at such points.

The *constraint problem* (*ii*) is of fundamental significance, and in the degenerate case presents the major obstacle to solving the equations (4.1). The generalized Hamilton equations are to be considered as evolution equations for the system, and hence must be *differential* equations. However, in order for the vectorfield X to be interpretable as a differential equation, it is necessary that X be "tangent" to $\overline{N}$ in the sense that $X \in \underline{T_N \overline{N}}$. In other words, if X is to describe the evolution of the system in phasespace, then it must generate a (local) flow. Since X is defined only along the g-submanifold N, it can (at best) give rise to a flow on N -- only if X is tangent to $\overline{N}$. Physically this has the interpretation that the motion of the system is constrained to lie in $\overline{N}$.

The existence and constraint problems will be the subjects of the next section, while (*iii*), the *uniqueness problem* -- which signals the presence of gauge degrees of freedom in the theory -- has been discussed elsewhere [8, 9].

The *integration problem* (*iv*) can be very severe for presymplectic systems as well. As discussed above, the interpretation of equations (4.1) as evolution equations requires that X be integrable, i.e., X must give rise to a well-defined (possibly local) flow. The demand that X be tangent to $\overline{N}$ gives a necessary, but certainly not sufficient, condition for X to be integrable. The difficulty is that X is not necessarily continuous (as it may not be defined globally; e.g., the Klein-Gordon field) so that the standard theorems on the existence and uniqueness of flows of vector fields are not applicable. Unlike difficulties (*i*) - (*iii*), the integration problem lies mainly outside the province of symplectic geometry and is better considered from the viewpoint of global analysis and the theory of partial differential equations [15]. Consequently, this problem will not be considered further here.

In this paper, techniques will be developed which will (eventually) enable one to "solve" problems (*i*) - (*iii*). In view of the first difficulty, however, the initial step in the "solution" must be to answer the question: "What does one mean by 'consistent equations of motion,' and how does one obtain and solve such equations?"

V. The Presymplectic Constraint Algorithm

In this section we present an improved version of our presymplectic constraint algorithm [7-9] which correctly handles the infinite dimensional case where the evolution may be defined only on a dense subset rather than globally. Given a presymplectic dynamical system $(M_1, \omega_1, \alpha_1)$, our procedure will be used to select a certain g-submanifold N of M_1 upon which one can define and solve "consistent equations of motion." More precisely, this technique will provide necessary and sufficient conditions for the existence of a g-submanifold N of M_1 such that the equations

$$i(X)\omega_1 = \alpha_1 \tag{5.1}$$

hold when restricted to N, i.e.,

$$[i(X)\omega_1 - \alpha_1]|N = 0,$$

with X *tangent* to $\overline{N}$.

Begin by noting that if α_1 is everywhere contained in the range of $\flat_1$, then the required solution (not necessarily unique) of the equations (5.1) is simply any smooth element of $\flat_1^{-1}\{\alpha_1\}$. In the generic case, however, this will not be so. But there may exist points of M_1 (such points being assumed to form a g-submanifold M_2 of M_1), for which $\alpha_1|M_2$ is in the range of $\flat_1|M_2$. One is thus led to try and solve equation (5.1) restricted to M_2, i.e.,

$$[i(X)\omega_1 - \alpha_1]|M_2 = 0. \tag{5.2}$$

Equation (5.2) evidently possesses solutions, but only in an *algebraic* sense. In accord with the discussion of the constraint problem in §IV, one must demand that X solve (5.2) in a *differential* sense, viz., that $X \in \underline{T_{M_2}\overline{M}_2}$, or else the equations of motion will try to evolve the system "off $\overline{M}_2$" into an unphysical domain.

This requirement will not necessarily be satisfied, forcing a further restriction of (5.1) to the g-submanifold M_3 of M_2 defined by

$$M_3 := \{m \in M_2 \text{ such that } \alpha_1(m) \in \underline{T\overline{M}_2}^{\flat}\},$$

with the shorthand notation $T_P\overline{P} =: \overline{T}P$. It must now be ensured that the solution to (5.1) restricted to M_3 is in fact tangent to $\overline{M}_3$; this will in general necessitate yet more restrictions.

It is now clear how the algorithm must proceed. A string of *constraint g-submanifolds* [22]

$$\ldots \rightarrow M_3 \xrightarrow{j_3} M_2 \xrightarrow{j_2} M_1$$

is generated, defined as follows:

$$M_{\ell+1} := \{m \in M_\ell \text{ such that } \alpha_1(m) \in \underline{T\overline{M}_\ell}^{\flat}\}.$$

Once the constraint algorithm so defined is set into motion, only one of four distinct possibilities may occur. They are:

Case 1: There exists a K such that $M_K = \phi$;

Case 2: Eventually, the algorithm produces a g-submanifold $M_K \neq \phi$ such that $\dim M_K = 0$;

Case 3: There exists a K such that $M_K = M_{K+1}$ with $\dim M_K \neq 0$; and

Case 4: The algorithm does not terminate.

In the first case, $M_K = \phi$ means that the generalized Hamilton equations (5.1) have no solutions at all in any sense. In principle, this means that $(M_1, \omega_1, \alpha_1)$ does not accurately describe the dynamics of any system.

The second possibility results in a constraint g-submanifold which consists of isolated points. The equations (5.1) are consistent, but the only possible solution is $X = 0$ and there is no dynamics.

For case three, one has a constraint g-submanifold M_K and completely consistent equations of motion on M_K of the form

$$[i(X)\omega_1 - \alpha_1]|M_K = 0, \tag{5.3}$$

with X tangent to $\overline{M}_K$. It is this g-submanifold M_K (the *final* constraint

g-submanifold) which corresponds to the g-submanifold N discussed in §III.

The situation described in case four is only possible for systems with an infinite number of degrees of freedom. In this circumstance, the final constraint g-submanifold can be taken to be the intersection M_∞ of all the g-submanifolds M_ℓ. One then recovers cases (1) - (3) depending upon whether $M_\infty = \phi$, $dim\ M_\infty = 0$, or $0 < dim\ M_\infty \leq \infty$.

If the algorithm terminates with some final constraint g-submanifold M_K $(1 \leq K \leq \infty)$, then *by construction* one is assured that at least one solution X to the canonical equations exists and furthermore that this solution is tangent to $\overline{M}_K$. Note that X need not be unique, for one can add to it any element of $ker\ \omega_1 \cap \underline{T\overline{M}_K}$. In addition, it is obvious, again by construction, that the final constraint g-submanifold is *maximal* in the following sense: if N is any other submanifold along which the equations (5.1) are satisfied, then $N \subseteq M_K$.

We have shown [7] that this constraint algorithm generalizes the local Dirac-Bergmann theory of constraints [6]. Indeed, it is possible to characterize the closures of the constraint g-submanifolds M_ℓ, as follows:

$$\overline{M}_\ell = \{m \in \overline{M}_{\ell-1} \text{ such that } \langle Z|\alpha_1\rangle(m) = 0 \text{ for all } Z \in T\overline{M}_{\ell-1}^\perp\}.$$

The constraint functions $\langle T\overline{M}_{\ell-1}^\perp|\alpha_1\rangle = 0$ which define $\overline{M}_\ell$ in $\overline{M}_{\ell-1}$ are none other than Dirac's *ℓ-ary constraints*.

This presymplectic constraint algorithm provides a geometrically intuitive and conceptually simple method for defining and solving consistent equations of motion on a presymplectic manifold. It provides a *constructive* solution to the existence and constraint problems of §IV, and is of very general applicability, requiring only that the phasespaces involved be Banach manifolds.

VI. *Special Presymplectic Manifolds*

Here, we broaden Tulczjew's notion of "special symplectic manifold" [1, 23] so as to encompass the presymplectic formalism necessary for the description of completely general dynamical systems.

A *special symplectic manifold* is a quintuple (P, p, M, λ, μ), where $p\colon P \to M$ is a fiber bundle, λ is a 1-form on P, and μ is a fiber-preserving diffeomorphism $P \to T^*M$ such that $\mu^*\Theta_M = \lambda$.

Essentially, one is transferring the symplectic structure on T^*M to P via μ. The 2-form $d\lambda$ on P is weakly nondegenerate, and strongly nondegenerate iff M is reflexive.

A *special presymplectic manifold* is obtained by relaxing the requirement that μ be a diffeomorphism. A special presymplectic manifold is therefore a degenerate "copy" of a cotangent bundle.

Example 1: If Q is the configuration space of a physical system, then momentum phasespace $(T^*Q,\ \pi_Q,\ Q,\ \Theta_Q,\ id_Q)$ is a special symplectic manifold.

Example 2: The Lagrangian system $(TQ, \tau_Q,\ Q,\ FL^*\Theta_Q,\ FL)$ is a special presymplectic manifold, where $L\colon TQ \to R$ is the Lagrangian. In a bundle chart $U \times F$ for TQ, one has

$$FL^*\Theta_Q(u,\ e) = \dot{D}L(u,\ e) \oplus 0.$$

[In finite-dimensions, $FL^*\Theta_Q = \frac{\partial L}{\partial \dot{q}^i}\, dq^i$]. The 2-form $dFL^*\Theta_Q$ is strongly (weakly) symplectic iff the velocity Hessian $\ddot{D}\ddot{D}L(u,\ e)$, viewed as a linear map $F \to F^*$, is strongly (weakly) nondegenerate [14].

Example 3: Let $(M,\ \omega)$ be a presymplectic manifold. Then $(TM,\ \tau_M,\ M,\ \flat_M{}^*\Theta_M,\ \flat_M)$ is a special presymplectic manifold, where $\flat_M$ is the map of TM to T^*M induced by ω. The presymplectic structure $d\flat_M{}^*\Theta_M$ on TM is denoted $\dot{\omega}$.

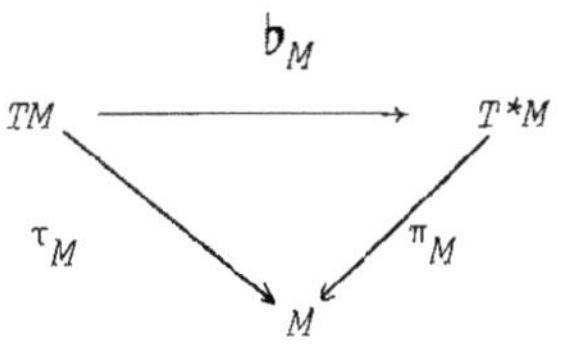

Consider the special case $M = T^*Q$. The local representative

$$\flat_{T^*U}\colon\ U \times F^* \times F \times F^* \to U \times F^* \times F^* \times F^{**}$$

of $\flat_{T^*Q}\colon\ T(T^*Q) \to T^*(T^*Q)$ is

$$\flat_{T^*U}\ (x,\ \sigma,\ e,\ \pi) = (x,\ \sigma,\ \pi,\ -e).$$

Consequently, one has

$$\lambda(x,\ \sigma,\ e,\ \pi) = (\ \pi,\ -e) \oplus (0,\ 0).$$

In a finite-dimensional natural bundle chart $(T(T^*U);\ q^i,\ p_i,\ \dot{q}^i,\ \dot{p}_i)$, this expression becomes

$$\lambda = \dot{p}_i dq^i - \dot{q}^i dp_i\ .$$

Example 4: If $U \subset F$ is a chart for Q, then

$$T(T^*U) = U \times F^* \times F \times F^*,$$

while

$$T^*(TU) = U \times F \times F^* \times F^*.$$

The map t: $U \times F^* \times F \times F^* \to U \times F \times F^* \times F^*$ given in charts by

$$t(x, \sigma, e, \pi) = (x, e, \pi, \sigma)$$

extends to a well-defined diffeomorphism $t\colon T(T^*Q) \to T^*(TQ)$ (for an intrinsic definition of t, see ref [2]). Since the diagram

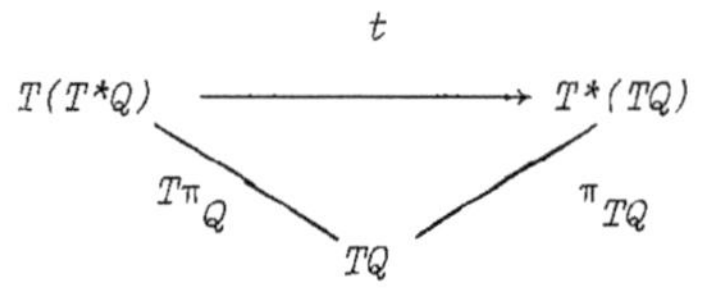

commutes, it follows that $(T(T^*Q), T\pi_Q, TQ, t^*\Theta_{TQ}, t)$ is a special symplectic manifold. Here,

$$\lambda(x, \sigma, e, \pi) = (\pi, 0) \oplus (\sigma, 0)$$

or, in finite-dimensions,

$$\lambda = \dot{p}_i dq^i + p_i\, d\dot{q}^i.$$

Combining examples (3) and (4), one sees that $T(T^*Q)$ can be realized as a special presymplectic manifold in two completely different ways. This fact is of fundamental significance for mechanics, since it provides the geometric link between the Hamiltonian and Lagrangian formalisms in terms of which the Legendre transformation is defined (cf §VIII). Note, however, that both *special* presymplectic structures on $T(T^*Q)$ give rise to the same *symplectic* structure, since

$$d\flat_{T^*Q}{}^*\Theta_{T^*Q} = \dot{\omega} = dt^*\Theta_{TQ}.$$

*****

Of particular importance for dynamics are the isotropic g-submanifolds of special presymplectic manifolds. Generalizing the construction at the end of §III, one has the following interpretation of such g-submanifolds in terms of generating forms.

Theorem [Śniatycki and Tulczyjew]: Let (P, p, M, λ, μ) be a special presymplectic manifold, $j_N\colon N \to M$ a g-submanifold of M, and α a closed 1-form on N. Define $\mathcal{D}(\alpha) = \{y \in p^{-1}(j_N(N)) \,|\, \langle Z|\lambda\rangle = \langle u|\alpha\rangle$ for all $Z \in T_yP$ and $u \in \underline{TN}$ with $Tp(Z) = u\}$. Then $\mathcal{D}(\alpha)$ is an isotropic g-submanifold of $(P, d\lambda)$ with inclusion $j_{\mathcal{D}}$, the map $p_{\mathcal{D}}$ defined by the commutative diagram

$$\begin{array}{ccc} \mathcal{D}(\alpha) & \xrightarrow{\;j_{\mathcal{D}}\;} & P \\ {\scriptstyle p_{\mathcal{D}}}\downarrow & & \downarrow{\scriptstyle p} \\ N & \xrightarrow[\;j_N\;]{} & M \end{array}$$

is a submersion, the fibers of $p_{\mathcal{D}}$ are connected, and $j_{\mathcal{D}}{}^*\lambda = p_{\mathcal{D}}{}^*\alpha$.

Conversely, suppose that $\mathcal{D}$ is an isotropic g-submanifold of $(P, d\lambda)$ with inclusion $j_{\mathcal{D}}$ such that $N := p\circ j_{\mathcal{D}}(\mathcal{D})$ is a g-submanifold of M and the induced projection $p_{\mathcal{D}}$ defined by the commutative diagram

$$\begin{array}{ccc} \mathcal{D} & \xrightarrow{\;j_{\mathcal{D}}\;} & P \\ {\scriptstyle p_{\mathcal{D}}}\downarrow & & \downarrow{\scriptstyle p} \\ N & \xrightarrow[\;j_N\;]{} & M \end{array}$$

is a submersion with connected fibers. Then there exists a unique closed 1-form α on M such that $j_{\mathcal{D}}{}^*\lambda = p_{\mathcal{D}}{}^*\alpha$. Furthermore, $\mathcal{D} \subseteq \mathcal{D}(\alpha)$.

The 1-form α is the *generating form* of $\mathcal{D}(\alpha)$, and $\mathcal{D}(\alpha)$ is said to be *generated* by α. Symbolically, we write

$$\mathcal{D}(\alpha) = \mu^{-1}\{\alpha(N) + \underline{TN}^{\vdash}\}.$$

If N happens to be a Banach submanifold of M, then $\mathcal{D}(\alpha)$ is actually Lagrangian.

The proof of the above result, given in [23] for submanifolds of special symplectic manifolds, in fact holds for generalized submanifolds of special presymplectic manifolds.

Example 5: Let $L\colon\ TQ \to \mathbb{R}$ be a Lagrangian. According to the above theorem, the Lagrangian submanifold $\mathcal{D}(dL)$ of $(T(T^*Q),\ T\pi_Q,\ TQ,\ t^*\Theta_{TQ},\ t)$ generated by dL is defined by $t^*\Theta_{TQ} = T\pi_Q{}^*(dL)$. In a natural bundle chart $(T(T^*U);\ q^i,\ p_i,\ \dot{q}^i,\ \dot{p}_i)$, this becomes

$$\dot{p}_i dq^i + p_i d\dot{q}^i = dL,$$

or, more suggestively,

$$\dot{p}_i = \frac{\partial L}{\partial q^i}\ ,\ p_i = \frac{\partial L}{\partial \dot{q}^i}\ .$$

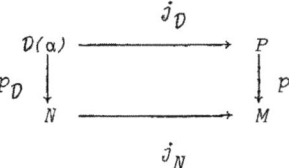

Example 6: Let N be a Banach submanifold of T^*Q, and $H: N \to \mathbb{R}$ a Hamiltonian. The exact 1-form $-dH$ generates an isotropic submanifold $D(-dH)$ of $(T(T^*Q), \tau_{T^*Q}, T^*Q, \flat_{T^*Q}{}^*\Theta_{T^*Q}, \flat_{T^*Q})$. In the natural bundle chart of Example [5], N may be described by the vanishing of certain functions $\phi^\alpha(q, p)$, $\alpha = 1, \ldots,$ codim N. $D(-dH)$ is then locally given by $\flat_{T^*Q}{}^*\Theta_{T^*Q} = -\tau_{T^*Q}{}^*dH$ subject to the constraints $\phi^\alpha = 0$. Using Lagrange multipliers, one has the local expressions

$$\dot{q}^i = \frac{\partial \bar{H}}{\partial p_i} + \lambda_\alpha \frac{\partial \phi^\alpha}{\partial p_i}$$

$$\dot{p}^i = -\frac{\partial \bar{H}}{\partial q^i} - \lambda_\alpha \frac{\partial \phi^\alpha}{\partial q^i} ,$$

where $\bar{H}$ is any extension of H to T^*Q. Physically, $N = FL(TQ)$ is Dirac's primary constraint submanifold, the ϕ^α are primary constraints, and the above two equations are the Dirac-Hamilton equations of motion (cf. [6], [9] and §VIII).

VII. *Generalized Constraint Algorithm*

Let $(M_1, \omega_1, \alpha_1)$ be a presymplectic dynamical system, and consider the generalized Hamilton equations

$$i(X)\omega_1 = \alpha_1. \tag{7.1}$$

We now restate the presymplectic constraint algorithm of §V, which provides the necessary and sufficient conditions for the solvability of (7.1), in terms of special presymplectic manifolds.

Construct the special presymplectic manifold $(TM_1, \tau_1, M_1, \flat_1{}^*\Theta_1, \flat_1)$, where $\flat_1$: $TM_1 \to TM_1{}^*$ is the map induced by ω_1. The closed 1-form α_1 on M_1 generates, according to the theorem in the last section, an isotropic g-submanifold

$$D_1 = \flat_1{}^{-1}\{\alpha_1(M_1)\}$$

of $(TM_1, \dot{\omega}_1)$. The secondary constraint g-submanifold

$$M_2 = \tau_1(D_1)$$

consists of those points of M_1 along which there exist algebraic solutions X of (7.1), viewed as smooth sections of D_1.

The g-submanifold D_1 will be a *differential equation* with respect to M_2 -- that is, vector fields X: $M_2 \to D_1$ will solve (7.1) in a differential sense -- iff the integrability conditions

$$D_1 \subseteq \underline{TM_2}$$

are satisfied. If this is not the case, then one must restrict attention to the subset

$$\mathcal{D}_2 = \mathcal{D}_1 \cap \underline{\overline{TM}_2}$$

of TM_1.

The motion of the system is thereby constrained to lie in the closure of the tertiary constraint g-submanifold.

$$M_3 = \tau_1(\mathcal{D}_2)$$

of M_1. Demanding that $\mathcal{D}_2$ be tangent to $\overline{M}_3$ (i.e., $\mathcal{D}_2 \subseteq \underline{\overline{TM}_3}$) may necessitate a further restriction to $\mathcal{D}_3 = \mathcal{D}_2 \cap \underline{\overline{TM}_3}$ etc.

Thus, the algorithm leads to a sequence of isotropic constraint g-submanifolds M_ℓ given by

$$M_\ell = \tau_1(\mathcal{D}_{\ell-1}), \tag{7.2}$$

where

$$\mathcal{D}_\ell = \mathcal{D}_{\ell-1} \cap \underline{\overline{TM}_\ell}, \tag{7.3}$$

and

$$\mathcal{D}_1 = \flat_1^{-1}\{\alpha_1(M_1)\}.$$

If the algorithm terminates with some non-empty final constraint g-submanifold M_K $(1 \leq K \leq \infty)$, then $\mathcal{D}_K = \mathcal{D}_{K+1} \subseteq \underline{\overline{TM}_K}$. Consequently, there exists at least one solution $X \in \underline{\overline{TM}_K}$ such that

$$[i(X)\omega_1 - \alpha_1]|M_K = 0. \tag{7.4}$$

The fact that $\mathcal{D}_K$ is not usually transverse to the fibers of $\underline{\overline{TM}_K}$ is indicative of the generic non-uniqueness of the evolution vectorfield X. Specifically, X is unique iff the fiber dimension of $\mathcal{D}_K \cap \underline{\overline{TM}_K}$ is everywhere unity, in which case the canonical system (M_1, ω_1, M_K) is second class.

There are two regularity conditions that must be satisfied for the successful application of the algorithm: (i) Each set $\tau_1(\mathcal{D}_{\ell-1})$ must be a g-submanifold of M_1, and (ii) The fibers of $\mathcal{D}_{\ell-1}$ over $\tau_1(\mathcal{D}_{\ell-1})$ must be isomorphic [24]. If, at the ℓth step of the algorithm, either of these two conditions fails to hold, then one must judiciously choose a g-submanifold M_ℓ' of $\tau_1(\mathcal{D}_{\ell-1})$ such that the fibers of $\mathcal{D}_{\ell-1}|M_\ell'$

are isomorphic and then proceed with the algorithm applied to $M_\ell{}'$. A proper treatment of such singularities, which is necessary for the correct physical interpretation of certain systems, will be given elsewhere [25] (see also [13]).

The above technique should be compared with that proposed by Menzio and Tulczyjew [4]. From the presymplectic standpoint, the integrability conditions

$$\mathcal{D}_\ell \subseteq \overline{T}[\underline{\tau_1(\mathcal{D}_\ell)}].$$

are applied during the course of the algorithm and consequently are <u>automatically satisfied</u> on the final constraint g-submanifold M_K, i.e., if M_K exists, then by construction

$$\mathcal{D}_K \subseteq \overline{T}[\underline{\tau_1(\mathcal{D}_K)}].$$

Therefore, integrability has no relation to the class of the canonical system (M_1, ω_1, M_K).

We note that this generalized constraint algorithm is applicable to any dynamical system determined by the specification of a submanifold $\mathcal{D}_1$ of "admissible" vector fields. Eqns. (7.2, 7.3) contain the essence of the Dirac constraint problem and are quite independent of the origin of $\mathcal{D}_1$.

VIII. *Applications*

(1) <u>The Lagrangian Formulation of Mechanics</u>

Let Q be the configuration space of a physical system, and TQ its velocity phase space. We want to include the case of field dynamics, where the Lagrangian may be only densely defined. Typically, one takes the domain of L to be the restriction T_CQ, where C is a manifold domain in Q.

For $w \in T_CQ$, we define the *energy* $E\colon\ T_CQ \to \mathbb{R}$ of L by

$$E(w) = \langle w | FL(w)\rangle - L(w),$$

where the Legendre transformation $FL\colon\ T_CQ \to T^*Q$ is given by (2.6). Pulling Ω_Q back to T_CQ, one obtains a generically presymplectic form $\Omega_L = FL^*\Omega_Q$. Our task is to define and solve consistent Lagrange equations of the form

$$i(X)\Omega_L = -dE. \tag{8.1}$$

Consider the special presymplectic manifold $(T(T_CQ), \tau_{T_CQ}, T_CQ, \flat_L{}^*\Theta_{T_CQ}, \flat_L)$, where $\flat_L\colon\ T(T_CQ) \to T^*(T_CQ)$ is induced by Ω_L. The 1-form $-dE$ on T_CQ generates an isotropic g-submanifold $\mathcal{D}_1 = \flat_L^{-1}\{-dE(T_CQ)\}$ of $(T(T_CQ), \dot{\Omega}_L)$. The constraint algorithm, applied to $\mathcal{D}_1$, then proceeds as in §VII, eventually (if the problem is solvable) producing a differential equation $\mathcal{D}_K \subseteq \mathcal{D}_1$ and a final constraint

g-submanifold $\tau_{T_CQ}(\mathcal{D}_K)$ of T_CQ. One is then assured of the existence of a section $X\colon \tau_{T_CQ}(\mathcal{D}_K) \to \mathcal{D}_K$ such that

$$[i(X)\Omega_L + dE]|\tau_{T_CQ}(\mathcal{D}_K) = 0. \tag{8.2}$$

(2) The Second-Order Equation Problem

The consistent Lagrange equations that follow from the constraint algorithm are typically a set of coupled first-order differential equations -- a feature of theories which are described mathematically by presymplectic geometries. Variational as well as physical considerations demand, however, that the Lagrange equations be a set of coupled second-order differential equations. Specifically, the equations of motion (8.1) will follow from a variational principle iff the *second-order equation* condition

$$\tau_{T_CQ}(X) = T(\tau_Q|T_CQ)(X) \tag{8.3}$$

holds at every point in the domain of X [26].

It is therefore necessary to find the conditions under which the Lagrange equations (8.2) admit solutions which are in fact second-order equations. Formally, special presymplectic techniques combined with the constraint algorithm allow us to easily solve this problem. Indeed, define

$$T_C^2Q = \{X \in T(T_CQ)\,|\,(8.3)\text{ holds}\}.$$

The isotropic g-submanifold

$$\mathcal{D}_1' = \flat_L^{-1}\{-dE(T_CQ)\} \cap T_C^2Q$$

consists of those vectors which satisfy both (8.1) and (8.3) along $\tau_{T_CQ}(\mathcal{D}_1')$. Applying the constraint algorithm to $\mathcal{D}_1'$, one obtains a second-order differential equation $\mathcal{D}_F'$ whose sections are solutions of the Lagrange equations along $\tau_{T_CQ}(\mathcal{D}_F')$.

Typically, however,

$$\tau_{T_CQ}(\mathcal{D}_F') \subset \tau_{T_CQ}(\mathcal{D}_K),$$

where $\mathcal{D}_K$ is as in example (1) above. Furthermore, it may happen that $\mathcal{D}_F' = \phi \neq \mathcal{D}_K$; and in the case $\mathcal{D}_F' \neq \phi$, there may not exist globally smooth sections $\tau_{T_CQ}(\mathcal{D}_F') \to \mathcal{D}_F'$ even though such sections of $\tau_{T_CQ}(\mathcal{D}_K) \to \mathcal{D}_K$ exist. Elsewhere [26] we have, subject to certain regularity conditions, proved the existence of, as well as classified, certain g-submanifolds of $\tau_{T_CQ}(\mathcal{D}_F')$ along which smooth sections exist.

(3) The Hamiltonian Formulation of Mechanics [27]

Given a Lagrangian L on the restricted velocity phasespace T_CQ, one may Legendre transform to the Hamiltonian description as follows: the 1-form dL on T_CQ generates an isotropic g-submanifold $\Lambda = t^{-1}\{dL(T_CQ)\}$ of the special symplectic manifold $(T(T^*Q),\ T\pi_Q,\ TQ,\ t^*\Theta_{TQ},\ t)$. However, $T(T^*Q)$ may be viewed as a special presymplectic manifold $(T(T^*Q),\ \tau_{T^*Q},\ T^*Q,\ \flat_{T^*Q}{}^*\Theta_{T^*Q},\ \flat_{T^*Q})$. The g-submanifold $\tau_{T^*Q}(\Lambda)$ of T^*Q is the primary constraint g-submanifold M_1 of the Dirac-Bergmann theory. Indeed,

$$\tau_{T^*Q}(\Lambda) = \tau_{T^*Q} \circ t^{-1} \circ dL(T_CQ)$$

$$= FL(T_CQ)$$

as may be verified in charts. These constructions are summarized in the following diagram:

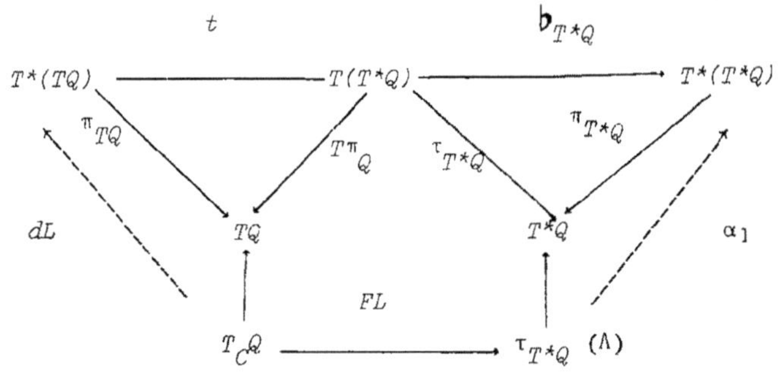

If the projection $\Lambda \to \tau_{T^*Q}(\Lambda)$ is a submersion whose fibers are connected [28], then Λ is generated by a unique closed 1-form α_1 on $M_1 = \tau_{T^*Q}(\Lambda)$ [29]. The form α_1 is the Hamiltonian 1-form for the presymplectic Hamiltonian system (M_1, ω_1), where $j_1\colon M_1 \to T^*Q$ is the inclusion and $\omega_1 = j_1{}^*\Omega_Q$.

There are two equivalent ways to proceed with a Hamiltonian analysis of the system. For example, one may apply the algorithm directly to Λ, effectively generating solutions of

$$i(X)\Omega_Q = \bar{\alpha}_1,$$

where $\bar{\alpha}_1$ is any extension of α_1 on M_1 to T^*Q. One thus obtains a symplectic version of the Dirac-Bergmann technique [6, 7]. Note, however, that this method only relies upon the existence of Λ, not the Hamiltonian 1-form $\bar{\alpha}_1$. Consequently, one has here a way to do Hamiltonian dynamics without ever mentioning Hamiltonians.

On the other hand, one may proceed more in the spirit of §VII by directly solving the Hamilton equations

$$i(X)\omega_1 = \alpha_1$$

associated to $(M_1, \omega_1, \alpha_1)$. In this case, the constraint algorithm is directly applied to the isotropic g-submanifold $\flat_1^{-1}\{\alpha_1(M_1)\}$ of the special presymplectic manifold $(TM_1, \tau_1, M_1, \flat_1{}^*\Theta_{M_1}, \flat)$.

*****

The Proca Field

As a concrete example of the generalized constraint algorithm applied to an infinite-dimensional second class system, we now work out the details for the Proca field in the Hamiltonian formulation.

The 3 + 1 decomposed Proca Lagrangian is

$$L(A,\dot{A}) = \tfrac{1}{2} \int_{\mathbb{R}^3} (\vec{\nabla}\dot{A}_\perp)^2 - 2(\vec{\nabla}A_\perp)\cdot\dot{\vec{A}} + \dot{\vec{A}} - (\vec{\nabla}\times\vec{A})^2 + m^2A_\perp^2 - m^2\vec{A}^2\}d\mu,$$

where the vector field A is decomposed $A = (A_\perp, \vec{A})$, $\mathbb{R}^3$ denotes a constant-time Cauchy surface in Minkowski spacetime and μ is some measure on $\mathbb{R}^3$.

One must first decide on a choice for velocity phasespace. The configuration space should be some Hilbert space of all four-vectors $(A_\perp, \vec{A})$. As L contains at most first spatial derivatives of A, an appropriate choice for configuration space is the manifold domain

$$C = H_\perp^1 \times \vec{H}^1$$

of

$$Q = L_\perp^2 \times \vec{L}^2,$$

with the obvious notational shorthand, where H^1 is the first Sobolev space on $\mathbb{R}^3$. Velocity phasespace, that is, the manifold of all $(A,\dot{A})$ is then the restriction of TQ to C:

$$T_CQ = (H_\perp^1 \times \vec{H}^1) \oplus (L_\perp^2 \times \vec{L}^2),$$

as no spatial derivatives of $\dot{A}$ appear in L. The measure μ can then be taken to be the ordinary L^2 measure on $\mathbb{R}^3$.

To Legendre transform to the Hamiltonian description a la example (3), we must calculate dL. For $(A,\dot{A}) \in T_CQ$ and $a \oplus \mathbf{b} \in T(T_CQ)$,

$$dL(A,\dot{A}):(a + b) = \int_{\mathbb{R}^3} \{(\vec{\nabla}A_\perp - \vec{\dot{A}})\cdot\vec{\nabla}a_\perp$$

$$+ (\vec{\dot{A}} - \vec{\nabla}A_\perp)\cdot\vec{b} - (\vec{\nabla}\times\vec{A})\cdot(\vec{\nabla}\times\vec{a})$$

$$+ m^2A_\perp a_\perp - m^2\vec{A}\cdot\vec{a}\}d\mu$$

Appealing to the theorem of §VI, one finds that the isotropic g-submanifold $dL(T_CQ) \subset T^*(TQ)$ consists of those points

$$(A,\dot{A}) \oplus (\sigma, \pi) \in (C\times Q) \oplus (Q^*\times Q^*)$$

such that

$$<a|\sigma> = \int_{\mathbb{R}^3}\{(\vec{\nabla}A_\perp-\vec{\dot{A}})\cdot\vec{\nabla}a_\perp + m^2A_\perp a_\perp$$

$$- (\vec{\nabla}\times\vec{A})\cdot(\vec{\nabla}\times\vec{a}) - m^2\vec{A}\cdot\vec{a}\}d\mu \tag{8.4a}$$

$$<b|\pi> = \int_{\mathbb{R}^3} (\vec{\dot{A}} -\vec{\nabla}A_\perp)\cdot\vec{b}\ d\mu \tag{8.4b}$$

for arbitrary $a \in C$ and $b \in Q$. Here, the natural pairing $< | >: TQ \times T^*Q \to \mathbb{R}$ is defined by

$$<(A,\dot{A})|(A,\pi)> = \int_{\mathbb{R}^3}\{\vec{\dot{A}}\cdot\vec{\pi} + \dot{A}_\perp\pi_\perp\}d\mu. \tag{8.5}$$

According to (8.5), (8.4b) implies that

$$\pi_\perp = 0. \tag{8.6}$$

Applying t^{-1}, we have that

$$\Lambda = t^{-1}\{dL(T_CQ)\}\subset T(T^*Q)$$

consists of those points

$$(A,\pi) \oplus (\dot{A},\sigma) \in (C\times\vec{L}^{2*}) \oplus (Q\times Q^*)$$

for which (8.4a, b) hold with $\pi = (0,\vec{\pi})$.

Viewing Λ as an isotropic g-submanifold of the special symplectic manifold $(T(T^*Q), \tau_{T^*Q}, T^*Q, \flat_{T^*Q}{}^*\Theta_{T^*Q}, \flat_{T^*Q})$, one finds that the primary constraint g-submanifold $M_1 = \tau_{T^*Q}(\Lambda)$ of T^*Q is

$$M_1 = C \times \vec{L}^{2*}.$$

The condition (8.6) is therefore a primary constraint.

The induced projection $\Lambda \to M_1$ is clearly a submersion whose fibers are connected. Thus, Λ is generated by a closed 1-form α_1 on M_1. In fact, $\alpha_1 = -dH_1$, where the Hamiltonian H_1 on M_1 is

$$H_1(A,\pi) = \tfrac{1}{2}\int_{R^3}\{\vec{\pi}^2 + 2(\vec{\nabla}A_\perp)\cdot\vec{\pi} + (\vec{\nabla}\times\vec{A})^2 - m^2A_\perp^2 + m^2\vec{A}^2\}d\mu$$

for $(A,\pi) \in M_1$ (cf. [29]).

We now apply the constraint algorithm to solve the field equations

$$i(X)\omega_1 = -dH_1 \tag{8.7}$$

of the presymplectic dynamical system (M_1, ω_1, dH_1), where $j_1\colon M_1 \to T^*Q$ is the inclusion and $\omega_1 = j_1{}^*\Omega_Q$.

The first step is to calculate the isotropic g-submanifold $\mathcal{D}_1 = \flat_1^{-1}\{-dH_1(M_1)\}$ of $(TM_1, \tau_1, M_1, \flat_1{}^*\Theta_{M_1^*}, \flat_1)$. If $(A,\pi) \in M_1$ and $b \oplus \tau \in TM_1$,

$$\begin{aligned} dH_1(A,\pi)\cdot(b\oplus\tau) = \int_{R^3} \{&\vec{\pi}\cdot\vec{\tau} + \vec{\nabla}b_\perp \cdot\pi \\ &+ (\vec{\nabla}A_\perp)\cdot\vec{\tau} + (\vec{\nabla}\times\vec{A})\cdot(\vec{\nabla}\times\vec{b}) \\ &- m^2A_\perp b_\perp + m^2\vec{A}\cdot\vec{b}\}d\mu. \end{aligned} \tag{8.8}$$

Writing $X(A,\pi) = a \oplus \sigma \in TM_1$, (8.7) becomes

$$\omega_1(a\oplus\sigma, b\oplus\tau)\,|\,(A,\pi) = -dH_1(A,\pi)\cdot(b\oplus\tau). \tag{8.9}$$

From the definition of ω_1, (2.4b) and (8.5),

$$\omega_1(a+\sigma, b+\tau)\,|\,(A,\pi) = \int_{R^3}\{\vec{b}\cdot\vec{\sigma} - \vec{a}\cdot\vec{\tau}\}d\mu\,.$$

Substituting this expression into (8.9), and then comparing with (8.8), one calculates

$$X(A,\pi) = (a_\perp, \vec{\pi} + \vec{\nabla}A_\perp) \oplus (0, \Delta\vec{A} - \vec{\nabla}(\vec{\nabla}\cdot\vec{A}) - m^2\vec{A}) \tag{8.10}$$

iff

$$\vec{\nabla}\cdot\vec{\pi} + m^2 A_\perp = 0 \,. \tag{8.11}$$

Note that these formal expressions are well-defined iff $\vec{A} \in \vec{H}^2$ and $\vec{\pi} \in \vec{H}^{1*}$. Thus, we

$$\mathcal{D}_1 = \{(A,\pi) \oplus (a,\sigma) \;\varepsilon\; [(H^1_\perp \times \vec{H}^2)\times\vec{H}^{1*}] \oplus [(H^1_\perp\times\vec{H}^1)\times\vec{L}^{2*}]$$

such that

$$a = (a_\perp,\ \vec{\pi} + \vec{\nabla}A_\perp),$$

$$\sigma = (0,\ \Delta\vec{A} = \vec{\nabla}(\vec{\nabla}\cdot\vec{A}) - m^2\vec{A}),$$

and

$$\vec{\nabla}\cdot\vec{\pi} + m^2 A_\perp = 0\}.$$

Proceeding with the algorithm, the secondary constraint g-submanifold $M_2 = \tau_1(\mathcal{D}_1)$ along which algebraic solutions X to (8.7) exist is

$$M_2 = \{(A,\pi) \;\varepsilon\; (H^1_\perp \times \vec{H}^2) \times \vec{H}^{1*} \mid \text{(8.11) holds}\}.$$

We now check the integrability conditions: is $\mathcal{D}_1 \subseteq \underline{\overline{TM}_2}$?

$$\underline{\overline{TM}_2} = \{(A,\pi) \oplus (a,\sigma) \;\varepsilon\; TM_1 \mid \text{(8.11) holds and } \vec{\nabla}\cdot\vec{\sigma} + m^2 a_\perp = 0\}.$$

From the definition of $\mathcal{D}_1$, however,

$$\vec{\nabla}\cdot\vec{\sigma} + m^2 a_\perp = \vec{\nabla}\cdot\{\Delta\vec{A} - \vec{\nabla}(\vec{\nabla}\cdot\vec{A}) - m^2\vec{A}\} + m^2 a_\perp$$

$$= m^2\,\{a_\perp - \vec{\nabla}\cdot\vec{A}\} \neq 0,$$

so that $\mathcal{D}_1 \not\subseteq \underline{\overline{TM}_2}$. Thus, we consider

$$\mathcal{D}_2 = \mathcal{D}_1 \cap \underline{\overline{TM}_2}$$

$$= \{(A,\pi) \oplus (a,\sigma) \;\varepsilon\; \mathcal{D}_1 \mid a_\perp - \vec{\nabla}\cdot\vec{A} = 0\}.$$

Calculating the tertiary constraint g-submanifold M_3, we have

$$M_3 = \tau_1(\mathcal{D}_2) = M_2,$$

and M_2 is the final constraint g-submanifold.

Thus, the constraint algorithm terminates, and we are assured that at least one solution X to (8.7) restricted to M_2 exists. From (8.10) and the expression for $\mathcal{D}_2$, one finds for $(A,\pi) \in M_2$

$$X(A, \pi) = (\vec{\nabla}\cdot\vec{A}, \vec{\pi} + \vec{\nabla}A_\perp) \oplus (0, \Delta\vec{A} - \vec{\nabla}(\vec{\nabla}\cdot\vec{A}) - m^2\vec{A}).$$

These are just the Proca equations:

$$dA_\perp/dt \equiv a_\perp = \vec{\nabla}\cdot\vec{A}$$

$$d\vec{A}/dt \equiv \vec{a} = \vec{\pi} + \vec{\nabla}A_\perp$$

$$d\pi_\perp/dt \equiv b_\perp = 0$$

$$d\vec{\pi}/dt \equiv \vec{b} = \Delta\vec{A} - \vec{\nabla}(\vec{\nabla}\cdot\vec{A}) - m^2\vec{A}.$$

Clearly, X is unique. That the Proca equations give rise to a well-defined flow on $\overline{M_2}$ follows from the hyperbolic version of the Hille-Yoshida Theorem [15].

The Proca canonical system is thus (M_1, ω_1, M_2) and is second class. Indeed, let $(A, \pi) \in M_2$ and $b \oplus \tau \in TM_1$. Then $b \oplus \tau \in TM_2^\perp$ iff

$$0 = \omega_1(a\oplus\sigma, b\oplus\tau)|(A,\pi)$$

$$= \int_{\mathbb{R}^3}\{\vec{b}\cdot\vec{\sigma} - \vec{a}\cdot\vec{\tau}\}d\mu.$$

for arbitrary $a\oplus\sigma \in \underline{TM_2}$. Taking $a\oplus\sigma = (0,\vec{a}) \oplus 0$, the above expression will vanish iff $\vec{\tau} = 0$. On the other hand, if $a\oplus\sigma = (-m^{-2}\vec{\nabla}\cdot\vec{\sigma}, \vec{0}) \oplus (0,\vec{\sigma})$, then this expression is zero iff $\vec{b} = 0$. Consequently, $b\oplus\tau \in TM_2^\perp$ iff $b \oplus \tau = (b, \vec{0}) + 0$. But such a $b \oplus \tau$ is an element of $\underline{TM_2}$ iff $b_\perp = 0$. Thus, $\underline{TM_2} \cap TM_2^\perp = \{0\}$ and the Proca canonical system is weakly symplectic.

Appendix

List of Symbols

$\alpha^*\Theta$ pullback of Θ by α

d exterior derivative

D Frechet derivative

$\bar{D}/\dot{D}$ partial Frechet derivative along the base/fiber of a fiber bundle

FL	Legendre transformation $TQ \to T^*Q$ induced by L
H^n	nth Sobolev space on $\mathbb{R}^3$
θ_Q	canonical 1-form on T^*Q
i	interior product
$\overline{N}$	topological closure of N
π_Q	cotangent bundle projection $T^*Q \to Q$
t	canonical diffeomorphism $T(T^*Q) \to T^*(TQ)$
T	tangent functor
Tf	pushforward, prolongation of f
T_CQ	restriction of TQ to $C \subseteq Q$, $TQ\|C$
TM	tangent bundle of M; set of all smooth vectorfields on M
$\underline{TN}$	image $Tj(TN)$ of TN in TM, where $j: N \to M$ is an inclusion
$TN^\flat$	image of TN in T^*M under $\flat$
$TN^\perp$	symplectic complement of $\underline{TN}$ in TM
$TN^\vdash$	annihilator of $\underline{TN}$ in T^*M
$\overline{T}N$	$T_N\overline{N}$
T^2Q	the diagonal in $T(TQ)$
T^*Q	cotangent bundle of Q; set of all smooth 1-forms on Q
τ_Q	tangent bundle projection $TQ \to Q$
ω, ω_1	presymplectic forms
$\dot{\omega}$	"special" presymplectic form on the tangent bundle of a presymplectic manifold
Ω_Q	canonical symplectic form on T^*Q
Ω_L	symplectic form on TQ; $FL^*\Omega_Q$
$\flat_M$	map $TM \to T^*M$ induced by the presymplectic form ω on M
$\langle\ \|\ \rangle$	dualization $TM \times T^*M \to \mathbb{R}$
$\cdot$	dualization $E \times E^* \to \mathbb{R}$ for Banach spaces
$\|$	restriction (not pullback)

Terminology and Conventions

All manifolds and maps appearing in this paper are assumed to be C^∞.

The symbol TM (T^*M) denotes both the tangent (cotangent) bundle of M and the space of all smooth vectorfields (1-forms) on M. Usually, lower-case italic letters will refer to tangent vectors, while upper-case italics will denote vectorfields.

Let Q be a manifold, τ_Q: $TQ \to Q$ its tangent bundle, and $(U;q^i)$ a chart on Q. For $w \in T_mQ$, the chart $(TU;q^i,\dot{q}^i)$ on TQ defined by

$$q^i(w) = q^i \circ \tau_Q(w)$$

$$\dot{q}^i(w) = \langle w | dq^i \rangle$$

is said to be a *natural bundle chart*. One can similarly define natural bundle charts on cotangent bundles and higher-order bundles.

Let $j: N \to M$ be a map of a Banach manifold N into a Banach manifold M. The pair (N,j) is said to be a

(*i*) *Banach submanifold* of M if j is an injective immersion (i.e., both j and Tj are injective and $Tj(TN)$ splits in $T\dot{M}$),

(*ii*) *manifold domain* of M if both j and Tj are injective and have dense range,

(*iii*) *submanifold domain* of M if (N, j) is a manifold domain of the injectively immersed submanifold N of M, and

(*iv*) *submersion* of N onto M if j and Tj are surjective and *ker* Tj splits in TN.

Throughout this paper, the term *generalized submanifold* ("*g*-submanifold") refers to any pair (N, j) which is a Banach submanifold, a manifold domain or a submanifold domain.

We now briefly explain how one calculates locally, following Refs. [15] and [20]. If $U \subset E$ is a chart on a manifold Q, then $T^*U = U \times E^*$ is a chart on T^*Q, and a point $m \in T^*Q$ has the local representation $m = (x,\sigma)$ where $x \in U$, $\sigma \in E^*$. A chart on $T(T^*Q)$ is $T(T^*U) = (U \times E^*) \oplus (E \times E^*)$. Thus a tangent vector X to T^*Q has the local representation $x(m) = (x,\sigma) \oplus (a, w)$ where $a \in E$ and $w \in E^*$. We will often suppress the base point (x,σ) and simply write this as $X =: a \oplus w$. Thus, for example, if α is a 1-form on T^*Q, the interior product $i(X)\alpha(m)$ is written locally as $\alpha(x,\sigma)\cdot(a \oplus \pi)$.

In general, we try to keep our notation and terminology consistent with that of references [14], [15] and [20].

Acknowledgements

The authors would like to express their appreciation to J. Arms, J. Marsden and W. Tulczyjew for stimulating and helpful conversations.

Notes and References

1. W. Tulczyjew, Symposia Mathematica 14, 247 (1974).
2. W. Tulczyjew, in Differential Geometric Methods in Mathematical Physics, Lecture Notes in Math., #570, 457, 464 (Springer-Verlag, Berlin, 1977).
3. W. Tulczyjew, Acta Phys. Polon., B8, 431 (1977).
4. M. Menzio and W. Tulczyjew, Ann. Inst. H. Poincaré, A28, 349 (1978).
5. J. Kijowski and W. Tulczyjew, A Symplectic Framework for Field Theories, to appear (Springer-Verlag)
6. P.A.M. Dirac, Lectures on Quantum Mechanics, Belfer Graduate School of Science Monograph Series #2 (1964).
7. M.J. Gotay, J.M. Nester and G. Hinds, J. Math. Phys. 19, 2388 (1978).
8. M.J. Gotay and J.M. Nester, Presymplectic Hamilton and Lagrange Systems, Gauge Transformations and the Dirac Theory of Constraints, to appear (Proc. of the VII Int'l. Colloq. on Group Theoretical Methods in Physics, Austin, 1978).
9. M.J. Gotay, Presymplectic Manifolds, Geometric Constraint Theory, and the Dirac-Bergmann Theory of Constraints, Dissertation, University of Maryland, 1979.

10. M.J. Gotay and J.M. Nester, Ann. Inst. H. Poincare, A30, 129 (1979).
11. J. Kijowski, Commun. Math. Phys., 30, 99 (1973).
12. J. Kijowski and W. Szczyrba, Commun. Math. Phys., 46, 183 (1976).
13. J. Sniatycki, Ann. Inst. H. Poincaré, A20, 365 (1974).
14. R. Abraham and J. Marsden, Foundations of Mechanics, second ed., (Benjamin-Cummings, New York, 1978).
15. P. Chernoff and J. Marsden, Properties of Infinite-Dimensional Hamiltonian Systems, Lecture Notes in Mathematics #425 (Springer-Verlag, Berlin, 1974).
16. J.-M. Souriau, Structures des Systemes Dynamiques, (Dunod, Paris, 1970).
17. H.P. Künzle, J. Math. Phys., 13, 739 (1972).
18. A. Lichnerowicz, C.R. Acad. Sci. Paris, A280, 523 (1975).
19. A Lagrangian subspace $\underline{TN}$ of TM is necessarily closed, so that if N is Lagrangian, then N must be a Banach submanifold of M.
20. S. Lang, Differential Manifolds, (Addison-Wesley, Reading, Mass., 1972).
21. A. Weinstein, Adv. Math., 6, 329 (1971).
22. It is assumed that all of the spaces appearing in this paper are in fact generalized submanifolds in the sense of the Appendix (cf. §VII).
23. J. Sniatycki and W. Tulczyjew, Indiana U. Math. J., 22, 267 (1972).
24. In finite-dimensions, condition (*ii*) is tantamount to requiring that $dim\ \{ker\ \omega_1 \cap TM_{\ell-1}\}$ be constant on M_ℓ.
25. J.M. Nester and M.J. Gotay, The Dynamics of Singular Presymplectic Systems, (work in progress).
26. J.M. Nester and M.J. Gotay, Presymplectic Lagrangian Systems II: The Second-Order Equation Problem, University of Maryland Preprint #79-141, (1979) (to be published).
27. See also Exercise 5.3*L* of [14].
28. This is equivalent to the almost regularity of L (cf. ref. [10]).
29. Equivalently, $FL^*\alpha_1 = -dE$.

* *Present Address: Department of Mathematics and Statistics, University of Calgary, Calgary, Alberta, Canada T2N 1N4*

DEFORMATIONS AND QUANTIZATION.

ANDRE LICHNEROWICZ

Lowell March 1979

It is possible to give a complete description of Classical Mechanics in terms of symplectic geometry and Poisson bracket. It is the essential of the hamiltonian formalism. In a common program with Bayen, Flato, Fronsdal, Sternheimer and J. Vey, we have study properties and applications of the deformations of the Poisson Lie algebra and of a trivial associative algebra. Such deformations give a new approach for Quantum Mechanics. I consider here only dynamical systems with a finite number of degrees of freedom, but the approach and a significative part of the results can be extended to physical fields. [9] .

Derivations and deformations of a Lie algebra process from a same cohomology of the Lie algebra, which is essential for our purpose. It is the cohomology with values in the Lie algebra itself and corresponding for the adjoint representation. This cohomology shall be called here the Chevalley cohomology of the Lie algebra.

1- LIE ALGEBRAS ASSOCIATED TO A SYMPLECTIC MANIFOLD.

a) Let (W,F) be a smooth symplectic manifold of dimension $2n$; the symplectic structure is defined on W by the closed 2-form F of rank $2n$ ($F^n \neq 0$ everywhere). We denote by $\mu : TW \to T^*W$ the isomorphism of vector bundles defined by $\mu(X) = - i(X)F$ (where $i(.)$ is the interior product); this isomorphism can be extended to tensors in a natural way. We denote by Λ the antisymmetric contravariant 2-tensor $\mu^{-1}(F)$. We put for simplicity $N = C^\infty(W;\mathbb{R})$.

A symplectic infinitesimal transformation (i.t.) is defined by a vector field X such that $\mathcal{L}(X)F = 0$ (where $\mathcal{L}$ is the Lie derivative); it is an infinitesimal automorphism of the structure. We denote by L the (infinite dimensional) Lie algebra of the symplectic vector fields; X belongs to L if and only if the 1-form $\mu(X)$ is closed ($d\mu(X) = 0$). If X, Y $\in$ L, we have

$$\mu([X,Y]) = d\, i(\Lambda)(\mu(X) \wedge \mu(Y)) \tag{1-1}$$

Let L^* be the subspace of L defined by the converse images of the exact 1-forms ($X_u = \mu^{-1}(du)$; $u \in N$). An element of L^* is a <u>hamiltonian vector field</u>. Consider the commutator ideal $[L,L]$ of L : each element of $[L,L]$ is, by definition, a finite sum of brackets of elements of L. It follows trivially from (1-1) that $[L,L] \subset L^*$. It has been proved (Arnold, Calabi, myself) that we have $[L,L] = L^*$; L/L^* is abelian and $\dim L/L^* = b_1(W)$, where $b_1(W)$ is the first Betti number of W for the homology with compact supports.

b) Let $\bar{N}$ be the space of the classes of elements of N modulo the additive constants; $\pi : u \in N \to \bar{u} \in \bar{N}$ is the projection of N onto $\bar{N}$. The natural isomorphism between the spaces L^* and $\bar{N}$ induces on $\bar{N}$ a structure of Lie algebra defined in the following way : if $\bar{u}, \bar{v} \in \bar{N}$, it follows from (1-1) that the function :

$$w = i(\Lambda)(d\bar{u} \wedge d\bar{v}) \tag{1-2}$$

defines a class $\bar{w}$ which is the bracket of $\bar{u}$ and $\bar{v}$. The function w is the Poisson bracket of $\bar{u}$, $\bar{v}$, or of two representants u, v in N. We put

$$w = \{u,v\} = P(u,v)$$

The Poisson bracket P defines on N itself a structure of Lie algebra; P is a bidifferential operator of order 1 in each argument, null on the constants; (N,P) is the <u>Poisson Lie algebra</u> of the manifold and we have a homomorphism of the Poisson Lie algebra L^* of the hamiltonian vector fields.

2- CLASSICAL DYNAMICS AND SYMPLECTIC MANIFOLDS.

a) Consider a dynamical system with time independent constraints and n degrees of freedom. The corresponding configuration space is an arbitrary differentiable manifold M of dimension n. It is well-known that the cotangent bundle T^*M admits a natural symplectic structure defined by the Liouville 2-form which may be written locally in terms of classical variables

$$F = \sum_a dp_\alpha \wedge dq^\alpha$$

For the hamiltonian formalism, a dynamical state of the system is nothing other as a point of $W = T^*M$, which is the usual phase space of the mechanicians and physicists. The analysis of the equations of Mechanics has showed, from a long time, that it is essential to may introduce changes of the variables (q^α, p_α) which doe not respect the cotangent structure of W. We are thus led to introduce as phase space a symplectic manifold (W,F) of dimension 2n.

Dynamics is determined by a function $H \in N$, the hamiltonian of the system, which defines a hamiltonian vector field X_H . A motion of the dynamical system is given, by an integral curve c(t) of the hamiltonian vector field X_H, the parameter t being the time.

Such is the geometrical meaning of the classical equations of Hamilton.

b) We can adopt another viewpoint. The space N admits the following two algebraic structures :

1) a structure of associative algebra defined by the usual product of functions which is here commutative)

2) a structure of Lie algebra defined by the Poisson bracket.

The derivations of the product are given by the vector fields; in particular, it follows from

$$\{u,v\} = \mathcal{L}(X_u)v$$

that we have :

$$(2\text{-}1) \qquad \{w,\ uv\} = \{w,u\}\ .\ v + u.\ \{w,v\}$$

Consider a family u_t of elements of N satisfying the differential equation :

$$(2\text{-}2) \qquad du_t/dt = \{H,\ u_t\}$$

and taking the initial value u_o at $t = 0$. It follows from (2-1) that the evolution in the time of u_t processes from the integral curves which appear in the first viewpoint; (2-2) may be considered as the intrinsic equation of Classical Dynamics.

c) We have completely described Classical Mechanics in terms of the two laws of composition defined on N and connected by (2-1). It is natural to study if it is possible to deform in a suitable way these two algebraic laws so that we obtain a model isomorphism to the Usual Quantum Mechanics. The first results are positive.

3- CHEVALLEY COHOMOLOGY AND DERIVATIONS.

a) The Chevalley cohomology of the Poisson Lie algebra (N,P) is defined in the following way : a p-cochain C of N is an alternate p-linear map of N^p into N, the 0-cochains being identified with the elements of N. The coboundary of the p-cochain C is (p+1)-cochain ∂C defined by :

$$(3\text{-}1) \qquad \partial C(u_o,\ldots,u_p) = \varepsilon^{\lambda_o\ldots\lambda_p}_{o\ldots p}\left(\frac{1}{p!}\{u_{\lambda_o},C(u_{\lambda_1},\ldots,u_{\lambda_p})\} - \frac{1}{2(p-1)!}C(\{u_{\lambda_o},u_{\lambda_1}\},u_{\lambda_2},\ldots,u_{\lambda_p})\right)$$

where ε is the skewsymmetrical Kronecker indicator and where $u_\lambda \in N$. The space of the 1-cocycles of (P,N) is the space of the <u>derivations</u> of the Lie algebra, the space of the exact 1-cocycles is the space of the <u>inner derivations</u>.

A p-cochain C is called <u>local</u> if, for each $u_1 \in N$ such that $u_1|_U = 0$ for a domain U, we have $C(u_1, .., u_p)|_U = 0$. If C is local, ∂C is local.

A p-cochain C is called <u>d-differential</u> ($d \geqslant 1$) if it is defined by a multidifferential operator of maximum order d in each argument. If C is d-differential, ∂C is d-differential also.

b) I have proved the following non trivial theorem ([3],[11])

<u>Theorem 1. If W is non compact, each 1-cochain T of N such that $C = \partial T$ is d-differential itself.</u>

<u>If W is compact, each exact d-differential 2-cochain C is the coboundary of a d-differential 1-cochain T.</u>

It is possible to deduce from this theorem the knowledge of all the derivations of (N,P). In the non compact case, the derivations are given by suitable first order differential operators; in the compact case there exist non local derivations. The most interesting result is the following

<u>Theorem 2 - The derivations $\mathcal{D}$ of the Poisson Lie algebra (N,P) which are null on the constants are given by</u> :

$$\mathcal{D}u = \mathcal{L}(X)u$$

<u>where $X \in L$. [1]</u> .

4- FORMAL DEFORMATIONS OF THE POISSON LIE ALGEBRA.

I will first recall and extend the main elements of the theory of Gerstenhaber concerning the deformations of the algebraic structures, in particular of the Lie algebras [2] .

a) Let $E(N;\lambda)$ be the space of the formal functions of $\lambda \in \mathbb{C}$ with coefficients in N; λ is said to be the parameter of deformation. Consider an alternate bilinear map $N \times N \to E(N;\lambda)$ which gives a formal series in λ :

(4-1) $$[u,v]_\lambda = \sum_{r=0}^{\infty} \lambda^r C_r(u,v) = \{u,v\} + \sum_{r=1}^{\infty} \lambda^r C_r(u,v)$$

where the $C_r (r \geqslant 1)$ are differential 2-cochains of (N,P). These cochains may be extended to $E(N;\lambda)$ in a natural way. If u, v, w $\in$ N, we have :

$$S\Big[[u,v]_\lambda ,w\Big]_\lambda = \sum_{t=1}^{\infty} \lambda^t D_t(u, v, w)$$

where S is the summation after circular permutation, and where D_t is the 3-cochain :

$$D_t(u,v,w) = S \sum_{r+s=t} C_r (C_s(u,v),w) \qquad (r,s \geqslant 0)$$

We say that (4-1) defines <u>a formal deformation of the Poisson Lie algebra</u> if the Jacobi identity is satisfied formally, that is if $D_t = 0$ $(t = 1,2,\ldots)$. We put :

$$E_t(u,v,w) = S \sum_{r+s=t} C_r(C_s(u,v),w) \qquad (r,s \geqslant 1)$$

and we have $D_t = E_t - \partial C_t$. If (4-1) is limited to the order q, we have <u>a deformation of order q</u> if the identity Jacobi is satisfied up to the order (q+1). If such is the case, E_{q+1} is automatically a 3-cocycle of (N,P). We can find a 2-cochain C_{q+1} satisfying $D_{q+1} = E_{q+1} - C_{q+1} = 0$ iff E_{q+1} is exact; E_{q+1} defines a cohomology class which is <u>the obstruction at the order (q+1)</u> to the construction of a formal deformation.

A deformation of order 1 is called an infinitesimal deformation. We have $E_1 = 0$ and so only $C_1 = 0$, that is C_1 is a 2-cocycle of (N,P).

b) Consider a formal series in

(4-2) $$T = \sum_{s=0}^{\infty} \lambda^s T_s = Id_N + \sum_{s=1}^{\infty} \lambda^s T_s$$

where the T_s are differential operators on N $(s > 1)$; T_λ acts naturally on $E(N;\lambda)$. Consider alos another alternate bilinear map $N \times N \to E(N;\lambda)$ corresponding to the formal series

(4-3) $$[u,v]'_\lambda = \{u,v\} + \sum_{r=1}^{\infty} \lambda^r C'_r (u,v)$$

where the C'_r are differential 2-cochains again. Suppose that (4-2), (4-3) are such that we have formally the identity

(4-4) $$T_\lambda [u,v]'_\lambda = [T_\lambda u, T_\lambda v]_\lambda$$

$[u,v]'_\lambda$ is transformed of $[u,v]_\lambda$ by means of T_λ. Using some universal formulas, we may prove by recursion the following

Proposition- The formal deformation (4-1) of the Poisson Lie algebra being given, each formal series (4-2) generates a unique bilinear map (4-3) satisfying Equation (4-4). This map is a new formal deformation which is said to be equivalent to (4-1). In particular a deformation is called trivial if it is equivalent to the identity deformation ($C_r = 0$ for every r).

If two deformations are equivalent at the order q, there appears a 2-cocycle, element of $H^2(N;N)$ which is the obstruction to the equivalence for the order (q+1); $H^2(N,N)$ is here the 2th space of Chevalley cohomology. In particular two infinitesimal deformations defined by the 2-cocycles C_1 and C'_1 are equivalent iff $(C'_1 - C_1)$ is exact.

5- FORMAL DEFORMATIONS OF THE ASSOCIATIVE ALGEBRA.

Derivations and deformations of an associative algebra process also form a same cohomology of the algebra which is called the Hochschild cohomology.

a) This cohomology is defined in the following way : a p-cochain Γ of (N,.) is a p-linear map of N^p into N, the 0-cochains being identified again with the elements of N. The coboundary of the p-cochain Γ is the (p+1)-cochain $\tilde{\partial}\Gamma$ defined by

$$(5\text{-}1)\qquad \tilde{\partial}\Gamma(u_o,\ldots,u_p) = u_o\,\Gamma(u_1,\ldots,u_p) - \Gamma(u_o u_1,u_2,\ldots,u_p) + \Gamma(u_o,u_1 u_2,\ldots,u_p) - \ldots + (-1)^p\,\Gamma(u_o,u_1,\ldots,u_{p-1}u_p) + (-1)^{p+1}\,\Gamma(u_o,\ldots,u_{p-1}).u_p$$

We have $\tilde{\partial}^2 = 0$ for $p \geqslant 1$. A 1-cocycle of (N,.) is a derivative of this algebra. The definitions concerning local and d-differential p-cochains are the same as for the Chevalley cohomology, but it is suitable to take $d \geqslant 0$. I have proved the following

Theorem 3 - If T is an 1-cochain of (N,.) such that $\Gamma = \tilde{\partial}T$ is d-differential ($d \geqslant 0$), T is (d+1)-differential itself [11]

In particular we see that each derivation of (N,.) is 1-differential and so is defined by an arbitrary vector field.

b) Let $E(N;\nu)$ be the space of the formal functions of $\nu \in \mathbb{C}$ with coefficients in N. An associative deformation of (N;.) is defined by a bilinear map $N \times N \to E(N;\nu)$ given by :

$$(5\text{-}2)\qquad u *_\nu v = u.v + \sum_{r=1}^{\infty} \nu^r\,\Gamma_r(u,v)$$

where the Γ_r are differential 2-cochains of (N,.) such that we have the associative identity

$$(5\text{-}3)\qquad (u *_\nu v) *_\nu w = u *_\nu (v *_\nu w)$$

The Hochschild cohomology plays the same role for the associative deformations as the Chevalley cohomology for the deformations of Lie algebras. We denote by $\tilde{H}^p(N;N)$ the pth space of Hochschild cohomology.

c) Let P be the Poisson operator of the symplectic manifold (W,F). It is easy to see that P defines a Hochschild 2-cocycle ($\tilde{\partial}P = 0$) and so

$$u *_\nu v = u.v + \nu P(u,v)$$

defines an infinitesimal deformation of the associative algebra. This deformation is never trivial. If P were exact, it would be the coboundary of a second order differential operator according to Theorem 3. But such a coboundary is never equal to P since it is symmetric in u, v and P is antisymmetric in u,v.

In the following part, we consider deformations of the associative algebra (N,.) - a $*_\nu$ -product defined by

$$(5\text{-}4) \qquad u *_\nu v = u.v + \nu P(u,v) + \sum_{r=2}^{\infty} \nu^r \Gamma_r(u,v)$$

where the Γ_r are null on the constants. Moreover we suppose that $\Gamma_r(u,v)$ is symmetric in u, v if r is even, antisymmetric if r is odd. We obtain by skewsymmetrization, with $\lambda = \nu^2$:

$$(5\text{-}5) \qquad [u,v]_\nu = (2\nu)^{-1} (u *_\nu v - v *_\nu u) = P(u,v) + \sum_{r=1}^{\infty} \lambda^r \Gamma_{2r+1}(u,v)$$

which is a deformation of the Poisson Lie algebra ($C_r = \Gamma_{2r+1}$) generated by (5-4).

Under these assumptions, I have proved that <u>there exists at most one associative deformation which generates a given Lie deformation</u> [11].

Now I consider a very important example of deformation described recently by Jacques Vey. My viewpoint is different from the viewpoint of Vey [5]..

6- THE FLAT CASE.

a) Let (W,F) be a symplectic manifold. Such a manifold admit atlases of charts for which F (or Λ) have constant components (<u>natural charts</u> $\{x^i\}$ $(i,j,.. = 1,\dots,2n)$.

<u>A symplectic connection</u> Γ is a linear connection without torsion such that $\nabla F = 0$, where ∇ is the operator of covariant differentiation defined by Γ. If $\{\Gamma^i_{jk}\}$ are the usual coefficients of a connection Γ in a <u>natural</u> chart $\{x^i\}$, we introduce the coefficients $\Gamma_{ijk} = F_{il}\,\Gamma^l_{jk}$. Such coefficients $\{\Gamma_{ijk}\}$ define a symplectic connection

iff they are completely symmetric for every natural chart. A symplectic manifold admits infinitely many symplectic connections; the difference between two symplectic connections is given by a symmetric covariant 3-tensor.

b) Suppose that (W,Λ) admits a symplectic connection <u>without curvature</u>; if such is the case, the manifold (W,Λ,Γ) is called <u>a flat symplectic manifold</u>. The simplest example is the cotangent bundle of $\mathbb{R}^n$, that is $\mathbb{R}^n \times \mathbb{R}^n$. Introduce on a flat symplectic manifold the bidifferential operators P^r of maximum order r on each argument, defined by the following expression on each domain U of an arbitrary chart $\{x^i\}$:

$$P^r(u,v)_{|U} = \Lambda^{i_1 j_1} \ldots \Lambda^{i_r j_r} \nabla_{i_1 \ldots i_r} u \nabla_{j_1 \ldots j_r} v \qquad (u,v \in N) \tag{6-1}$$

We put $P^o(u,v) = u.v$. For $r = 1$, we obtain the Poisson bracket operator P.

Given a formal function $f(z)$ with constant coefficients such that $f(o) = 1$, substitute P^r to z^r in the expansion of $f(\nu z)$; we obtain a bilinear map $(u,v) \in N \times N \to u \star_\nu v = f(\nu P)(u,v) \in E(N;\nu)$. We wish to choose f so that we define thus a deformation of the associative algebra $(N,.)$. The answer is given by the following :

<u>Proposition. If (W,Λ,Γ) is a flat symplectic manifold, there is only one formal function of the Poisson bracket P (up to a constant factor and a linear change of the deformation parameter ν) that generates a formal deformation of the associative $(N,.)$: it is the exponential function.</u>

We have :

$$u \star_\nu v = \sum_{r=0}^{\infty} (\nu^r/r!)\, P^r(u,v) = \exp(\nu P)(u,v) \tag{6-2}$$

which generates <u>the deformation of the Poisson Lie algebra</u> ($\lambda = \nu^2$)

$$[u,v]_\lambda = \sum_{r=0}^{\infty} (\lambda^r/(2r+1)!)\, P^{2r+1}(u,v) = \nu^{-1} \sinh(\nu P)(u,v) \tag{6-3}$$

It is remarkable that, for $\nu = i\hbar/2$, we deduce from (6-3) a bracket $\frac{2}{\hbar}\sin(\frac{\hbar}{2}P)$

given in 1949 by Moyal in the context of the Hermann Weyl-Wigner quantization ([7], [8]).

Consider the term P^3 of (6-3). If this cocycle were exact in the Chevalley cohomology, it would be the coboundary of a 1-cochain, which can be assumed 3-differential, according to Theorem 1. But it is easy to see that such a coboundary has no term of bidifferential type (3,3). It is possible to prove that, for a flat symplectic manifold, the second space $H^2(N;N)$ of Chevalley cohomology has the dimension 1; P^3 defines a cohomology 2-class β which is a generator for this space. We see that the deformation (6-3) is non trivial, even for the order 1.

7- GENERALIZATIONS.

It is natural to study if the deformations (6-2) and (6-3) may be generalized to non flat symplectic manifolds. It is easy to see that we doe not obtain generalizations if we extend the formula (6-1) to the case where ∇ corresponds to an arbitrary connection Γ.

a) If $u \in N$, denote by $\mathcal{L}(X_u)\Gamma$ the symmetric covariant 3-tensor defined by means of the Lie derivative of the symplectic connection Γ by the hamiltonian vector field X_u. The 2-cochain S^3_Γ defined by :

$$(7\text{-}1) \qquad S^3_\Gamma(u,v)_{|U} = \Lambda^{i_1 j_1} \Lambda^{i_2 j_2} \Lambda^{i_3 j_3} (\mathcal{L}(X_u)\Gamma)_{i_1 i_2 i_3} (\mathcal{L}(X_v)\Gamma)_{j_1 j_2 j_3}$$

admits the same principal symbol as P^3. According to the properties of the Lie derivative, we have $\partial S^3_\Gamma = 0$. The same argument as for the flat case shows that the 2-cocycle S^3_Γ is non exact. If we change the symplectic connection, S^3_Γ is changed by additive of a coboundary. We see that the cohomology 2-class β of (N,P) defined by this 2-cocycle depends only upon the symplectic structure of the manifold.

b) Introduce now the following notation : we denote by Q^r a bidifferential operator of maximum order r on each argument, null on the constants and such that its principal symbol coincides with the principal symbol of P^r; Q^r is supposed symmetric in u, v if r is even, antisymmetric if r is odd. We take in particular $Q^o(u,v) = u.v$, $Q^1(u,v) = P(u,v)$ and $Q^3 \in \beta$; J. Vey has recently proved by a long and fine cohomology study, the following

Theorem 4- (Vey). Let (W,F) be a symplectic manifold such that the third Betti number $b_3(W)$ is null. There exist formal deformations of the Poisson Lie algebra of the manifold such that

$$(7\text{-}2) \qquad [u,v]_\lambda = \sum_{r=0}^{\infty} (\lambda^r/(2r+1)!)\, Q^{2r+1}(u,v)$$

General explicit forms for Q^{2r+1} are not known. For the 2-cocycle Q^3, I have proved the following result : there is a unique symplectic connection Γ such that :

$$(7\text{-}3) \qquad Q^3 = S^3_\Gamma + \partial K$$

where K is a differential operator of order $\leqslant 2$ such that K(1) = const.

c) I shall say that we have a $\star$ -products (or twisted product) on the symplectic manifold (W,F) if there are Q^r's such that

$$(7\text{-}4) \qquad u \star_\nu v = \sum_{r=0}^{\infty} (\nu^r/r!)\, Q^r(u,v)$$

is associative. The general problem of the existence of such $\star_\nu$-products on (W,F) is much more difficult than the problem solved by Vey and the answer is unknown. I have obtained however construction processes of such $\star_\nu$-products for large classes of cotangent bundles of classical groups and homogeneous spaces.

I will limit myself to the simplest example. Consider the flat symplectic manifold defined by the cotangent bundle of the space $\mathbb{R}^n - \{0\}$, that is the manifold $E = (\mathbb{R}^n - \{0\}) \times \mathbb{R}^n$. The solvable group G_2 of dimension 2 acts on E in the following way :

$$(x,y) \in E = (\mathbb{R}^n - \{0\}) \times \mathbb{R}^n \rightarrow (x' = e^{\rho}x,\ y' = e^{-\rho}(y + \sigma x) \qquad (\rho,\sigma \in \mathbb{R})$$

The group G_2 leaves the natural symplectic structure of E and the flat connection invariant. It follows that it preserves the P^r defined by (6-1). The space of the orbits of E by this group is isomorphic to T^*S^{n-1}, where $S^{n-1} = SO(n)/SO(n-1)$ is the sphere of dimension (n-1). We deduce from the $*_\nu$-product invariant under G_2 defined on E by the P^r, a natural $*_\nu$-product on T^*S^{n-1}; this product is invariant under SO(n). We may deduce from this method the existence of natural $*_\nu$-products for example for the cotangent bundles of the Stiefel manifolds and of the Grassamnn manifolds. Twisted products may be defined also on the symplectic manifolds determined by the orbits of a Lie group for the coadjoint representation, according to the classical theorem of Kirilov-Kostant-Souriau.

8- INTRODUCTION TO A SPECTRAL THEORY AND QUANTIZATION [10].

a) Come back to the flat symplectic manifold $\mathbb{R}^n \times \mathbb{R}^n$. Under suitable assumptions, Hermann Weyl has defined in this case, in terms of Fourier transform, a map Ω (the Weyl map) which associates with each element u of a large class of classical functions or distributions an operator $\hat{u}$ of a Hilbert space and conversely. The usual quantization processes in terms of these operators. But the $*_\nu$-product defined by (6-1) corresponds by Ω to the product of operators (for $\nu = i\hbar/2$). If

$$u * v = \exp((i\hbar/2)P)(u,v) \qquad \text{we have} \qquad \Omega(u * v) = \Omega(u) \cdot \Omega(v)$$

The Moyal bracket is the image by Ω^{-1} of the natural commutator of operators. We note that if u or v has a compact support, we have :

(8-1) $$\int_W (u * v)\eta = \int_W u\, v\, \eta$$

where η is the symplectic volume element. Moreover, it is possible to prove that the change of ordering for operators may be translated in terms of equivalent twisted products.

It appears as possible to develop directly Quantum Mechanics in terms of ordinary functions or distributions and $*$-products, without reference to some Ω and to operators, in a complete and autonomous way.

b) Consider a symplectic manifold (W,F) admitting a $*_\nu$-product; we put $N^c = C^\infty(W;\mathbb{C})$. Let H be the classical hamiltonian of our problem. If we consider the value $\nu = i\frac{\hbar}{2}$ of the parameter of deformation suggested by the Moyal product, we are led to translate the dynamical Heisenberg equation by :

$$\frac{du_t}{dt} = \frac{2\nu}{i\hbar}[H,u_t]_{\nu^2} \qquad (u_t \in E(N^c\,;\nu\times\mathbb{R}) \tag{8-2}$$

If we put $\tilde{H} = H/i\hbar$, we have :

$$\frac{du_t}{dt} = 2\nu[\tilde{H},u_t]_{\nu^2} = \tilde{H} *_\nu u_t - u_t *_\nu \tilde{H} \tag{8-3}$$

Introduce the $*$-powers of $\tilde{H}$ ($\tilde{H}^{(*)p} = H^{(*)p-1} *_\nu H$). It is easy to see that it follows from the symmetry properties of the Γ_r that $\tilde{H}^{(*)p}$ depends only upon the even powers of ν. We can define the $*_\nu$-exponential of $\tilde{H}t$ in the following way :

$$\mathrm{Exp}_*(\tilde{H}t) = \sum_{p=0}^{\infty} \frac{t^p}{p!}\tilde{H}^{(*)p} \tag{8-4}$$

If $u_o \in E(N^c;\nu)$, define u_t formally by :

$$u_t = \mathrm{Exp}_*(\tilde{H}t) *_\nu u_o *_\nu \mathrm{Exp}_*(-\tilde{H}t) \tag{8-5}$$

(8-5) gives the formal solution of (8-3) taking the value u_o at $t = 0$.

c) We now consider <u>the viewpoint of the mathemetical analysis</u> and give to ν the value $i\hbar/2$. Assume that H is such that, for t in a complex neighborhood of the origin, the right-side of (8-4) converges to a distribution denoted by $\mathrm{Exp}_*(\tilde{H}t)$ again and that $\mathrm{Exp}_*(\tilde{H}t)$ admits a unique Fourier-Dirichlet expansion :

(8-6) $$\mathrm{Exp}_*(\tilde{H}t) = \sum_{\tilde{\lambda} \in I} e^{\tilde{\lambda} t}\, \Pi_\lambda$$

where I is a set of C and $\Pi_\lambda \in N^c$. This expansion is similar to the spectral decomposition of an operator. It is easy to see that

(8-7) $$\Sigma\, \Pi_\lambda = 1, \quad \Pi_\lambda * \Pi_{\lambda'} = \delta_{\tilde{\lambda}\tilde{\lambda}'} \Pi_\lambda, \quad H * \Pi_\lambda = \Pi_\lambda * H = (i\hbar\tilde{\lambda})\Pi_\lambda, H = \Sigma(i\hbar\tilde{\lambda})\, \Pi_\lambda$$

We are led to the following definition :

<u>Definition - If H satisfies (8-6), $i\hbar I$ is said to be the spectrum of H; $\lambda = (i\hbar\tilde{\lambda})$ $i\hbar I$ is an eigenvalue of H and Π_λ is the corresponding eigenprojector</u>.

Come back, just for a moment, to the formal viewpoint and suppose that the parameter ν of deformation is subject to the condition to be <u>purely imaginary</u>. If $u,v \in N^c$, the property of symmetry of the Γ_r can be translated by the identity :

(8-8) $$\overline{u *_\nu v} = \bar{v} * \bar{u}$$

We say that our $*_\nu$-product is <u>symmetric</u>.

A $*_\nu$-product is sayed to be <u>nondegenerate</u>, if, for any $u \in N^c$, $\bar{u} *_\nu u = 0$ on a domain implies $u = 0$ on this domain. It follows from (8-1) that the Moyal $*$-product and the $*$ - products deduced by quotient are nondegenerate.

Consider a nondegeenrate symmetric $*$ - product. It is possible to prove that <u>the spectrum of each real-valued function H admitting a spectral expansion in the sense of (8-6) is real and that the corresponding Π_λ are real-valued</u>.

Define now N_λ by

(8-9) $$N_\lambda = \int_W \Pi_\lambda\, \tilde{\eta}$$

where $\tilde{\eta} = \eta/(2\pi\hbar)^n$, if η is the symplectic volume element. If the integral (8-9) does not converge, we say that N_λ is infinite. If N_λ is finite, a normalized state ρ_λ is defined by $\rho_\lambda = \Pi_\lambda/N_\lambda$. It is easy to verify that N_λ is the multiplicity of the state ρ_λ in the usual sense of Quantum Mechanics. In all the cases, we say that N_λ is the multiplicity of the eigenvalue λ of H.

More generally, we may consider the Fourier transform in the sense of the distributions :

(8-10) $$\mathrm{Exp}_*(\tilde{H}t) = \int e^{\tilde{\lambda}t} \, d\mu\,(\tilde{\lambda},x)$$

In general the support of $d\mu(\tilde{\lambda})$ will be referred to as the spectrum of $\tilde{H}$. It is the spectrum of $\mathrm{Exp}_*(\tilde{H}t)$ as a distribution in t in the sense of Schwartz. We obtain the spectrum of H by product by $i\hbar$.

d) A state ρ is here a real (pseudo probability) distribution on phase space normalized by the condition :

(8-11) $$\int_W \rho\,\tilde{\eta} = 1$$

and such that :

(8-12) $$\rho * \rho = \frac{1}{N}\rho$$

where N is the multiplicity. We have in the Moyal case for N = 1

(8-13) $$\int_W \rho^2\,\tilde{\eta} = 1$$

The measurable value $\langle u\rangle_t$ of the observable u at time t for the state ρ is given by :

(8-14) $$\langle u\rangle_t = \int_W (u_t * \rho)\,\tilde{\eta}$$

e) The previous algorithm directly applied to the flat case gives for the n-dimensional harmonic oscillator the energy levels $E_m = \hbar(m + \frac{n}{2})$ with the correct multiplicities. For the Hydrogen Atom, we may consider T^*S^3 as the phase space and we introduce the corresponding $*_\nu$-product invariant under SO(4) (Fock). We obtain then the complete spectrum, that is the negative discrete spectrum and the positive continuous spectrum ($E_m = -\frac{1}{2}(\hbar m)^{-2}$ with multiplicities m^2) (see [10]) .

REFERENCES

[1] A.Avez and A. Lichnerowicz C.R. Acad. Sci. Paris t275, A(1972), p.113-117.

[2] M. Gerstenhaber Ann. of Math. 79, (1964), 59-103.

[3] M. Flato, A. Lichnerowicz, D. Sternheimer C.R. Acad. Sci. Paris t283, A(1976), p.19-24.

[4] J.E. Moyal Proc. Cambridge Phil. Soc. 45, (1949), p.99-124.

[5] J. Vey Comm . Math. Helv. 50, (1975), p.421-454.

[6] A. Lichnerowicz Journ. Geom. diff. Liège 1976.

[7] H. Weyl The Theory of Groups and Quantum Mechanics, Dover New-York 1931.

[8] E.P. Wigner Phys. Rev. 40 (1932), p. 749.

[9] I.E. Segal Sympos. Mathematica t.14, p.99-117 Academic Press New-York 1974. and references quoted here

[10] F. Bayen, M. Flato, C. Fronsdal, A. Lichnerowicz, D. Sternheimer Lett. in Math. Phys. 1 (1977), p.521-530; Deformation Theory and Quantization, Ann. of Physics 111, (1978), p. 61-152.

[11] A. Lichnerowicz C.R. Acad. Sci. Paris t.286, A(1978), p.49-53; Sur les algèbres formelles associées par déformation à une variété symplectique. Ann. di Matem. pura e appl. (to appear).

HOLOMORPHIC GAUGE THEORY

Gerald Kaiser
Mathematics Department
University of Lowell
Lowell, Massachusetts 01854

ABSTRACT

A new invariant way of obtaining interactions from gauge freedom is explored. No use is made of Lagrangians. Instead, the starting point is a scalar quantity of immediate physical interest: the probability density ρ of the particle in phase space, as defined in references [3-6]. This theory is based not on space-time R^4 but on the <u>forward tube</u> T, which is interpreted as an extended classical phase space. The probability density ρ is a positive function on T which can be expressed as the fiberwise inner product $\langle f,f \rangle$ of the wave function f with itself. Here f is a holomorphic section of the trivial holomorphic vector bundle $T \times C^s$, and the inner product is with respect to a fiber metric h: $\langle f,f \rangle = f^*hf$. Conservation of probability, combined with holomorphy, leads to an equation for f which is closely related to the Klein-Gordon equation for a particle minimally coupled to a Yang-Mills field. The Yang-Mills potential is uniquely determined as the canonical connection of type (1,0) defined by h.

1. INTRODUCTION

I shall outline here an attempt to construct a radically new formalism for relativistic quantum theories, based upon the following general idea:

Consider an isolated system of quantum fields and/or particles with a Hilbert space H. Relativistic invariance requires that H carry a unitary representation U of the restricted Poincaré group $P_+^\uparrow$. The generators P_μ of space-time translations are interpreted as the total momentum and energy observables of the system, hence are required to satisfy the spectral condition: their joint spectrum must be contained in the closure of the forward light cone V_+. Consequently, if y is any vector in V_+, the operator $yP \equiv y^\mu P_\mu$ is non-negative and the group of space-time translations $U(x) = \exp(-ixP)$ can be extended to a (holomorphic) semi-group

$$U(z) = \exp(-izP) = \exp(-yP)\; U(x), \tag{1}$$

where $z=x-iy$ belongs to the forward tube

$$T=\{x-iy \in C^4 \mid y \in V_+\}, \tag{2}$$

which may be regarded as a complexification of the spacetime associated with the theory. This extension has some important and useful consequences, such as the holomorphy of the Wightman functions [1] and the existence of quantum fields in the Euclidean region [2]. However, the extended objects (complexified space-time T, holomorphic Wightman functions, etc.) are usually not given a direct physical interpretation but are, rather, regarded as auxiliary technical devises. The "real" physics is believed to take place in the "real" space-time R^4, which mathematically plays the role of a "distinguished boundary" of T. Our primary aim will be to show that T can, in fact, have a direct physical significance: it can be interpreted as an extended classical phase space for particles associated with the theory. This interpretation can then be used to motivate the construction of a new formalism based on T rather than R^4.

The plan of the paper is as follows: In Section 2 we review some recent results [3-6] which establish the phase-space formalism for single free massive scalar particles. In Section 3 we propose a generalization of this formalism to the case of a particle in an external electromagnetic or Yang-Mills field. In section 4 we indulge in a little speculation.

2. FREE PARTICLES

A free scalar particle of mass m>o is described by a wave-function f(x) satisfying the Klein-Gordon equation

$$(\Box + m^2) f(x) = o. \tag{3}$$

This means that the Fourier transform of f is supported on $\Omega \cup (-\Omega)$, where Ω is the positive mass shell $(p^2=m^2)$. Since only positive-energy states are of interest to us, we consider only those f(x) with Fourier transform on Ω. They are given by

$$f(x) = (2\pi)^{-3/2} \int_\Omega e^{-ixp} \hat{f}(p)\, d\Omega(p) \tag{4}$$

where

$$xp = x^\mu p_\mu = x_o p_o - \vec{x}\cdot\vec{p}$$
$$d\Omega(p) = dp_1 dp_2 dp_3 / p_o$$

and $\hat{f}(p) \epsilon L^2(\Omega)$. Note that if $y \epsilon V_+$, then $yp = y_o p_o - \vec{y}\cdot\vec{p} > o$ for all $p \epsilon \Omega$. Hence, the replacement of x by x-iy can only help the above integral to converge. In fact, this results in a holomorphic function in $z = x - iy \epsilon T$:

$$f(z) = (2\pi)^{-3/2} \int_\Omega e^{-izp} \hat{f}(p)\, d\Omega(p) . \tag{5}$$

For each $z \epsilon T$ define

$$e_z(p) = (2\pi)^{-3/2} e^{i\bar{z}p} \tag{6}$$

where $\bar{z}=x+iy$. Then for $w\in T$,

$$<e_z|e_w>_{L^2(\Omega)} = (2\pi)^{-3}\int_{\Omega} e^{-i(z-\bar{w})p} d\Omega(p) \tag{7}$$

$$= -2i\Delta_+(z-\bar{w})$$

where Δ_+ is the two-point Wightman function for the free scalar field of mass m [1]. In particular, each e_z belongs to $L^2(\Omega)$ and

$$f(z) = <e_z|\hat{f}>_{L^2(\Omega)}. \tag{8}$$

This means that the vector space $K = \{f(z)|\hat{f}\in L^2(\Omega)\}$, with inner product

$$<f|g> \equiv <\hat{f}|\hat{g}>_{L^2(\Omega)}, \tag{9}$$

is a reproducing-kernel Hilbert space [7]. The wave-packets e_z play a very special role: they are evaluation maps for K (eq. (8)). This has an immediate consequence of physical interest:

Theorem 1. Fix $z\in T$. Then the function

$$\tilde{e}_z(w) \equiv <e_w|e_z>/\|e_z\|$$

(where $\|e_z\|^2 \equiv <e_z|e_z>$)is the unique solution (up to a phase factor) to the following variational problem: Find $f\in K$ such that $\|f\|=1$ and $|f(z)|$ is a maximum.

The proof merely consists of noting that

$$|f(z)| = |<e_z|\hat{f}>| \leq \|e_z\| \, \|f\| = \|e_z\|,$$

equality holding iff $\hat{f}=ce_z$, so $f(w) = c<e_w|e_z>$.

In this sense, the $\tilde{e}_z$ are "optimal" wave packets in K. The physical interpretation of T can now be established. Let

$$X_k = i\left(\frac{\partial}{\partial p_k} - \frac{p_k}{2p_o{}^2}\right), k=1,2,3,$$

be the Newton-Wigner position operators [8] on $L^2(\Omega)$. Then it is easily computed that in the state e_z, $z=x-iy$ with $x^o=o$, we have

$$<X_k>\equiv<e_z|X_k e_z>/\|e_z\|^2=x_k. \tag{10}$$

A more involved calculation gives

$$<P_\mu>\equiv<e_z|P_\mu e_z>/\|e_z\|^2=A(\lambda)y_\mu, \tag{11}$$

where P_μ is the energy-momentum operator (multiplication by p_μ in $L^2(\Omega)$) and A depends on z only through

$$\lambda\equiv\sqrt{y_\mu y^\mu}\equiv\sqrt{y^2}.$$

It follows that for each $\lambda>o$, the submanifold

$$P_\lambda=\{z=x-iy\varepsilon T|x^o=o,y^2=\lambda^2\} \tag{12}$$

can be identified as a classical phase space for the particle. A possible physical interpretation of λ will be discussed in Section 4.

We have defined the inner product in the space of solutions K by using the Fourier transform (eq. (9)). This obviously won't work when an external field is present, hence we must find an expression for $<f|g>$ using only the function values f(z) and g(z). This will further clarify the role of classical phase spaces in quantum mechanics. The following result has been obtained [4]:

Theorem 2. Let

$$d\mu_\lambda(z)=C_\lambda dx_1 dx_2 dx_3 dy_1 dy_2 dy_3$$

(with a certain choice of C_λ). Then

$$<f|g>=\int_{P_\lambda}\overline{f(z)}g(z)\,d\mu_\lambda(z) \tag{13}$$

for all $f,g\varepsilon K$.

Using eq. (8), this gives a "continuous resolution of the identity" in terms of the e_z:

$$\int_{P_\lambda}|e_z><e_z|\,d\mu_\lambda(z)=\mathbb{1}\ .$$

Now $L^2(\Omega)$ carries an irreducible unitary representation of the restricted Poincaré group $P_+^\uparrow$, under which the e_z are covariant: $U(a,\Lambda)e_z=e_{\Lambda z+a}$. This gives rise to an equivalent representation of $P_+^\uparrow$ on K:

$$(U(a,\Lambda)f)(z)=<e_z|U(a,\Lambda)f>=<U(a,\Lambda)^*e_z|f>=f(\Lambda^{-1}(z-a)).$$

Clearly P_λ is not invariant under the action of $P_+^\uparrow$ on T, hence the expression (13), though invariant, is not "manifestly" invariant. To rectify this, we will now define a large class $\mathcal{S}$ of "phase spaces" such $<f|g>$ can be expressed in terms of any $\sigma\varepsilon\mathcal{S}$.

Let $s:R^4\to R$ be a C^∞ function such that $S\equiv\{x\varepsilon R^4|s(x)=o\}$ is a 3-dimensional submanifold. Fix $\lambda>o$ and let

$$\begin{aligned}\sigma&=\{x-iy\varepsilon T|s(x)=o,y^2=\lambda^2\}\\ &\equiv S-i\Omega_\lambda.\end{aligned} \tag{14}$$

We wish to define a "covariant" symplectic structure [9] on σ. Begin with the Poincaré-invariant symplectic form on T:

$$\alpha=dy_\mu\wedge dx^\mu \tag{15}$$

and ask: under what conditions is the pullback α_σ of α to σ a symplectic form? Clearly it suffices for α_σ to be non-degenerate, i.e., for α_σ^3 to be a volume form on σ. But

$$\alpha^3 = 3!\, d\hat{y}^\mu \wedge d\hat{x}_\mu$$

where

$$d\hat{y}^\mu = (-)^\mu dy_o \wedge \ldots dy_{\mu-1} \wedge dy_{\mu+1} \wedge \ldots \wedge dy_3$$

$$d\hat{x}_\mu = (-)^\mu dx^3 \wedge \ldots \wedge dx^{\mu+1}\ dx^{\mu-1} \wedge \ldots \wedge dx^o .$$

From

$$\alpha^3 \wedge ds \wedge d(y^2 - \lambda^2) = 3! \frac{\partial s}{\partial x^\mu} (2y^\mu) dy_o \wedge \ldots \wedge dy_3 \wedge dx^3 \wedge \ldots \wedge dx^o$$

it now follows that α_σ is non-degenerate if and only if

$$y^\mu \frac{\partial s}{\partial x^\mu} \neq o \text{ on } \sigma, \tag{16}$$

and we may assume the left-hand side to be positive. For fixed x, (16) must hold for all $y \varepsilon \Omega_\lambda$, hence for all $y \varepsilon V_+$. It follows that the 4-vector $\frac{\partial s}{\partial x^\mu}$ must belong to the closure of the cone dual to V_+. Since V_+ is self-dual (with c=1) this means that

$$\frac{\partial s}{\partial x^\mu} \frac{\partial s}{\partial x_\mu} \geq o . \tag{17}$$

Thus we have proved:

Theorem 3. α_σ is a symplectic form if and only if s(x) satisfies eq. (17), i.e., S is non-timelike.

We denote by $\mathcal{S}$ the class of all such σ's with all $\lambda > o$. These will be our phase spaces. $\mathcal{S}$ contains P_λ and is clearly invariant under $P_+^\uparrow$. In fact, the invariance of α implies that $P_+^\uparrow$ maps σ's onto one another by canonical transformations. To generalize theorem 2, define

$$<f|g>_\sigma = \int_\sigma \overline{f(z)} g(z) \mu_\sigma(z) \tag{18}$$

where μ_σ is the volume form obtained by restricting

$$\mu = \frac{1}{3!} C_\lambda \alpha^3 = C_\lambda d\hat{y}^\mu \wedge d\hat{x}_\mu \tag{19}$$

to σ. (We orient σ so that $<f|f>_\sigma \equiv \|f\|_\sigma^2 \geq o$.) Note that $s(x)=x^o$ gives $\sigma = P_\lambda$ and $\mu_\sigma = d\mu_\lambda$, hence $<f|g>_{P_\lambda} = <f|g>$.

Theorem 4. Let $\sigma \epsilon \mathcal{S}$. Then

$$<f|g>_\sigma = <f|g>$$

for all $f, g \epsilon K$.

We sketch the proof (see [5] for a full account). Clearly it suffices to set $g=f$, i.e., prove $\|f\|_\sigma = \|f\|$. Set

$$B_\lambda = \{y \epsilon V_+ | y^2 > \lambda^2\} \tag{20}$$

so that $\Omega_\lambda = -\partial B_\lambda$ (ignoring the part at $y=\infty$, which turns out not to contribute). Now

$$\|f\|_\sigma^2 = C_\lambda \int_\sigma |f(z)|^2 d\hat{y}^\mu \wedge d\hat{x}_\mu$$

$$= C_\lambda \int_S \left(\int_{\Omega_\lambda} |f(x-iy)|^2 d\hat{y}^\mu \right) d\hat{x}_\mu$$

$$\equiv\int_S j^\mu(x)\,d\hat{x}_\mu, \tag{21}$$

with

$$j^\mu(x)=C_\lambda\int_{\Omega_\lambda} |f(x-iy)|^2 d\hat{y}^\mu$$

$$=-C_\lambda\int_{B_\lambda} \frac{\partial|f|^2}{\partial y_\mu} d^4y$$

by Stokes' theorem. But

$$\frac{\partial j^\mu}{\partial x^\mu}=-C_\lambda\int_{B_\lambda}\frac{\partial^2|f|^2}{\partial x^\mu \partial y_\mu}\, d^4y \tag{22}$$

and it follows from the Klein-Gordon equation plus holomorphy that the integrand on the right-hand side vanishes; hence j^μ is a conserved current. A lengthy analysis shows that if S and S' are any two non-timelike 3-dimensional submanifolds of R^4, then

$$\int_S j^\mu d\hat{x}_\mu=\int_{S'} j^\mu d\hat{x}_\mu,$$

i.e., there is no leakage to spacelike infinity. This proves the theorem, since we can choose $\sigma'=P_\lambda$.

3. HOLOMORPHIC GAUGE THEORY

We now wish to "perturb" the preceding formalism so as to describe a particle in an external field. In the usual (Lagrangian) approach, this is done by exploiting gauge invariance [10,11]. It turns out that in the phase-space framework, a similar construction is possible without the use of a Lagrangian, as described below.

We begin by developing a gauge theory appropriate to the phase-space formalism. Very roughly speaking, the role of Lagrangian will now be played by $|f(z)|^2$. This is invariant under the "local gauge transformation "

$$f'(z)=e^{i\phi(z)}f(z), \tag{23}$$

where ϕ is a C^∞ real-valued function on T. However, such ϕ is clearly non-holomorphic (unless ϕ=constant, which is excluded from consideration), hence neither is f'. On the other hand, if ϕ is holomorphic, then so is f' but $|f(z)|^2$ cannot be invariant since ϕ is not real-valued. To retain holomorphy, yet preserve the invariance of the inner product, we introduce a positive weight function h(z): For an arbitrary $\sigma \in \mathcal{S}$ and an arbitrary holomorphic f(z), define

$$\|f\|_\sigma^2=\int_\sigma \overline{f(z)}h(z)f(z)\mu_\sigma(z), \tag{24}$$

which we assume finite, and require that along with (23), h transform as

$$h'(z)=e^{2\mathrm{Im}\,\phi(z)}h(z), \tag{25}$$

where ϕ is now assumed holomorphic.

More generally, let $f:T\to C^s$ (s=1,2,...) be a holomorphic section of the (trivial) holomorphic vector bundle $T\times C^s$. We assume given a Hermitian fiber metric [12] h on $T\times C^s$. That is, h(z) is a C^∞ function on T whose values are positive-definite s×s matrices. The fiberwise inner product of two sections f(z) and g(z) is

$$\langle f(z),g(z)\rangle=f(z)^*h(z)g(z)$$

$$\equiv\langle f,g\rangle(z). \tag{26}$$

We now define a holomorphic gauge transformation as a holomorphic change of frame on $T\times C^s$, i.e.,

$$\left.\begin{aligned} f'(z) &= (X(z))^{-1} f(z) \\ h'(z) &= X(z)^* h(z) X(z) \end{aligned}\right\} \tag{27}$$

where X is a holomorphic function whose values are non-singular matrices. Clearly, (26) is invariant under (27), hence, for fixed $\sigma\varepsilon\mathcal{S}$, so is the norm

$$\|f\|_\sigma^2 = \int_\sigma \langle f,f\rangle \mu_\sigma \tag{28}$$

which we assume to be finite.

Note that so far, we have put no constraints on f except for holomorphy and $\|f\|_\sigma<\infty$. To obtain dynamics, we work <u>backwards</u>: require that f be such that $\|f\|_\sigma$ be "conserved," that is independent of S (where $\sigma=S-i\Omega_\lambda$). Since $\Omega_\lambda = -\partial B_\lambda$ (see proof of theorem 4),

$$\begin{aligned} \|f\|_\sigma^2 &= -C_\lambda \int_S \Big(\int_{\partial B_\lambda} \langle f,f\rangle d\hat{y}^\mu \Big) \, d\hat{x}_\mu \\ &= -C_\lambda \int_S \Big(\int_{B_\lambda} \frac{\partial \langle f,f\rangle}{\partial y_\mu} \, d^4y \Big) \, d\hat{x}_\mu \, . \end{aligned} \tag{29}$$

Therefore, to conserve $\|f\|_\sigma$ it is sufficient (and, with minor changes, also necessary) that f satisfy the "continuity equation"

$$\frac{\partial^2}{\partial x^\mu \partial y_\mu} \langle f,f\rangle = o. \tag{30}$$

To get the derivatives in (30) inside the bracket, i.e., to differentiate the section f, we need a <u>connection</u> compatible with h [12]. In simple terms, this means that we must find an sxs matrix θ of 1-forms on T such that

$$d\langle f,g\rangle \equiv d(f^* h g)$$

$$= (df)^*.hg+f^*.dh.g+f^*h.dg$$

$$= (df+\theta f)^*.hg+f^*h.(dg+\theta g)$$

$$\equiv \langle Df,g\rangle+\langle f,Dg\rangle \tag{31}$$

where

$$Df = df+\theta f. \tag{32}$$

Thus θ must satisfy the "compatibility condition "

$$dh = \theta^*h+h\theta. \tag{33}$$

Equation (33) determines only the "h-Hermitian" part of θ, i.e. its Hermitian part relative to the metric h. The anti-Hermitian part is still arbitrary. Write

$$\theta = \theta_\mu dx^\mu+\theta_{\dot\mu} dy^\mu \tag{34}$$

(i.e., a dot denotes the coefficient of dy^μ) and

$$Df = (D_\mu f)\ dx^\mu + (D_{\dot\mu} f)\ dy^\mu \tag{35}$$

where

$$D_\mu f = \frac{\partial f}{\partial x^\mu} + \theta_\mu f$$

$$D_{\dot\mu} f = \frac{\partial f}{\partial y^\mu} + \theta_{\dot\mu} f. \tag{36}$$

Then (30) inplies

$$\langle D_\mu D^{\dot\mu} f,f\rangle+\langle D_\mu f,D^{\dot\mu} f\rangle+\langle D^{\dot\mu} f,D_\mu f\rangle+\langle f,D_\mu D^{\dot\mu} f\rangle=0. \tag{37}$$

To reduce (30) to a linear equation in f (dynamics), we require that the second and third terms in (37) cancel. For this it suffices to have

$$iD_{\dot{\mu}}f=\eta D_{\mu}f \tag{38}$$

where $\eta\equiv\eta\,(z)$ is an sxs h-Hermitian matrix: $\eta^*h=h\eta$. But holomorphy of f means that

$$i\,\frac{\partial f}{\partial y^{\mu}} = \frac{\partial f}{\partial x^{\mu}}, \tag{39}$$

hence $\eta(z)$ must be the identity, and (38) implies that

$$i\theta_{\dot{\mu}} = \theta_{\mu}, \tag{40}$$

so that by (34),

$$\theta=\theta_{\mu}(dx^{\mu}-idy^{\mu})=\theta_{\mu}dz^{\mu}. \tag{41}$$

That is, θ is of <u>type (1,0)</u>! This, together with (33), is sufficient to determine θ uniquely. For

$$\begin{aligned} dh &= \frac{\partial h}{\partial x^{\mu}}\, dx^{\mu} + \frac{\partial h}{\partial y^{\mu}}\, dy^{\mu} \\ &= 1/2\,\left(\frac{\partial h}{\partial x^{\mu}} + i\frac{\partial h}{\partial y^{\mu}}\right) dz^{\mu} + 1/2\,\left(\frac{\partial h}{\partial x^{\mu}} - i\,\frac{\partial h}{\partial y^{\mu}}\right) d\overline{z^{\mu}} \\ &\equiv \frac{\partial h}{\partial z^{\mu}}\, dz^{\mu} + \frac{\partial h}{\partial \bar{z}^{\mu}}\, d\bar{z}^{\mu} \\ &\equiv \partial h + \bar{\partial} h \end{aligned} \tag{42}$$

(see ref. [12], where $i\to -i$ because $z=x+iy$), hence

$$\partial h + \bar{\partial} h = \theta^* h + h\theta. \tag{43}$$

Equating the terms of type (1,0), we have $\partial h = h\theta$, or

$$\theta = h^{-1}\partial h = h^{-1}\,\frac{\partial h}{\partial z^{\mu}}\, dz^{\mu}$$

$$= 1/2\ h^{-1}\ (\frac{\partial h}{\partial x^\mu} + i\ \frac{\partial h}{\partial y^\mu})\ dz^\mu . \tag{44}$$

This is a familiar object in complex-manifold theory, called the Hermitian (or canonical) connection on TxC^s determined by h [12,13]. Inserting $iD_{\dot\mu} f = D_\mu f$ into (37), we have

$$\langle D_\mu D^\mu f, f\rangle = \langle f, D_\mu D^\mu f\rangle . \tag{45}$$

This will be satisfied if we require that f obey the linear equation

$$D_\mu D^\mu f = \xi f \tag{46}$$

where $\xi \equiv \xi(z)$ is an h-Hermitian matrix:

$$\xi^* h = h\xi . \tag{47}$$

Equation (46) is our dynamical equation. As explained below, it is formally similar to the Klein-Gordon equation of a particle minimally coupled to a Yang-Mills field. To make a detailed comparison with the latter, return to the scalar case s=1. Then

$$h(z) = e^{2\psi(z)} \tag{48}$$

for some real-valued function ψ, and (44) gives

$$(\frac{\partial}{\partial x^\mu} + \psi_\mu + i\psi_{\dot\mu})^2 f(z) = \xi(z) f(z) \tag{49}$$

where the subscripts on ψ denote derivatives with respect to x^μ and y^μ. This is to be compared with the usual Klein-Gordon equation for a particle in an electromagnetic field with four-potential $A_\mu(x)$:

$$(\frac{\partial}{\partial x^\mu} + ieA_\mu(x))^2\ f(x) = -m^2 f(x) . \tag{50}$$

The major formal differences are:

(a) In (50), f(x) is a distribution over space-time, whereas in (49), f(z) is a holomorphic function on T.

(b) Letting

$$eA_\mu(x)=\lim_{y\to o} \psi_{\dot{\mu}}(x-iy), \tag{51}$$

we note that (49) has an "extra" potential ψ_μ.

(c) The conserved norm for (49),

$$\|f\|_\sigma^2 = \int_\sigma e^{2\psi}|f|^2\mu_\sigma, \tag{52}$$

depends explicitely on ψ, whereas the conserved norm for (50),

$$\|f\|_S^2 = -\int_S \mathrm{Im}(\bar{f}\frac{\partial f}{\partial x^\mu})\ d\hat{x}^\mu \tag{53}$$

does not depend on A_μ.

(d) The right-hand side of (49) is more general than that of (50).

To a large extent, these differences can be reconciled as follows: (d) simply means that the particle is coupled to a scalar field $\xi(z)$ as well as the field defined by $\psi(z)$. We may choose $\xi(z)=-m^2$. As for (b), set

$$\tilde{f}(z)=e^{\psi(z)}\ f(z) \tag{54}$$

(which amounts to a <u>non-holomorphic</u> gauge transformation). Then (49) becomes

$$(\frac{\partial}{\partial x^\mu}+i\ \psi_{\dot{\mu}})^2\ \tilde{f}(z) = \xi(z)\tilde{f}(z), \tag{55}$$

which by (51) corresponds to (50). The transformation (54) also has a bearing on (c) since (52) becomes

$$\|\tilde{f}\|_\sigma^2 \equiv \int_\sigma |\tilde{f}|^2\mu_\sigma \tag{56}$$

which no longer depends on ψ explicitely. A possible resolution of (a) is suggested by (51), (54) and (55): we expect that solutions of (50) are related to boundary values of solutions of (49) by

$$f(x) \sim \lim_{y\to o} \tilde{f}(x-iy)$$

$$\sim \lim_{y\to o} (e^{\psi(x-iy)} f(x-iy)). \tag{57}$$

This shows that formally, the scalar theory is closely related (though probably not equivalent) to the conventional theory of a scalar particle minimally coupled to an electro-magnetic field. Similarly, for arbitrary s we will think of $\theta = h^{-1}\partial h$ as a potential of the Yang-Mills type [10,11] by regarding it as a 1-form valued in the Lie algebra gl (s,C) (under commutators). This is in agreement with the fact that our structure group is GL (s,C), as evidenced by (27) since X(z) can be any non-singular matrix. The usual structure group is U(s) or SU(s), hence smaller than ours. On the other hand, our gauge transformations are more restricted than the usual ones, being holomorphic.

Incidentally, (51) provides a rough model for h in the scalar case, if the conventional vector potential A_μ is given:

$$h(z) = \exp (ey^\mu A_\mu(x)). \tag{58}$$

This has an interesting interpretation: Let $A_k = o$ (k=1,2,3) and $eA_o(x) \equiv V(x)$. Then

$$\|f\|^2_\sigma = \int_\sigma \exp (y^o V(x)) |f(z)|^2 \mu_\sigma(z), \tag{59}$$

and the factor exp $(y^o V(x))$ may be viewed as specifying the accessibility of the phase-space point $z \varepsilon \sigma$. Thus if we normalize $\|f\|_\sigma = 1$ and if V is large and positive in some region of σ, the particle is effectively repelled from that region. Although the model (58) is probably too crude (see below), a similar interpretation of h may be kept in mind for the general case (eq. (28)). This gives a direct, intuitive picture for our "Yang-Mills" fields.

An important question, to which I presently have no answer, is: For what choices of h and ξ do non-zero holomorphic solutions to (46) exist? Since we are looking for holomorphic solutions to an equation with non-holomorphic coefficients, it is likely that h and ξ must satisfy some strong conditions. For example, one cannot expect $h(z)=\exp(y^{0}V(x))$, with V the Coulomb potential, to give rise to any holomorphic solutions. That is, the model (58) is probably too crude. Singularities such as that of the Coulomb potential should probably be confined to the distinguished boundary R^4, h being at least C^{∞} on T. It may even happen that (46) must be regarded as a system of coupled, non-linear equations for f and h together, subject to proper initial or boundary conditions, before holomorphic solutions can be found. This is in line, of course, with the usual Yang-Mills approach where the "gauge fields" take an active part in the dynamics.

One of the principal novelties of the present theory is that our potential θ can itself be derived from a function h: $\theta=h^{-1}\partial h$. In the usual theory, this would mean that the Yang-Mills field, defined as the curvature of the connection θ, is "pure gauge," i.e., vanishes. In our case this is not so because h is not holomorphic (since $h^*=h$). The curvature is

$$\begin{aligned} F &\equiv d\theta + \theta\wedge\theta \\ &= \bar{\partial}\theta + (\partial\theta + \theta\wedge\theta). \end{aligned} \tag{60}$$

Note that the first term is of type (1,1) whereas the second is of type (2,0). Now $\theta=h^{-1}\partial h$ means that θ must satisfy an "integrability condition":

$$\begin{aligned} \partial\theta &= \partial(h^{-1})\wedge\partial h \\ &= -(h^{-1}\partial h h^{-1})\wedge\partial h \end{aligned}$$

$$= -\theta\wedge\theta. \tag{61}$$

That is, the (2,0) part of the curvature vanishes, so our "Yang-Mills" field is $F = \bar{\partial}\theta$, which is of pure type (1,1) and need not vanish. Furthermore, note that due to (61), the relation between F and θ is linear; this could prove to be a useful feature in the solution process, if F were brought into the dynamics.

Let us compute F explicitely for the scalar case with h given by (48):

$$2iF = 2i\bar{\partial}\theta = 2i\bar{\partial}((\psi_{\mu}+i\psi_{\dot{\mu}})dz^{\mu})$$

$$= i(\frac{\partial}{\partial x}\nu - i\frac{\partial}{\partial y}\nu)(\psi_{\mu}+i\psi_{\dot{\mu}})d\bar{z}^{\nu}\wedge dz^{\mu}$$

$$= (\psi_{\dot{\nu}\mu}-\psi_{\dot{\mu}\nu})(dx^{\nu}\wedge dx^{\mu}+dy^{\nu}\wedge dy^{\mu})$$

$$+(\psi_{\mu\nu}+\psi_{\dot{\mu}\dot{\nu}})(dx^{\mu}\wedge dy^{\nu}+dx^{\nu}\wedge dy^{\mu}), \tag{62}$$

where subscripts mean partials with respect to x^{μ} and y^{μ} as before. Note that the first tensor is anti-symmetric and the second is symmetric. In view of (51), the pullback of 2iF to space-time ($y\to o$) is

$$\lim_{y\to o}\ (\psi_{\dot{\nu}\mu}-\psi_{\dot{\mu}\nu})dx^{\nu}\wedge dx^{\mu} = e(A_{\nu,\mu}-A_{\mu,\nu})dx^{\nu}\wedge dx^{\mu} \tag{63}$$

which coincides with the usual electromagnetic field defined by A_{μ}.

To see how holomorphic gauge transformations are related to the usual ones (for scalars, again), set

$$f' = e^{i\phi}f$$

$$h' = e^{2\mathrm{Im}\phi}h$$

as in (23) and (25). Then $\psi'=\psi+\mathrm{Im}\phi$ in (48), hence by (51),

$$eA'_\mu(x)=\lim_{y\to o}\frac{\partial}{\partial y^\mu}\psi'(x-iy)=eA_\mu(x)+\lim_{y\to o}\frac{\partial}{\partial y^\mu}\mathrm{Im}\phi(x-iy)$$

$$=eA_\mu(x)-\lim_{y\to o}\frac{\partial}{\partial x^\mu}\mathrm{Re}\phi(x-iy)$$

$$=eA_\mu(x)-\frac{\partial}{\partial x^\mu}(\lim_{y\to o}\mathrm{Re}\phi(x-iy)) \qquad (64)$$

since ϕ is holomorphic in $z=x-iy$. We recognize in (64) the usual sort of gauge transformation.

To complete the parallel with Yang-Mills theory, we need the counterpart of the inhomogeneous Maxwell-Yang-Mills equation, to determine h. This problem is under investigation.

4. REMARKS

(a) I have tried to illustrate some ways in which the complexification of space-time, as based on the spectral condition, may be used to reformulate physical theories. Whether the constructions of the last two sections prove to be useful depends, of course, on their ability to describe systems of physical interest. In view of the instabilities which usually plague relativistic one-particle theories [14], it may turn out that ours also proves to be unsatisfactory. In that case it may still be useful as a stepping stone to quantum field theory. Here, I believe, lies the true calling of the phase-space approach. For whereas holomorphy may not be physically justifiable in the external field problem (strictly speaking, the spectral condition only holds for isolated systems), it is a basic ingredient in quantum field theory [1]. Such a theory, based on T instead of R^4, can be expected to be much more regular than the usual one. Furthermore, the "regularization" $x\to x-iy$ has the advantage of being physical: y has presumably got a direct physical significance (see below), and there is no need to let $y \to o$ at the end of calculations. The free-particle formalism of section 2 suggest that such a theory would deal directly with extended particles rather than point particles. It is therefore possible that some of the famous divergences [15] of the theory may disappear in the phase-space framework.

(b) We have seen that in the free-particle theory, $|f(x-iy)|^2$ represents the probability density relative to μ_σ that at time x^o the particle has position $\vec{x}$ and velocity $\vec{v}=\vec{y}/y_o$. Note that $\lambda \equiv \sqrt{y^2}$ has been a free parameter in the theory. We now ask: what significance can y (and, in particular, λ) have in quantum field theory, where there is no fixed type or number of particles? I can presently only offer a somewhat vague (but, I think, intriguing) suggestion. It is known [16,17] that there exists a close "analogy" between quantum field thoery and quantum statistical mechanics, based on the substitution $t \to -i\beta, \beta = 1/kT > o$. Here t is time, k is Boltzmann's constant and T is the absolute temperature. Under this substitution, the evolution operator $\exp(-itH)$ (H is the Hamiltonian) turns into the mixed state $\rho = \exp(-\beta H)$ which describes the system in thermodynamic equilibrium at temperature T (canonical ensemble). Now equilibrium is obviously a frame-dependent condition. If the system is in equilibrium in some frame, its state in a frame moving with velocity $-\vec{v}$ relative to the first will be (using $H=cP_o$) $\rho' = \exp(-y^\mu P_\mu)$, where

$$y^o = \beta c(1-v^2/c^2)^{-1/2}$$

$$\vec{y} = \beta\vec{v}(1-v^2/c^2)^{-1/2}.$$

Thus, y is a covariant version of β: its length is

$$\lambda \equiv \sqrt{y^2} = \beta c = c/kT \equiv \hbar c/kT \tag{65}$$

and its direction $\vec{y}/y_o = \vec{v}$ is the velocity of the given frame relative to that in which the system is in equilibrium.

So much for statistical mechanics. But how do these considerations apply to quantum field theory? Usually one works either with one ($z^0=-i\beta$) or the other ($z^0=t$) but not both. The analytic continuation between the two is viewed merely as a (powerful) method of carrying results from one theory to the other [2,17]. In the spirit of the approach advocated here, one might hope that the two theories can be unified into a single one using T as a base space. But what significance can temperature have in the presence of time evolution? We conjecture as follows: whereas the operator $\exp(-ixP)$ translates field observables in space-time, the operator $\exp(-yP)$ "cools" them to a local equilibrium at temperature $T=\hbar c/k\lambda$ in the frame moving at the velocity $\vec{v}=\vec{y}/y_0$ relative to the given one. To make this precise, one must specify how expectation values, etc., are obtained in the new theory.

Another, related interpretation of λ is that it provides a typical local scale for space- and time-phenomena in the theory. For example, λ measures the spatial extent of the optimal wave packets e_z (section 2) at the time x^0 [4], and the time-interval λ/c governs their decay (see [6], where $\lambda=c=1$). Thus for large λ we expect a situation in which all typical distances and times are large, the high-frequency phenomena having been suppressed. This corresponds intuitively to low temperatures! For a review of the relation between special relativity and classical statistical mechanics, see [18].

(c) The complexification of space-time discussed here is made possible only as a result of quantum theory. Yet, it welds together two important classical geometries: the Lorentzian geometry of space-time and the symplectic geometry of phase space. The synthesis is an (indefinite) Kählerian geometry [12,13] with metric

$$ds^2=g_{\mu\nu}d\bar{z}^{\mu}dz^{\nu}, \tag{66}$$

the Kähler form (that is, the anti-symmetric part) of which is a covariant version of the classical symplectic form. It would be natural to attempt a generalization of the phase-space approach by replacing T with a general (curved) indefinite

Kähler manifold M. Presumably such a theory would bring gravity into the picture as well. Furthermore, rather than working with separate metrics g (on the base space) and h (on the fibers) for holomorphic vector bundles over M, it would make sense to begin with a single (Kähler) metric on the entire bundle.

ACKNOWLEDGEMENTS

It is a pleasure to thank Stanley Deser, Lloyd Kannenberg, André Lichnerowicz, Jerrold Marsden, Richard Palais, Tudor Ratiu and Phillip Yasskin for helpful discussions.

REFERENCES

1. R. F. Streater and A. S. Wightman, PCT, Spin and Statistics and All That, (Benjamin, New York, 1964).

2. E. Nelson, in Constructive Quantum Field Theory, G. Velo and A. S. Wightman, editors (Springer, 1973); K. Osterwalder and R. Schrader, Phys. Rev. Letters 29, 1423 (1972).

3. G. Kaiser, Thesis, University of Toronto, 1977.

4. G. Kaiser, J. Math. Phys., 18, 952 (1977).

5. G. Kaiser, J. Math. Phys., 19, 502 (1978).

6. G. Kaiser, Lett. Math. Phys. 3, 61 (1979).

7. H. Meschkowski, Hilbertsche Räume mit Kernfunktion, (Springer, 1962).

8. T. D. Newton and E. P. Wigner, Rev. Mod. Phys. 21, 400 (1949).

9. R. Abraham and J. E. Marsden, Foundations of Mechanics - second edition (Benjamin/Cummings, 1978).

10. E. S. Abers and B. W. Lee, Gauge Theories, Physics Reports 9, no. 1 (1973).

11. W. Drechsler and M. E. Mayer, Fiber Bundle Techniques in Gauge Theories, (Springer, 1977).

12. R. O. Wells, Differential Analysis on Complex Manifolds, (Prentice Hall, 1973).

13. S. Kobayashi and K. Nomizu , Foundations of Differential Geometry, vols. I and II (Interscience, 1963 and 1969).

14. Invariant Wave Equations, G. Velo and A. S. Wightman, editors (Springer, 1978).

15. J. D. Bjorken and S. D. Drell, Relativistic Quantum Mechanics (McGraw-Hill, 1964); Relativistic Quantum Fields (McGraw-Hill, 1965).

16. L. P. Kadanoff and G. Baym, Quantum Statistical Mechanics (Benjamin, 1962).

17. F. Guerra, L. Rosen and B. Simon, Annuals of Math. 101, 111 (1975).

18. D. Ter Haar and H. Wergeland, Thermodynamics and Statistical Mechanics in the Special Theory of Relativity, Physics Reports 1, no. 2 (1970).

A GEOMETRIC VARIATIONAL FORMALISM FOR THE THEORY OF NONLINEAR WAVES

Robert Hermann
Division of Applied Sciences
Harvard University

1. INTRODUCTION

The hybrid, interdisciplinary subject called "nonlinear wave theory" has revitalized an old subject, the differential-geometry of differential equations. This was extensively developed in the 19th century; much of it has been lost and forgotten, but is highly pertinent to contemporary developments. Darboux's classic treatise, Theorie des Surfaces, for example, is right in the ball park, e.g., with the theory of Bäcklund transformations, Sine-Gordon equation, etc. Analytical mechanics and Hamilton-Jacobi theory are also major 19th century topics which are essential background to the current work. Finally, it has been recognized [1] that the 1920's work by Burchnall and Chaundy [2] (which is very close in spirit to the major 19th century work on linear ordinary differential equations with complex analytic coefficients) contains the algebraic essence of much of the modern work relating the linear Sturm-Liouville and nonlinear Korteweg-de Vries equations.

So far, the most striking results have been found for a relatively small number of equations--the Korteweg-de Vries and Sine-Gordon, of course, pre-eminent--in two independent, one dependent, variables. Some of us who work in this field feel that there is a potential for widespread application to differential equations in more dependent and independent variables; we want to push on to refine and generalize the methodology. This paper is pointed in that direction.

Right now, there are two leading contenders for a general theory encompassing the many special phenomena encountered in the study of the Korteweg-de Vries, Sine-Gordon et al.--the prolongations of Estabrook and Wahlquist [3,4] and the generalized

---

*This work was supported by a grant from Ames Research Center (NASA), NSG-2252 and the National Science Foundation grant MCS 78-06000.

Hamiltonian structure of Gelfand, Dikii, Manin and Kupershmidt [5,6]. It is not at all clear what are the relations between the two approaches, indeed, figuring out these relations will probably be a major advance in the theory. For example, the Koreteweg-de Vries equation has two major structures--an "inverse-scattering-Bäcklund" structure and on "infinite-number-of-conservation-laws-which-are-in-involution-complete-integrability-structure". Although we more-or-less understand how one of these structures generates the other for the special case of Korteweg-de Vries, and a few other equations, we do not know the relation in reasonable generality. Of course, we are even further from any systematic knowledge of <u>why</u> equations have <u>one</u> of the structures. It would be enormously significant for physics if possession of these structures could be traced back to the physical genesis of such equations, but we are not yet in that position. Note also that Manin and Kupershmidt [6,7] have discovered an equation in three independent variables (the Benney equation which derives from hydrodynamics) which admits an "infinite number of conservation laws in involution", but (so far) no prolongation-inverse scattering structure. Kupershmidt has discovered (private communication) other classes of equations which admit related structures.

The aim of these notes is to outline certain general ideas which are in the process of development. I would like to thank Frank Estabrook, Hugo Wahlquist, and Boris Kupershmidt, with whom I have had many fruitful discussions about these topics.

The context of these notes is the theory of differential and integral calculus on manifolds, particularly the theory of differential forms and vector fields. Notation will be that of [8,9]. The main technical thrust is to work as systematically as possible with <u>differential forms</u>.

2. HAMILTONIAN EQUATIONS IN GENERALITY

Let Γ be real vector spaces with positive definite symmetric real-valued inner product

$$(\gamma_1,\gamma_2) \to \langle\gamma_1,\gamma_2\rangle$$

Let t be a real parameter, varying, say, over $a \le t \le b$. Consider curves in V,

$\underline{\gamma}: t \to \gamma(t)$ which can be differentiated with respect to t. Let

$$A: \Gamma \to \Gamma$$

be a linear operator and let

$$h: \Gamma \to R$$

be a real-valued function. We suppose that A is <u>skew-symmetric</u> with respect to the form $< , >$, i.e.,

$$<A\gamma_1,\gamma_2> + <\gamma_1,A\gamma_2> = 0 \tag{2.1}$$

for $\gamma_1,\gamma_2 \in \Gamma$.

We then associate with each curve $\underline{\gamma}: t \to \gamma(t)$, $a \leq t \leq b$, in Γ the following <u>action function</u>

$$L(\underline{\gamma}) = \int_a^b \left[\left\langle A\gamma(t), \frac{d}{dt}\gamma(t)\right\rangle - h(\gamma(t))\right] dt \quad . \tag{2.2}$$

We now construct <u>first variations</u> of the function $\underline{\gamma} \to L(\underline{\gamma})$ in the usual classical calculus of variations manner.

Calculate

$$\frac{d}{dt}\underline{h}(\underline{\gamma}+\varepsilon\delta\underline{\gamma})\Big|_{\varepsilon=0} = \int_a^b \left[\left\langle A\delta\gamma, \frac{d\underline{\gamma}}{dt}\right\rangle + \left\langle A\gamma, \frac{d}{dt}\delta\gamma\right\rangle - \nabla h(\gamma),\delta\gamma\right] dt \tag{2.3}$$

where $\gamma \to \nabla h(\gamma)$ is the gradient of h with respect to the form $< , >$. We now integrate by parts on the right hand side of (2.3) (assuming that the variations vanish at end points, i.e., $\delta\underline{\gamma}(a) = \gamma\underline{\delta}(b) = 0$), obtaining the following <u>first variational formula</u>

$$\underline{\nabla h}(\underline{\gamma},\delta\underline{\gamma}) = \int_a^b \left\langle \delta\underline{\gamma}, -2A\frac{d\underline{\gamma}}{dt} - \nabla h(\underline{\gamma})\right\rangle dt \tag{2.4}$$

The curve $\underline{\gamma}: t \to \gamma(t)$ is then an <u>extremal</u> if it satisfies the following differential equations:

$$2A\frac{d\underline{\gamma}}{dt} = -\nabla\underline{h}(\underline{\gamma}) \quad .$$

This is called <u>Hamilton's differential equation.</u> $\underline{h}$ is the <u>Hamiltonian function.</u>

3. PARTIAL DIFFERENTIAL EQUATIONS OF HAMILTONIAN TYPE

In Section 2, Γ and A were general objects. Let us now specialize them so that Eqs. (2.5) become partial differential equations for functions $\gamma(x,t)$ of the "time" variable t and the additional "space" variables x. For the moment, local considerations suffice. Let x denote an element of R^n. Set:

Γ = vector space of C^∞ maps $\gamma: R^n \to R^m$, which are rapidly decreasing (in the sense of Schwartz)

Differential operators (linear or nonlinear) are defined in the usual way. (Alternately, the Ehresmann jet spaces $J^r(R^n,R^m)$ may be defined. The r-jet of each $\gamma \in \Gamma$ is a map $\partial^r\gamma: R^n \to J^r(R^n,R^m)$. Follow this with a fiber-preserving map $\sigma: J^r(R^n,R^m) \to R^p$ to obtain mappings $\gamma \to \partial^r\gamma \overset{\sigma}{\to} \partial^r\gamma \equiv D(\gamma)$, which are the r-th order differential operators.)

$$(\gamma_1,\gamma_2) \to \langle\gamma_1,\gamma_2\rangle$$

is defined as the usual pre-Hilbert space inner product

$$\langle\gamma_1,\gamma_2\rangle = \int_{R^n} \gamma_1(x)^T \gamma_2(x)\, dx \quad . \tag{3.2}$$

(Consider $\gamma_2(x)$ as a column vector; $\gamma_2(x)^T$ is its transpose.) Suppose now that

$$A: \Gamma \to \Gamma \tag{3.3}$$

is a linear differential operator and $h: \Gamma \to R$ is defined as follows

$$h(\gamma) = \int_{R^n} D(\gamma)(x)\, dx \quad . \tag{3.4}$$

where

$$D: \Gamma \to \mathcal{F}(R^n)$$

is a (possibly nonlinear) differential operator. ($\mathcal{F}(R^n) \equiv C^\infty$ for real-valued functions on R^n.) Thus,

$$< \nabla h(\gamma), \delta\gamma > \;=\; \frac{d}{d\varepsilon} h(\gamma + \varepsilon\delta\gamma)\Big|_{\varepsilon=0}$$

$$= \int \frac{d}{d\varepsilon} D(\gamma + \varepsilon\delta\gamma)\Big|_{\varepsilon=0} dx \tag{3.5}$$

This formula defines the "gradient" ∇h as a differential operator.

With these special choices for A, $< , >$ and h, we see that Eqs. (2.5) become a set of nonlinear partial differential equations to be solved for maps $(x,t) \to \gamma(x,t)$ of $R^n \times R \to R^m$. Here are some examples:

Example 1. Sine-Gordon

$$n = 1 = m$$

$$A = \partial_x$$

$$h(\gamma) = -\int \cos \gamma(x)\, dx$$

$$< \nabla h(\gamma), \delta\gamma > \;=\; -\int \frac{d}{d\varepsilon} \cos(\gamma + \varepsilon\delta\gamma)\Big|_{\varepsilon=0} dx$$

$$= \int \sin(\gamma)\, \delta\gamma\, dx$$

$$= < \sin \gamma, \delta\gamma >$$

i.e.,

$$\nabla h(\gamma) = \sin \gamma \quad .$$

Equations (2.5) take the following form in this case

$$2 \frac{d}{dt} (\partial_x \underline{\gamma}) = \sin \underline{\gamma} \quad ,$$

where $t \to \underline{\gamma}(t)$ is a curve in Γ. We can write

$$\underline{\gamma}(t)(x) = \gamma(x,t)$$

where γ: $R^2 \to R$ is a function of two variables and the equation is, of course, in the usual notation:

$$2\partial_x\partial_t = \sin \gamma \quad .$$

Example 2. Korteweg-de Vries

Let Γ and A be as in Example 1, but choose h and D as follows:

$$D(\gamma) = (\partial_x \gamma)^3 + \frac{1}{2} (\partial_x^2 \gamma)^2$$

$$h(\gamma) = \int D(\gamma)\, dx$$

$$\langle \nabla h(\gamma), \delta\gamma \rangle = \int \left[3(\partial_x \gamma)^2 \partial_x(\delta\gamma) + (\partial_x^2 \gamma)\, \partial_x^2(\delta\gamma) \right] dx$$

$$= \int \left[-6(\partial_x \gamma)(\partial_x^2 \gamma) + \partial_x^4 \gamma \;\delta\gamma \right] dx \quad ,$$

i.e.,

$$\nabla h(\gamma) = -6(\partial_x \gamma)(\partial_x^2 \gamma) + \partial_x^4 \gamma \quad .$$

The generalized Hamilton equations (2.5) then take the following form:

$$-2\partial_x \partial_t \gamma = -6(\partial_x \gamma)(\partial_x^2 \gamma) + \partial_x^4 \gamma \quad ,$$

which is the Korteweg-de Vries equation after the change

$$u = -2\partial_x \gamma \quad .$$

4. THE HAMILTONIAN FORMALISM AS A SPECIAL SORT OF INTEGRAL VARIATIONAL PROBLEM

Continue with Γ as the space of C^∞, rapidly decreasing maps: $R^n \to R^m$,

$$A: \Gamma \to \Gamma \quad , \qquad D: \Gamma \to \mathcal{F}(R^n)$$

differential operators (with A linear)

$$h(\gamma) = \int D(\gamma)\, dx \quad .$$

We have considered the one-variable variational problem

$$L(\underline{\gamma}) = \int_a^b \left[\left\langle A\underline{\gamma}, \frac{d\underline{\gamma}}{dt} \right\rangle - h(\underline{\gamma}(t)) \right] dt \qquad (4.1)$$

We can also regard each curve $t \to \underline{\gamma}(t)$ in Γ as a map

$$\phi : X \times T \to R^n \quad ,$$

i.e.,

$$\phi(x,t) = \underline{\gamma}(t)(x) \quad .$$

When the integrand in (4.1) is made explicit in terms of integrals over R^n, we obtain

$$L(\phi) = \int\int_{R^n \times R} F(\partial^r \phi) \, dx \, dt \tag{4.2}$$

where $(x,t) \to \partial^r \phi(x,t) \in J^2(R^{n+1}, R^m)$ is the r-jet of the map ϕ. F is a map $J^2(R^{n+1}, R^m) \to R$. Thus, extremizing (4.1) is equivalent to extremizing the multiple integral (4.2). However, the integrand function F in (4.2) is not the most general function on the jet-space; it is a function of only the first time derivatives. I will now present a formalism for handling variational problems of this type in terms of differential forms.

5. A GENERAL VARIATIONAL FORMALISM [8-14]

Let Z be a manifold. Let $\mathscr{D}(Z)$ be the graded differential algebra of C^∞ differential forms on Z. The algebra is exterior multiplication, denoted

$$(\theta_1, \theta_2) \to \theta_1 \wedge \theta_2 \quad ,$$

which maps bilinearly

$$\mathscr{D}^r(Z) \times \mathscr{D}^s(Z) \to \mathscr{D}^{r+s}(Z)$$

and the "differential" is exterior differentiation

$$d: \mathscr{D}^r(Z) \to \mathscr{D}^{r+1}(Z) \quad .$$

$\mathscr{V}(Z)$ denotes the Lie algebra (under Jacobi bracket) of vector fields on Z, and

$$(\theta, V) \to \mathscr{L}_V(\theta) \quad ,$$

mapping bilinearly

$$\mathcal{D}(Z) \times \mathcal{V}(Z) \to \mathcal{D}(Z)$$

is Lie derivative.

Contraction

$$(V,\theta) \to V \lrcorner \theta$$

is a bilinear map $\mathcal{V}(Z) \times \mathcal{D}(Z) \to \mathcal{D}(Z)$ which lowers degrees by one. The basic formula for the calculus of variations is the Cartan family identity

$$\mathcal{L}_V(\theta) = V \lrcorner d\theta + d(V \lrcorner \theta) \quad . \tag{5.1}$$

An exterior differential system, denoted as $\mathcal{E}$, is a subset of $\mathcal{D}(Z)$ which is an ideal relative to the algebra defined by exterior multiplication and which is closed under exterior differentiation.

Let such an $\mathcal{E}$ be chosen. A vector field $V \in \mathcal{V}(Z)$ is an infinitesimal symmetry of $\mathcal{E}$ if

$$\mathcal{L}_V(\mathcal{E}) \subset \mathcal{E} \quad . \tag{5.2}$$

The set of $V \in \mathcal{V}(Z)$ which satisfy (5.2) forms a Lie subalgebra of $\mathcal{V}(Z)$, denoted as

$$\mathcal{S}(\mathcal{E}) \quad .$$

An integral submanifold of $\mathcal{E}$ is a submanifold map

$$\phi: Y \to Z$$

such that

$$\phi^*(\mathcal{E}) = 0 \quad .$$

For a given manifold Y, $IM(Y)$ denotes the set of such integral submanifolds.

Let θ be an n-form on Z with

$$n = \dim Y \quad .$$

Suppose that Y is orientable and that an orientation is chosen. θ then defines a map

$$\underline{\theta}: IM(Y) \to R$$

by the following formula:

$$\underline{\theta} = \int_Y \phi^*(\theta) \quad . \tag{5.3}$$

The <u>first variational formula</u> can now be readily computed by means of formula (5.1). Choose a vector field $V \in \mathscr{S}(\mathscr{E})$. Let

$$t \to \exp(tV)$$

be the one-parameter group generated by V. Then,

$$\begin{aligned} \frac{\partial}{\partial t}(\exp(tV)^*(\theta)) &= \mathscr{L}_V(\exp(tV)^*(\theta)) \\ &= \exp(tV)^*(\mathscr{L}_V(\theta)) \end{aligned} \tag{5.4}$$

for $V \in \mathscr{V}(Z)$, $\theta \in \mathscr{D}(Z)$.

Given $\phi \in IM(Y)$, set

$$\phi_t = \exp(tV)\phi \quad ,$$

i.e., $t \to \phi_t$ is the one parameter family of submanifolds obtained by moving ϕ around via the one-parameter group generated by V. Then,

$$\frac{d}{dt}\int_Y \phi_t^*(\theta) = \frac{d}{dt}\int \exp(tV)\phi)^*(\theta)$$

$$= \int \frac{\partial}{\partial t}(\phi^* \exp(tV)^*\theta)$$

$$= \int \phi^*(\mathscr{L}_V(\exp(tV)^*(\theta))$$

$$= \int_Y \phi^*(d(V \lrcorner \exp(tV)^*(\theta) + V \lrcorner\, d \exp(tV)^*(\theta))$$

= , using Stoke's formula and assuming that Y is a manifold with boundary ∂Y,

$$\int_{\partial Y} \phi^*(V \lrcorner \exp(tV)^*(\theta)) + \int_Y \phi^*(V \lrcorner\, d \exp(tV)^*(\theta)) \tag{5.5}$$

This is the <u>first variation formula</u>.

<u>Definition</u>. The submanifold $\phi: Y \to Z$ with boundary ∂Y is an <u>extremal</u> for the variational problem $(\theta, \mathscr{E})$ if

$$\int_Y \phi^*(V \lrcorner\, d\theta) = 0 \tag{5.6}$$

for all $V \in \mathscr{S}(\mathscr{E})$ such that

$$V(\partial\phi) = 0 .$$

Of course, this definition does not directly define the extremals as a solution of a set of <u>differential equations</u>. To do this requires further knowledge about θ and $\mathscr{E}$. Here is a method which seems to work very elegantly for the classical cases [9]. We shall see how it works for the Hamiltonian situations in the next section.

Notice that in the definition of "variational problem" given above, θ is not uniquely determined; forms in $\mathscr{E}$ can be added to it without affecting the action function. The method presented here involves choosing θ so that $d\theta$ has a sort of "canonical form" with respect to $\mathscr{E}$.

Let us suppose that $\mathscr{E}$ is generated by a set of forms

$$\theta^1,\dots,\theta^m .$$

It may be possible to choose θ so that there are forms $\omega_1,\dots,\omega_m$ so that:

$$d\theta = \omega_1 \wedge \theta^1 + \cdots + \omega_m \wedge \theta^m . \tag{5.7}$$

(This may require that the space Z on which θ and $\mathscr{E}$ were originally defined be "prolonged"; this means that a fiber space (i.e., a submersion map)

$$\pi: Z' \to Z$$

is constructed and everything is pulled back to Z'. In classical notation, this means that "additional variables" (e.g., "Lagrange multipliers") are added to the variables that were originally present in Z)

Having obtained the formula (5.7), it is natural to define the exterior differential system $\mathscr{E}'$ as that generated by the $\theta^1,\dots,\theta^m,\omega_1,\dots,\omega_m$ and <u>define</u> the extremals as the integral submanifolds of $\mathscr{E}'$. If $\phi: Y \to Z$ is such an integral manifold, then an <u>arbitrary</u> $V \in \mathscr{V}(Z)$ satisfies:

$$\phi^*(V \lrcorner d\theta) = 0 \quad .$$

6. THE GELFAND-DIKII HAMILTONIAN FORMALISM

In this section we shall construct the form θ with $d\theta$ in the "canonical form" (5.7) for the type of Hamiltonian structure in two independent variables of the simplest type; this leads to the Korteweg-de Vries and Sine-Gordon equations and, in the work of Kodama and Wadati [15], to the Bäcklund transformation for these equations.

Consider the two-independent variables variational problem

$$\iint [v_x v_t - h(v, v_x, v_{xx}, \ldots)] \, dx \, dt \quad . \tag{6.1}$$

We use the classical notation; v is a scalar-valued function $v(x,t)$ of two real variables, subscript denotes partial derivatives. Choosing h appropriately, we get as extremal equation Sine-Gordon, Korteweg-de Vries, higher Korteweg-de Vries and other interesting nonlinear partial differential equations.

Let Z be the space of variables

$$(x, t, v, v_x, v_t, v_{xx}, \ldots) \quad .$$

Let $\mathscr{E}$ be the exterior differential system generated by the two-forms:

$$\begin{aligned} &dv - v_x dx - v_t dt \\ &(dv_x - v_{xx} dx) \wedge dt \\ &\quad \vdots \end{aligned} \tag{6.2}$$

Set:

$$\theta = (v_x v_t - h) dx \wedge dt + (\lambda_1 dx + \lambda_2 dt) \wedge (dv - v_x dx - v_t dt) + \lambda_3 (dv_x - v_{xx} dx) \wedge dt$$

The $\lambda_1, \lambda_2, \ldots$ are functions possibly depending on additional variables which we are <u>free to choose</u> so that $d\theta$ can be written in form (5.7).

Now,

$$d\theta = (v_t dv_t + v_x dv_t) \wedge dx \wedge dt - (h_{v_x} dv_x t \ldots) \wedge dx \wedge dt$$

$$+ (d\lambda_1 \wedge dx + d\lambda_2 \wedge dt) \wedge (dv - v_x dx - v_t dt)$$

$$+ (\lambda_1 dx + \lambda_2 dt) \wedge (dv_x \wedge dx + dv_t \wedge dt) + d\lambda_3 \wedge (dv_x - v_{xx} dx) \wedge dt$$

$$- \lambda_3 dv_{xx} \wedge dx \wedge dt + \cdots \quad .$$

Let us assume, for simplicity, that h is a function of v_x, v_{xx} alone. First, choose λ_3 so that the terms in dv_{xx} vanish. This forces the following condition

$$\lambda_3 = -h_{v_{xx}} \quad .$$

Now, choose λ_1 so that the terms in dv_t vanish. This requires the following condition:

$$v_x = \lambda_1 \quad .$$

With these choices, the canonical form is achieved. λ_2 is unconstrained, and <u>will be chosen as a new variable</u>

$$d\theta = v_t (dv_x - v_{xx} dx) \wedge dx \wedge dt - h_{v_x} (dv_x - v_{xx} dx) \wedge dx \wedge dt$$

$$+ dv_x \wedge dx \wedge (dv - v_x dx - v_t dt) + d\lambda_2 \wedge dt \wedge (dv - v_x dx - v_t dt)$$

$$+ \lambda_2 dt \wedge dv_x \wedge dx - dh_{v_{xx}} \wedge (dv_x - v_{xx} dx) \wedge dt$$

$$= (dv_x \wedge dx + d\lambda_2 \wedge dt) \wedge (dv - v_x dx - v_t dt)$$

$$+ (v_t dx \wedge dt - h_{v_x} dx \wedge dt - \lambda_2 dx \wedge dt) \wedge (dv_x - v_{xx} dx) \quad .$$

Thus we see that $d\theta$ can be written in canonical form as follows:

$$d\theta = \omega_1 \wedge \theta^1 + \omega_2 \wedge \theta^2 \tag{6.5}$$

with

$$\omega_1 = dv_x \wedge dx + d\lambda_2 \wedge dt$$

$$\theta^1 = dv - v_x dx - v_t dt$$

$$\omega_2 = \lambda_4 dt + (v_t - hv_x - \lambda_2)dx + dh_{v_{xx}} \tag{6.6}$$

$$\theta^2 = dt \wedge (dv_x - v_{xx}dx) \quad .$$

Let $\mathcal{E}'$ be the exterior differential system generated by the forms (6.6). (λ_4 is a new variable which is independent of the others.) A two-dimensional integral manifold (which implies $dx \wedge dt \neq 0$, i.e., which can be parameterized by (x,t)), then is determined by a function $v(x,t)$. We see that:

$$\lambda_2 = \left(h_{v_{xx}}\right)_x - h_{v_x} - v_t$$

$$\lambda_4 = \left(h_{v_{xx}}\right)_t$$

$$\left(h_{v_{xx}}\right)_{xx} - \left(h_{v_x}\right)_x - (v_t)_x = -v_{xt} \quad .$$

These are just classical extremal equations of the variational problem with which we started.

With this special case in hand, let us study certain general features.

7. A GENERALIZATION OF SYMPLECTIC STRUCTURE

One of the most useful ideas in differential geometry and analytical mechanics has been that introduced by E. Cartan in his book Lecons sur les Invariants Integraux; assign to each single variable calculus of variation problem a closed two-form Ω in such a way that the extremal curves of the variational problem are the Cauchy characteristic curves of the form Ω. The set of extremals has a symplectic structure; the Cauchy characteristics form a symplectic foliation. We see that there is a generalization that works for some multiple integral variational problem:

> Assign a closed (n+1)-form Ω. (n = number of independent variables.) Look for sets $(\omega_1, \theta^1, \ldots, \omega_m, \theta^m)$ of differential forms so that $\Omega = \omega_1 \wedge \theta^1 + \cdots + \omega_m \wedge \theta^m$. Construct the

exterior differential system $\mathscr{E}$ generated by the forms $\omega_1,\dots,\omega_m$, $\theta^1,\dots,\theta^m$. The n-dimensional integral submanifolds of $\mathscr{E}$ are then called the characteristics of Ω. (They are no longer of "Cauchy" type.)

If $n = 1$, and if the forms $\omega_1,\dots,\theta^m$ are linearly independent, the one-dimensional integral manifolds of $\mathscr{E}$ are just the Cauchy characteristics of Ω, i.e., the orbits of vector fields V such that

$$V \lrcorner \Omega = 0 \quad .$$

Thus, we have formulated what is probably the most natural and/or simplest way of generalizing to field theories the "symplectic structure" material that has proved so fruitful in the study of particle mechanics. The next step is to formulate the natural notion of "Poisson bracket". This has already been done in Lie Algebras and Quantum Mechanics [10]; here is a more polished algebraic version.

8. THE POISSON BRACKET ALGEBRA ASSOCIATED TO A CLOSED (n+1)-FORM

Let Z be a manifold and let $\Omega \in \mathscr{D}^{n+1}(Z)$ be a differential form of degree $(n+1)$ such that

$$d\Omega = 0 \quad .$$

Let $\mathscr{P}$ be the set of pairs

$$(V,\theta) \in \mathscr{V}(Z) \times \mathscr{D}^{n-1}(\Omega)$$

such that:

$$d\theta = V \lrcorner \Omega \quad .$$

Define an operation by the following formula:

$$\left\{(V,\theta),(V',\theta')\right\} = \left([V,V'], \tfrac{1}{2}\left(\mathscr{L}_V(\theta') - \mathscr{L}_{V'}(\theta)\right)\right) \tag{8.1}$$

Theorem 8.1. $\left\{\mathscr{P},\mathscr{P}\right\} \subset \mathscr{P}$. This operation (which does not satisfy the Jacobi identity) defines an algebra. The quotient by its center is a Lie algebra.

Proof. First,

$$\mathscr{L}_V(\Omega) = V \lrcorner d\Omega + d(V \lrcorner \Omega)$$

$$= 0 + dd\theta$$

$$= 0 \quad .$$

Hence,

$$d(\mathscr{L}_V(\theta') - \mathscr{L}_{V'}(\theta)) = \mathscr{L}_V(d\theta') - \mathscr{L}_{V'}(d\theta)$$

$$= \mathscr{L}_V(V' \lrcorner \Omega) - \mathscr{L}_{V'}(V \lrcorner \Omega)$$

$$= [V,V'] \lrcorner \Omega - [V',V] \lrcorner \Omega$$

$$= 2[V,V'] \lrcorner \Omega$$

This proves that

$$\{\mathscr{P},\mathscr{P}\} \subset \mathscr{P} \quad .$$

We can say a bit more about the algebraic structure of $\mathscr{P}$. Let $\mathscr{S}$ = Lie algebra of vector fields V such that

$$\mathscr{L}_V(\Omega) = 0 \quad .$$

The map

$$(V,\theta) \to V$$

$$\pi: \mathscr{P} \to \mathscr{S}$$

is a Lie algebra homomorphism. It will be onto if the n-th Betti number of Z vanishes. The kernel of π consists of the (n-1)-forms θ such that

$$d\theta = 0 \quad .$$

It is in the center of $\mathscr{P}$. Thus, $\mathscr{P}$ is a central extension of the Lie subalgebra $\pi(\mathscr{P})$ of $\mathscr{S}$. It is well known that, algebraically, there is an element of the second Lie algebra cohomology class of $\pi(\mathscr{P})$ (with coefficients in the trivial representation) which determines this extension up to isomorphism. The study of this Lie algebra--which has barely begun--is the key to treatment of what physicists call "Gauge Theories".

BIBLIOGRAPHY

1. I.M. Krichever, "Methods of Algebraic Geometry in the Theory of Nonlinear Equations", Russian Math. Surveys 32 (1977), 125-213.

2. J.L. Burchnall and T.W. Chaundy, Proc. London Math. Soc. 21 (1922), 420-440.

3. H. Wahlquist and F. Estabrook, "Prolongation Structures of Nonlinear Evolution Equations", J. Math. Phys. 16 (1975), 1-7.

4. R. Hermann, The Geometry of Nonlinear Differential Equations, Bäcklund Transformations and Solitons, Parts A and B, Math Sci Press, Brookline, MA.

5. I.M. Gelfand and L.A. Dikii, "Resolvents and Hamiltonian Systems", Functional Anal. Appl. 11 (1977), 93-105.

6. Yu. Manin, "Algebraic Aspects of Nonlinear Differential Equations" (in Russian), Modern Problems in Mathematics 11 (1978), 5-152, Viniti, Moscow.

7. Yu. Manin and B. Kupershmidt, "Long-Wave Equation with Free Boundaries", Funct. Anal. Appl. 11 (1978), 188.

8. R. Hermann, Yang-Mills, Kaluza-Klein and the Einstein Program, Interdisciplinary Mathematics, Vol. 19, Math Sci Press, Brookline, MA. 1968.

9. R. Hermann, Differential Geometry and the Calculus of Variations, 2nd Edition, Interdisciplinary Mathematics, Volume 17, Math Sci Press, Brookline, MA.

10. R. Hermann, Lie Algebras and Quantum Mechanics, W.A. Benjamin, 1970.

11. R. Hermann, "The Second Variation for Variational Problems in Canonical Form", Bull. Amer. Math. Soc. 71 (1965), 145-148.

12. R. Hermann, "The Second Variation for Minimal Submanifolds", J. Math. and Mech. 16 (1966), 473-492.

13. P. Dedecker, "On the Generalization of Symplectic Geometry to Multiple Integrals in the Calculus of Variations", in Springer Math. Lecture Notes No. 520, K. Bleuler and A. Reetz (eds.), Springer-Verlag, 1977.

14. W.F. Shadwick, "The Hamilton-Cartan Formalism for Higher Order Conserved Currents", Preprint, Kings College, London.

15. Y. Kodama and M. Wadati, "Theory of Canonical Transformations for Nonlinear Evolution Equations", I & II, Prog. Theor. Phys. 56 (1976), 1740 and 57 (1977), 1900.

16. J. Moser (ed.), Lecture Notes in Physics, Vol. 38, Springer-Verlag, 1975.

17. R.M. Miura (ed.), Lecture Notes in Mathematics, Vol. 15, Springer-Verlag, 1976.

18. G. Whitham, Linear and Non-Linear Waves, Wiley, New York, 1974.

19. M.J. Ablowitz, D.J. Kaup, A.C. Newell, and H. Segur, Studies in Applied Math 53 (1974), 249.

20. H. Morris, "Prolongation Structures and Nonlinear Evolution Equations in Two Spatial Dimensions", J. Math. Phys. 17 (1976), 1870-1872.

21. R. Hermann, "The Pseudopotentials of Estabrook and Wahlquist, the Geometry of Solitons and the Theory of Connections", Phys. Rev. Lett. 36 (1976), 835.

22. J. Corones, "Solitons, Pseudopotentials and Certain Lie Algebras", J. Math. Phys. 18 (1977), 163-164.

23. R. Hermann, Toda Lattices, Cosymplectic Manifolds, Bäcklund Transformations and Kinks, Parts A and B, Math Sci Press, Brookline, MA.

24. R. Hermann, "'Modern' Differential Geometry in Elementary Particle Physics", Proceedings of the VII GIFT Summer School, Lecture Notes in Physics, J. Azcarraga (ed.), Springer-Verlag.

25. R. Hermann, "The Lie-Cartan Geometric Theory of Differential Equations and Scattering Theory", to appear, Proceedings of the 1977 Park City (Utah) Conference on Differential Equations, P. Bynes (ed.)

26. R. Hermann, "Prolongation, Bäcklund Transformations and Lie Theory as Algorithms for Solving and Understanding Nonlinear Differential Equations", in Solitons in Action, K. Lonngren and A. Scott (eds.), Academic Press, New York, 1978.

27. M. Crampin, F.A.E. Pirani and D.C. Robinson, "The Soliton Connection" (to appear, Letters in Math. Phys.).

28. R. Hermann (ed.), The 1976 Ames Research Center (NASA) Conference on the Geometric Theory of Nonlinear Waves (articles by Estabrook, Wahlquist, Hermann, Morris, Corones, R. Gardner, and Scott), Math Sci Press, Brookline, MA.

29. F. Estabrook and H. Wahlquist, "Prolongation Structures, Connection Theory and Bäcklund Transformations", Proceedings of the International Symposium on Nonlinear Evolution Equations..., F. Calogero (ed.), Research Notes in Math., Pitman, London, 1978.

30. S. Chern and C. Terng, "Analogue of the Bäcklund Theorem for Affinely Connected Manifolds", Preprint, Math. Dept., University of California, Berkeley, 1977.

31. R. Hermann, Cartanian Geometry, Nonlinear Waves, and Control Theory, Part A, Interdisciplinary Mathematics, Vol. 20, Math Sci Press, Brookline, MA. 1979.

GEOMETRY OF JET BUNDLES AND THE STRUCTURE OF LAGRANGIAN AND HAMILTONIAN FORMALISMS

B. A. Kupershmidt

Massachusetts Institute of Technology
Department of Mathematics
Cambridge, Massachusetts 02139

PREFACE

It is widely recognized that jet language is the natural way to speak with the local problems of differentiable mathematics. To cite a few examples, one can refer to differential equations ([1], [4], [6], [8], [9], [16]), singularities ([7]), calculus of variations and field theory ([5], [10], [11] - [14]).

The main topic of this paper is Lagrangian formalism in the calculus of variations and Hamiltonian formalism in field theory.

The presentation begins by developing some basic facts about the geometry of jet bundles which are necessary for the rest of the paper (Chapter I). There, this geometry is calculated on both a finite and on an infinitesimal level, and main philological events are described. Among them is the geometric interpretation of evolution equations, which is the much needed analog of the famous proverb: "ODE = a vector field on a manifold."

Chapter II begins with the definition in §1 of Lagrangian formalism, e.g., the formula for the first variation. The main result, proved in §3, states that this formula exists (globally). In §4, the Hamilton-Cartan principle about the connection between the main and additional (through the Legendre transformation) variational problems, is stated and proved. In §5 is proved formal version of direct and inverse Noether theorem. In §6,7,8 two different resolvents are constructed for the Euler-Lagrange operator ∂ (= "functional derivative"); this gives the full description of the Image and Kernel of this operator. Note that in §7 it is considered the simplest case when, roughly speaking, the density of Lagrangian is differential form itself.

With the tool of Chapter II, the Hamiltonian form of field theory - "Mechanics $\leftrightarrow T^*M$, Field Theory $\leftrightarrow$?" - is established in Chapter III. The main concern there is Jacoby identity for Poisson bracket, r.h.s. of which turns out to be rather long.

At last , in Chapter IV a short outline is given for Hamiltonian formalism in general, as it's understood now. An example shown there indicates how far away the notion of Hamiltonian structure has diverged from the classical point of view.

The main results of Chapters I and II were announced in [4], [5] correspondingly. The language here is improved.

I would like to take this opportunity to express my gratitude to Yu.I. Manin who insisted that this paper be published; to A.M. Vinogradov with whom the matter of the paper was discussed; to D. Coray for help in preparing the manuscript, and to G. Kaiser for hospitality during the conference.

TABLE OF CONTENTS

CHAPTER I. GEOMETRY OF JET BUNDLES

§1. Jet Bundles, Main Definitions

1.1 In this section we shall briefly introduce main definitions and arrange notions and notations. For additional information see [8,9]. From here on all manifolds, bundles and maps will be smooth (=C^∞), π and ν denote fixed bundles $\pi: E\to M$, $\nu: F\to M$, $\dim M=m$.

Let $\Gamma(\pi)$ denote the set of all local sections of π and let $\gamma\in(\pi)$, $x\in M$. The equivalent class of all sections of $\Gamma(\pi)$ which are tangent to γ in the point x with order $\geq k$ is called k-jet of γ in the point x and is denoted by $[\gamma]_x^k$ or γ_x^k. The set $\bigcup_{x\in M,\gamma\in\Gamma(\pi)}\{[\gamma]_x^k\}$ naturally possesses smooth structure and is denoted $J^k\pi$; projection $\pi_k: J^k\pi\to M$, $\pi_k([\gamma]_x^k)=x$, is a smooth bundle map; $J^k\pi$ is called k-jet manifold (of bundle π). Let $j_k=j_k(\pi)$ denote the map $\Gamma(\pi)\to\Gamma(\pi_k)$, $(j_k\gamma)(x)=[\gamma]_x^k$. When $k\geq \ell$, $\pi_{k,\ell}: J^k\pi\to J^\ell\pi$ is natural projection. Note that $J^0\pi=E$. Let denote: $J^\infty\pi \overset{\text{def}}{=} \lim\text{proj}\ J^k\pi$ with natural projections $\pi_{\infty,k}: J^\infty\pi\to J^k\pi$, $\pi_\infty: J^\infty\pi\to M$; $K=C^\infty(J^\infty\pi)\overset{\text{def}}{=}\lim\text{ind}\ F(J^k\pi)$, where $F(\cdot)\overset{\text{def}}{=}C^\infty(\cdot)$. Similarly $\Lambda^s(K)\overset{\text{def}}{=}\lim\text{ind}\ \Lambda^s(J^k\pi)$.

The map $\Delta:\Gamma(\pi)\to\Gamma(\nu)$ is called differential operator of order $\leq k$, if there exists the map $\phi_\Delta: J^k\pi\to F$ (over M) such that $\Delta=\phi_\Delta\cdot j_k(\pi)$. $\forall s\geq 0$, ϕ_Δ generates map $\phi_\Delta^s: J^{k+s}\pi\to J^s\nu$ by universal property $\phi_\Delta^s\cdot j_{k+s}(\pi)=j_s(\nu)\cdot\Delta$. In particular $\phi_\Delta^0=\phi_\Delta$. Since $\nu_{s,r}\cdot\phi_\Delta^s=\phi_\Delta^r\cdot\pi_{k+s,\,k+r}$, is defined the map $\bar\phi\overset{\text{def}}{=}\lim\text{proj}\ \phi_\Delta^s: J^\infty\pi\to J^\infty\nu$.

1.2 Let $\phi: C\to B$, $\psi: B\to A$, $\chi=\psi\cdot\phi: C\to A$ be bundles. Additive maps $X: F(B)\to F(C)$ with property $X(fg)=\phi^*(f)\,X(g)+\phi^*(g)\,X(f)$, $\forall f,g\in F(B)$ are called fields. These fields form $F(C)$-module $D(\phi)$. Denote $D(B):=D(1_B)$.

Denote: $\Lambda_\circ^k(\phi)$ be the set of all k-forms on C which are horizontal over B; when $k\geq\ell$ let $\Lambda_\ell^k(\phi)=\{\omega\in\Lambda^k(C)\,|\,(X_1\lrcorner\ldots\lrcorner X_\ell\lrcorner\,\omega)\in\Lambda_\circ^{k-\ell}(\phi),\ \forall X_i\in D(C):X_i\cdot\phi^*=0\}$; $\Lambda_\ell^k(X;\phi)\overset{\text{def}}{=}\Lambda_\ell^k(X)\cap\Lambda_\circ^k(\phi)$.

1.3 Let us introduce special local coordinates in $J^k\pi$. Let (x_i), $i = \overline{1,m}$ be local coordinates in M; (x_i, q^a), $a = \overline{1,n}$ - in E. Then (x_i, q^a_σ), where $\sigma = (i_1,\ldots,i_{|\sigma|})$, $i_\cdot = \overline{1,m}$, $0 \leq |\sigma| \leq k$, are such coordinates in $J^k\pi$ that $\forall\gamma\in\Gamma(\pi)$, $(j_k\gamma)^*(q^a_\sigma) =$

$$:= \frac{\partial^{|\sigma|}(\gamma^*(q^a)}{\partial x_\sigma} \overset{\text{def}}{:=} \frac{\partial^{|\sigma|}(\gamma^*(q^a)}{\partial x_{i_1}\ldots\partial x_{i_{|\sigma|}}}. \quad \text{Let } \sigma + i := (i_1,\ldots,i_{|\sigma|}, i).$$

When we calculate in local coordinates we mean summation over repreated indices.

§2. Lift of exterior differentiations

2.1 Among all sections $\theta\in\Gamma(\pi_k)$ the sections of form $j_k\gamma$, $\gamma\in\Gamma(\pi)$, are naturally isolated. We start with the following problem: among all submanifolds in $J^k\pi$ which have the form $\theta(M)$ to separate those of form $j_k(\gamma)(M)$ (here and in the following we will write for simplicity $\theta(M)$, etc. instead of $\theta(U)$, where $U\subset M$ is region of difinition for θ). The last will be characterized as integral submanifolds of some distribution on $J^k\pi$. In §2 and §3 we shall prepare instruments for making this distribution.

2.2 Let's consider bundle $\phi: B\to M$ and induced bundles $\phi^*(\pi_k):\phi^*(J^k\pi)\to B$, $\phi(\pi_k):\phi^*(J^k\pi)\to J^k\pi$.

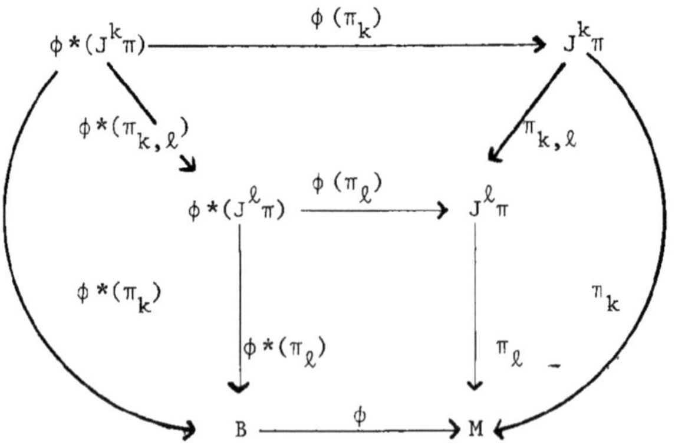

(2.1)

When $k > \ell$ we can identify $\phi^*(J^k\pi)$ with $\phi(\pi_\ell)^*(J^k\pi)$ (see (2.1)). Hence it defined map $\phi^*(\pi_{k,\ell}):\phi^*(J^k\pi) \to \phi^*(J^\ell\pi)$ with properties $\phi(\pi_\ell)\cdot\phi^*(\pi_{k,\ell}) = \pi_{k,\ell}\cdot\phi(\pi_k)$, $\phi^*(\pi_\ell)\cdot\phi^*(\pi_{k,\ell}) = \phi^*(\pi_k)$. Let $\tilde{\phi}_k := \pi_{k+1,k}\cdot\phi(\pi_{k+1}):\phi^*(J^{k+1}\pi) \to J^k\pi$.

Let $X \in D(\phi)$. $\forall\ k \geq 0$ define operator $X^k \in D(\tilde{\phi}_k)$ by universal property: $\forall\,\gamma \in \Gamma(\pi)$, $[\phi^*(j_{k+1}\gamma)]^* \cdot X^k = X \cdot (j_k\gamma)^*$. Speaking locally, if $f \in F(J^k\pi)$, then the value of function $X^k f$ in the point $(b, j_{k+1}\gamma_{\phi(b)}) \in \phi^*(J^{k+1}\pi)$, $b \in B$, equals the value of function $X[(j_k\gamma)^*(f)]$ in point $b \in B$. As $j_{k+1}\gamma_{\phi(b)}$ defines 1-jet of function $(j_k\gamma_{\phi(b)})^*(f)$ and X=differential operator of first order, so function $X[(j_k\gamma_{\phi(b)})^*(f)](b)$ and operator X^k are correctly defined.

Note. In definitions and in proofs we often will use, although not mention the fact that for equality of two functions on manifold of form $\phi^*(J^k\pi)$ is sufficient to check that restrictions of these functions coincide on submanifolds of form $\phi^*(j_k\gamma)$, $\gamma\in\Gamma(\pi)$.

Obviously, X^k is additive. Let's check that $X^k \in D(\tilde{\phi}_k)$. $\forall f,g \in F(J^k\pi)$ we have $[\phi^*(j_{k+1}\gamma)]^*X^k(fg) = X[(j_k\gamma)^*(fg)] = X[(j_k\gamma)^*(f)\cdot(j_k\gamma)^*(g)] = \phi^*[(j_k\gamma)^*(f)]X[(j_k\gamma)^*(g)] + \phi^*[(j_k\gamma)^*(g)]X[(j_k\gamma)^*(f)] = \phi^*[(j_{k+1}\gamma)^*(\pi_{k+1,k}^*f)]$ $[\phi^*(j_{k+1}\gamma)]^*X^k g + \phi^*[(j_{k+1}\gamma)^*(\pi_{k+1,k}^*g)]$ $[\phi^*(j_{k+1}\gamma)]^*X^k f = [\phi^*(j_{k+1}\gamma)]^*[\tilde{\phi}_k^*(f)X^k g + \tilde{\phi}_k^*(g)X^k(f)]$, that is $X^k(fg) = \tilde{\phi}_k^*(f)X^k(g) + \tilde{\phi}_k^*(g)X^k(f)$.

2.3 The **following lemmas** describe the basic properties of fields X^k and of corresponding map $\Pi^k = \Pi^k(\phi):D(\phi) \to D(\tilde{\phi}_k)$, $X \to X^k$.

Lemma 2.1. The fields X^k agree with projections $\pi_{k,\ell}$ and $\phi^*(\pi_{k,\ell})$, i.e. $X^k \cdot \pi_{k,\ell}^* = [\phi^*(\pi_{k+1,\ell+1})]^* \cdot X^\ell$.

<$\forall f \in F(J^\ell\pi)$, $[\phi^*(j_{k+1}\gamma)]^*X^k[\pi_{k,\ell}^*f] = X[(j_k\gamma)^*(\pi_{k,\ell}^*f)] = X[(j_\ell\gamma)^*(f)] = [\phi^*(j_{\ell+1}\gamma)]^*X^\ell(f) = [\phi^*(\pi_{k+1,\ell+1})\cdot\phi^*(j_{k+1}\gamma)]^*X^\ell(f) = [\phi^*(j_{k+1}\gamma)]^*[\phi^*(\pi_{k+1,\ell+1})]^*$ $X^\ell(f)$ >

The **fields** X^k belong to $F(\phi^*(J^{k+1}\pi))$-module $D(\tilde{\phi}_k)$. Projection $\phi^*(\pi_k)$ defines homomorphism of rings $[\phi^*(\pi_k)]^*:F(B) \to F(\phi^*(J^{k+1}\pi))$, so we can consider $D(\tilde{\phi}_k)$ as $F(B)$-module too.

<u>Lemma 2.2.</u> Π^k is homomorphis of $F(B)$-modules.

< $\forall h \in F(B)$, $[\phi^*(j_{k+1}\gamma)]^*(hX)^k = (hX)\cdot(j_k\gamma)^* = h\cdot X\cdot(j_k\gamma)^* = h[\phi^*(j_{k+1}\gamma)]^*\cdot X^k =$ $[\phi^*(j_{k+1}\gamma)]^*[\phi^*(\pi_{k+1})]^*(h)\cdot X^k$ >

Let's consider natural properties of homomorphism $\Pi^k(\phi)$. Let $\psi:C\to B$ be another bundle, $\chi:\phi\cdot\psi:C\to M$. The map $\psi^*:D(\phi)\to D(\chi)$, $X\to\psi^*\cdot X$, induces some map $\Pi^k(\phi)\to\Pi^k(\chi)$. Let's see what is this map. Note that we can identify $\chi^*(J^k\pi)$ with $\psi^*(\phi^*(J^k\pi))$, so $\phi(\pi_k)\cdot\psi(\phi^*(\pi_k)) = \chi(\pi_k)$ and $\psi(\phi^*(\pi_\ell))\cdot\chi^*(\pi_{k,\ell}) =$ $\phi^*(\pi_{k,\ell})\cdot\chi(\phi^*(\pi_k))$.

<u>Lemma 2.3.</u> $\Pi^k(\chi)\cdot\psi^* = [\psi(\phi^*(\pi_k))]^*\cdot\Pi^k(\phi)$, i.e. $(\psi^*\cdot X)^k =$ $[\psi(\phi^*(\pi_{k+1}))]^*\cdot X^k$, $\forall X\in D(\phi)$.

< $\chi^*(j_{k+1}\gamma) = \psi^*(\phi^*(j_{k+1}\gamma))$ and $\psi\cdot\chi^*(\pi_{k+1}) = \phi^*(\pi_{k+1})\cdot\psi(\phi^*(\pi_{k+1}))$, hence $[\chi^*(j_{k+1}\gamma)]^*\cdot(\psi^*\cdot X)^k = \psi^*\cdot X\cdot(j_k\gamma)^* = \psi^*\cdot[\phi^*(j_{k+1}\gamma)]^*\cdot X^k = [\chi^*(j_{k+1}\gamma)]^*$ $\cdot\,[\psi(\phi^*(\pi_{k+1}))]^*\cdot X^k$ >

§3. <u>Lift of interior differentiations</u>

3.1 Here we consider the most important example of lifting of differentiations when the role of ϕ of §2 will play $\pi_s:J^s\pi\to M$.

Let $X\in D(\pi_s)$, $X^k\in D(\pi_{k+1,k}\cdot\pi_s(\pi_{k+1}))$, $[s,k]:=\max(s,k)$. Define immersion $i_{s,k} = i_{s,k}(\pi): J^{[s,k]}\pi\to\pi_s^*(J^k\pi)$ by formulae $i_{s,k}(j_{[s,k]}\gamma_x) = (j_s\gamma_x, j_k\gamma_x)$. It's easy to see that $\pi_s(\pi_k)\cdot i_{s,k} = \pi_{s,k}$ when $s\geq k$; $\pi_s^*(\pi_k)\cdot i_{s,k} = \pi_{k,s}$ when $s\leq k$; and

$$\pi_s^*(\pi_{k,\ell})\cdot i_{s,k} = i_{s,\ell}\,,\quad s\geq k\geq \ell\ ; \tag{3.1}$$

$$\pi_s^*(\pi_{k,\ell})\cdot i_{s,k} = i_{s,\ell}\cdot\pi_{k,s}\,,\quad k\geq s\geq \ell\ ; \tag{3.2}$$

$$\pi_s^*(\pi_{k,\ell})\cdot i_{s,k} = i_{s,\ell}\cdot\pi_{k,\ell}\,,\quad k\geq \ell\geq s\ . \tag{3.3}$$

These inclusions allow us to define lift operators $\overline{X}^k\in D(\pi_{[s,k+1],k})$ which act on jet-bundles of π. Namely, let $\overline{X}^k := i_{s,k+1}^*\cdot X^k$. Then $\overline{X}^k\in D(\pi_{k+1,k})$ when $k+1\geq s$ and $\overline{X}^k\in D(\pi_{s,k})$ when $k+1\leq s$. Lemmas of §2 which describe the properties of fields X^k are transformed in such manner.

Lemma 3.1. Fields $\bar{X}^k$ agree with projections $\pi_{k,\ell}$, i.e. $\bar{X}^k \cdot \pi^*_{k,\ell}$ is: $\bar{X}^\ell$, $s \geq k+1$; $\pi^*_{k+1,s} \cdot \bar{X}^\ell$, $k+1 \geq s \geq \ell+1$; $\pi^*_{k+1,\ell+1} \cdot \bar{X}^\ell$, $\ell+1 \geq s$.

< By lemma 2.1, $X^k \cdot \pi^*_{k,} = [\pi^*_s(\pi_{k+1,\ell+1})]^* \cdot X^\ell$. Applying from the left operator $i^*_{s,k+1}$ we get $\bar{X}^k \cdot \pi^*_{k,\ell} = [\pi^*_s(\pi_{k+1,\ell+1}) \cdot i_{s,k+1}]^* \cdot X$ which coincides with: $i^*_{s,\ell+1} \cdot X^\ell = \bar{X}^\ell$, $s \geq k+1$, by (3.1); $(i_{s,\ell+1} \cdot \pi_{k+1,s})^* \cdot X^\ell = \pi^*_{k+1,s} \cdot \bar{X}^\ell$, $k+1 \geq s \geq \ell+1$, by (3.2); $(i_{s,\ell+1} \cdot \pi_{k+1,\ell+1})^* \cdot X^\ell = \pi^*_{k+1,\ell+1} \cdot \bar{X}^\ell$, $\ell+1 \geq s$, by (3.3) >

Lemma 3.2. Let $h \in F(J^s\pi)$. Then $\overline{hX}^k = h\bar{X}^k$, $s \geq k+1$; $\overline{hX}^k = \pi^*_{k+1,s}(h)\,\bar{X}^k$, $s \leq k+1$.

< By lemma 2.2, $(hX)^k = [\pi_s{*}(\pi_{k+1})]^*(h) \cdot X^k$. Applying from the left operator $i^*_{s,k+1}$ we get $\overline{hX}^k = [\pi^*_s(\pi_{k+1}) \cdot i_{s,k+1}]^*(h) \cdot X^k$. But $\pi^*_x(\pi_{k+1})\, i_{s,k+1}$ equals: $1_{J^s\pi}$, $s \geq k+1$; $\pi_{k+1,s}$, $s \leq k+1$ >

If $X \in D(\pi_s)$ then $\pi^*_{r,s} \cdot X \in D(\pi_r)$, $r > s$. Relation between lifts of X and $\pi^*_{r,s} \cdot X$ describes

Lemma 3.3. $\overline{\pi^*_{r,s} \cdot X}^k$ equals: $\pi^*_{r,s} \cdot \bar{X}^k$, $s \geq k+1$; $\pi^*_{r,k+1} \cdot \bar{X}^k$, $r \geq k+1 \geq s$; $\bar{X}^k$, $k+1 \geq r$.

< By lemma 2.3, $(\pi^*_{r,s} \cdot X)^k = [\pi_{r,s}(\pi^*_s(\pi_{k+1}))]^* \cdot X^k$, so $i^*_{r,k+1} \cdot (\pi^*_{r,s} \cdot X)^k = \pi^*_{r,s} \cdot X^k = [\pi_{r,s}(\pi^*_s(\pi_{k+1})) \cdot i_{r,k+1}]^* \cdot X^k$. But $\pi_{r,s}(\pi^*_s(\pi_{k+1})) \cdot i_{s,k+1}$ equals: $i_{s,k+1} \cdot \pi_{r,s}$, $s \geq k+1$; $i_{s,k+1} \cdot \pi_{r,k+1}$, $r \geq k+1 \geq s$; $i_{r,k+1}$, $k+1 \geq r$ >

3.2. Fields $\bar{X}^k$ may be described in pure "inner" terms.

Lemma 3.4. Let $X \in D(\pi_s)$. Define the field $\tilde{X}^k \in D(\pi_{[s,k+1],k})$ by the universal property $(j_{[s,k+1]}\gamma)^* \cdot \tilde{X}^k = (j_s\gamma)^* \cdot X \cdot (j_k\gamma)^*$, $\forall \gamma \in \Gamma(\pi)$. Then $\tilde{X}^k = X^k$.

< Evidently, X^k is correctly defined. Then $(j_{[s,k+1]}\gamma)^* \cdot \bar{X}^k = (j_{[s,k+1]}\gamma)^* \cdot i^*_{s,k+1} \cdot X^k = [i_{s,k+1} \cdot j_{[s,k+1]}\gamma]^* \cdot X^k = (j_s\gamma\,,\, j_{k+1}\gamma)^* \cdot X^k = (j_s\gamma)^* \cdot X^k = (j_s\gamma)^* \cdot X \cdot (j_k\gamma)^*$, i.e. $\bar{X}^k = \tilde{X}^k$ >

Let's calculate $\overline{X}^k$ in special local coordinates. For this let's note that every field $X \in D(\pi_s)$ can be represented (locally) as $X = \alpha_i \partial/\partial x_i$ where $\alpha_i \in F(J^s\pi)$. Next, $\overline{X}^k \in D(\pi_{[s,k+1],k})$ so it's sufficient to know how $\overline{X}^k$ acts on the coordinate's functions $\{x_i, q^a_\sigma, |\sigma| \leq k\}$ in J^k. Let $\gamma \in \Gamma(\pi)$, $\gamma:(x_i) \to (u^a(x), x_i)$. Then $(j_s\gamma)^* \cdot X \cdot (j_k\gamma)^*(q^a_\sigma) = (j_s\gamma)^* \cdot \alpha_i \partial/\partial x_i$; $\dfrac{\partial^{|\sigma|}u^a}{\partial x_\sigma} = (j_s\gamma)^*(\alpha_i) \dfrac{\partial^{|\sigma|+1}u^a}{\partial x_{\sigma+1}} = (j_s\gamma)^*(\alpha_i)\ (j_{k+1}\gamma)^*(q^a_{\sigma+1}) = [j_{[s,k+1]}\gamma]^*(\pi^*_{[s,k+1],s}(\alpha_i) \cdot q^a_{\sigma+1}) = (j_s\gamma)^* \cdot X \cdot (j_k\gamma)^*(x_j) = (j_s\gamma)^*(\alpha_j)$. Hence, by lemma 3.4, $\overline{X}^k = (\pi^*_{[s,k+1],s}(\alpha_i))\left(\dfrac{\partial}{\partial x_i} + q^a_{\sigma+1} \dfrac{\partial}{\partial q^a_\sigma}\right)$, $|\sigma| \leq k$.

3.3. Denote $\overline{X} = (X, \overline{X}^0, X^1, \ldots)$. From lemmas 3.1 and 3.3 we conclude that $\overline{X}$ can be considered as a differentiation of the ring K. In addition, lemma 3.2 tells us that the set of differentiations $\overline{X}$ is K-module elements of whom can be represented locally by the form $\alpha_i \dfrac{\overline{\partial}}{\partial x_i}$, where $\alpha_i \in K$, $\dfrac{\overline{\partial}}{\partial x_i} := \dfrac{\partial}{\partial x_i} + q^a_{\sigma+1} \dfrac{\partial}{\partial q^a_\sigma}$. Lemma 3.2 says also that map $\overline{\Pi} : D(\pi_\infty) \to D(K)$, $X \to \overline{X}$, is homomorphism of K-modules. We remark that field $\dfrac{\overline{\partial}}{\partial x_i}$ is nothing else but a "full derivative" (in a classical sense) corresponding to field $\partial/\partial x_i$ in base M. Indeed, for such field the statement of lemma 3.4 can be written in the form $(j_{s+1}\gamma)^* \cdot \overline{\partial}/\partial x_i = \partial/\partial x_i \cdot (j_s\gamma)^*$, $\forall s \geq 0$.

The homomorphism $\overline{\Pi}$ of K-modules $D(\pi_\infty)$ and $D(K)$ can be considered as "connection" for ring homomorphism $\pi^*_\infty : F(M) \to K$. Indeed, usually one means by connection if the bundle $\pi : E \to M$ the homomorphism of $F(M)$-modules $t : D(M) \to D(E)$ such that $t(X) \cdot \pi^* = \pi^* \cdot X$, $\forall X \in D(M)$. This is equivalent to the homomorphism of $F(E)$-modules $\overline{t}:D(\pi) \to D(E)$ such that $\overline{t}(X) \cdot \pi^* = X$, $\forall X \in D(\pi)$. In addition we have also the homomorphism of $F(M)$-modules $\Pi:D(M) \to D(K)$ which is composition of $\pi^*_\infty:D(M) \to D(\pi_\infty)$ and $\overline{\Pi}$; homomorphism Π is an analog of the usual meaning of connection homomorphism t. In the following both Π and $\overline{\Pi}$ will be called canonical formal connection of bundle π.

<u>Lemma 3.5.</u> $\Pi:C(M) \to D(K)$ is homomorphism of Lie algebras.

Let $X, Y \in D(M)$. By lemma 3.4, fields $\overline{X}^k$ and $\overline{Y}^k$ are uniquely defined by equalities $(j_{k+1}\gamma)^* \cdot \overline{X}^k = X \cdot (j_k\gamma)^*$, $(j_{k+1}\gamma)^* \cdot \overline{Y}^k = Y \cdot (j_k\gamma)^*$. Consider operator $(\bar{X}^{k+1} \cdot \bar{Y}^{k} - \bar{Y}^{k+1} \cdot \bar{X}^{k}) \in D(\pi_{k+2,k})$. We have $(j_{k+2}\gamma)^* \cdot (\bar{X}^{k+1} \cdot \bar{Y}^{k} - \bar{Y}^{k+1} \cdot \bar{X}^{k}) = X \cdot (j_{k+1}\gamma)^* \cdot \bar{Y}^{k} - Y(j_{k+1}\gamma)^* \cdot \bar{X}^{k} = (XY-YX) \cdot (j_k\gamma)^* \; \overline{[X,Y]}^{k}$, hence $\bar{X}^{k+1} \cdot \bar{Y}^{k} - \bar{Y}^{k+1} \cdot \bar{X}^{k} = \pi^*_{k+2,k+1} \cdot \overline{[X,Y]}^{k}$, i.e. $[\bar{X},\bar{Y}] = \overline{[X,Y]}$ >

<u>Theorem 3.6.</u> $\overline{D(\pi_\infty)}$ is Lie algebra and K-module. The curvature of canonical formal connection vanishes.

< Let $X,Y \in D(\pi_\infty)$ and locally $X=f_i Z_i$, $Y=g_i Z_j$ where $f_i, g_i \in K$, $Z_i \in D(M)$. Then $[\bar{X},\bar{Y}] = [f_i\bar{Z}_i, g_j\bar{Z}_j] = f_i g_j [\bar{Z}_i,\bar{Z}_j] + f_i\bar{Z}_i(g_j)\bar{Z}_j - g_j\bar{Z}_j(f_i)\bar{Z}_i = \overline{f_i g_j [Z_i,Z_j] + f_i Z_i(g_j) Z_j - g_j Z_j(f_i) Z_i} = \overline{[X,Y]|_{F(M)}}$ >

By the way, the canonical formal connection is natural. To give a sense to this statement consider some bundle morphism $\alpha : E \to F$. Map α defines morphism of bundles $\alpha^k : J^k\pi \to J^k\nu$ (over M).

Let $X \in D(\nu_s)$, then $\alpha^{s*} \cdot X \in D(\pi_s)$.

<u>Lemma 3.7.</u> Operator Π is natural, i.e. $\alpha^{[s,k+1]*} \cdot \bar{X}^k_\nu = \overline{\alpha^{s*} \cdot X}^k_\pi \cdot \alpha^{k*}$.

< By lemma 3.4, $\forall \gamma \in \Gamma(\pi)$, $[j_{[s,k+1]}(\pi)\gamma]^* \cdot \overline{\alpha^{s*} \cdot X}^k_\pi \cdot \alpha^{k*} = [j_s(\pi)\gamma]^* \cdot \alpha^{s*} \cdot X \cdot [j_k(\pi)\gamma]^* \cdot \alpha^{k*} = [j_s(\nu)(\alpha \cdot \gamma)]^* \cdot X \cdot [j_k(\nu)(\alpha \cdot \gamma)]^* = [j_{[s,k+1]}(\nu)(\alpha \cdot \gamma)]^* \cdot \bar{X}^k_\nu = [j_{[s,k+1]}(\pi)\gamma]^* \cdot \alpha^{[s,k+1]*} \cdot \bar{X}^k_\nu$ >

<u>Corollary 3.8.</u> Canonical formal connection is natural, i.e. $\bar{\alpha}^* \cdot \Pi_\nu = \Pi_\pi \cdot \bar{\alpha}^*$.

<u>§4. Cartan distribution, its properties and symmetries</u>

In this section we shall define the above mentioned distribution -- Cartan distribution -- and calculate the geometry which $J^k\pi$ is provided by this distribution.

Let $\phi : B \to M$ be bundle, $\omega \in \Lambda^p_0(\phi)$, $p > 0$, $Y_1, Y_2 \in D(B)$ and $Y_1 \cdot \phi^* = Y_2 \cdot \phi^* = X \in D(\phi)$. As $(Y_1 - Y_2) \lrcorner \omega = 0$ so expression $X \lrcorner \omega := Y_1 \lrcorner \omega = Y_2 \lrcorner \omega$ is well defined. Note that for every $X \in D(\phi)$ we can always locally find such $Y \in D(B)$ that $Y \cdot \phi^* = X$ (also locally); as such, the form (or function when p=1) $X \lrcorner \omega$ is correctly defined.

For fixed $X \in D(\phi)$ let define his annihilator $\mathrm{Ann}(X) := \{\omega \in \Lambda_0^1(\phi) \mid \omega(X)=0\}$. Clearly $\mathrm{Ann}(X)$ is $F(B)$-submodule in $\Lambda^1(B)$ and $\mathrm{Ann}(fX) = \mathrm{Ann}(X)$, $\forall f \in F(B)$, $f \neq 0$. If $\psi: C \to B$ is another bundle, then $\psi^* \cdot X \in D(\phi \cdot \psi)$ and is evident that $\mathrm{Ann}(\psi^* \cdot X)$ is generated by $\psi^*(\mathrm{Ann}(X))$ in $F(C)$-module $\Lambda_0^1(\phi \cdot \psi)$. At last, if $X' \in D(\psi)$ and $X' \cdot \phi^* = \psi^* \cdot X$ then $\psi^*(\mathrm{Ann}(X)) \subset \mathrm{Ann}(X')$.

Let's see what distribution $\mathrm{Ker}\,\mathrm{Ann}(X)$ in B is like. Let $b \in B$, $x = \phi(b) \in M$. Let us define vector $X_b \in T_x(M)$ by equality $X_b(f) = (X(f))(b)$, $\forall f \in F(M)$. Let $\mathrm{Ann}X_b := \{\theta \in T_x^*(M) \mid \theta(X_b)=0\}$.

<u>Lemma 4.1.</u> $(\mathrm{Ann}(X))|_b = \phi_b^*(\mathrm{Ann}\, X_b)$.

< $\Lambda_0^1(\phi)|_b \approx \phi_b^*(T_x^*(M))$ and $\forall \omega \in \Lambda_0^1(\phi)$,

$$\omega(X)|_b \approx (\phi_b^{*-1}(\omega|_b))(X_b) \quad > \tag{4.1}$$

4.2. Now let us specialize this to our basic situation: bundles $\pi: E \to M$ and $\pi_{k,\ell}: J^k\pi \to J^\ell\pi$. Let $X \in D(M)$, $\overline{X}^k \in D(\pi_{k+1,k})$. Recall k-th Cartan submodule $I_k := \bigcap_{X \in D(M)} \mathrm{Ann}(\overline{X}^k) \subset \Lambda_0^1(\pi_{k+1,k})$. From the above notes and lemmas 3.1-3.3 it follows that $\pi_{k+1,\ell+1}^* I_\ell \subset I_k$, $\mathrm{Ann}\,\Pi(D(M)) = \mathrm{Ann}\,\overline{\Pi}(D(\pi_\infty)) = \lim\,\mathrm{ind}\, I_k := I^1 \subset \Lambda^1(K)$, $\mathrm{Ker}\, I^1 = \overline{\Pi}(D(\pi_\infty))$. K-submodule I^1 in Λ^1 is called Cartan submodule.

Distribution $\Delta_k := \mathrm{Ker}\, I_k$ in $J^{k+1}\pi$ is called k-th Cartan distribution and integral manifolds of this distribution are called R-manifolds (R-resolve). Note that I_k contains $\pi_{k+1,k}$-horizontal forms, hence fibers of projection $\pi_{k+1,k}$ are R-manifolds. From inclusion $\pi_{k+1,\ell+1}^* I_\ell \subset I_k$ follows

<u>Lemma 4.2.</u> Let $b \in J^{k+1}\pi$, $\xi \in \Delta_k|_b$. Then $\pi_{k+1,\ell+1}(\xi) \in \Delta_\ell|\pi_{k+1,\ell+1}(b)$.

4.3. We have received Cartan distribution by algebraic considerations. Let's check the geometrical meaning of this distribution.

Let $x \in M$, $\gamma \in \Gamma(\pi)$, $b = [\gamma]_x^{k+1} \in J^{k+1}\pi$, $X \in D(M)$, $0 \neq \xi = X|_x \in T_x(M)$.

<u>Lemma 4.3.</u> $(j_{k+1}\gamma)_x(\xi) \in \Delta_k|_b$.

< $(j_{k+1}\gamma)^* \cdot \overline{X}^k = X \cdot (j_k\gamma)^*$, hence, $\forall f \in F(J^k\pi)$, we have $\overline{X}_b^k(f) = \overline{X}^k(f)(b) = (j_{k+1}\gamma_x)^*\overline{X}^k(f)(b) = (j_{k+1}\gamma_x)^*\overline{X}^k(f) = X|_x((j_k\gamma)^*(f)) = \xi((j_k\gamma)^*(f) = (j_k\gamma)_x(\xi)(f)$,

i.e. $\bar{X}^k_b = (j_K\gamma)_x(\xi)$ and

$$\bar{X}^k_{[\gamma]^{k+1}_x} = (j_k\gamma)_x(X|_x) \ . \tag{4.2}$$

I_k is horizontal **over** $J^k\pi$ so $I_k|_b((j_{k+1}\gamma)_x(\xi)) = (\pi^{*-1}_{k+1,k}|_b I_k|_b)((j_k\gamma)_x(\xi)) = (\pi^{*-1}_{k+1,k}|_b\ I_k|_b)(\bar{X}^k_b) = I_k(\bar{X}^k)(b) = 0$ >

So, tangent spaces to graphs of jets of sections belong to Cartan distribution. The inverse is also true:

Theorem 4.4. Cartan distribution is spanned by tangent spaces to graphs of jets of sections of bundle π. (Note. In this manner Cartan distribution was defined by A.M. Vinogradov. He also introduced the terminology "Cartan distribution" and "R-manifolds").

< $I_k|_b = \bigcap_{X \in D(M)} (\mathrm{Ann}\ \bar{X}^k)|_b = \pi^*_{k+1,k}|_b(\bigcap \mathrm{Ann}\ \bar{X}^k_b) = \pi^*_{k+1,k}|_b(\mathrm{Ann}((j_k\gamma)_x(T_x(M))),$ i.e. $\Delta_k|_b$ is linear span of those vectors $\eta \in T_b(J^{k+1}\pi)$ that $\pi_{k+1,k}|_b(\eta) \in (j_k\gamma)_x(T_x(M))$. But the subspace of such η differs from $(j_{k+1}\gamma)_x(T_x(M))$ by $\pi_{k+1,k}$ - vertical part. This part, in turn, belongs to the linear span of $(j_{k+1}\gamma)_x(T_x(M))$ when $j_{k+2}\gamma_x$ varies >

From this proof also follows

Lemma 4.5. If $\eta \in \Delta_k|_b$ then $\pi_{k+1,k}\eta \neq 0 \leftrightarrow \pi_{k+1}\eta \neq 0$.

4.4. Let's see what the local generators of I_k in special local coordinates are. Let coordinates of point b be (x, q^a_σ) , $|\sigma| \leq k+1$, $c = \pi_{k+1,k}(b)$. If $X|_x = \alpha_i \frac{\partial}{\partial x_i}$ then $\bar{X}^k_b = \alpha_i(\frac{\partial}{\partial x_i} + q^a_{\sigma+i}\ \frac{\partial}{\partial q^a})$, $|\sigma| \leq k$. By lemma 4.1, annihilator of all fields $\bar{X}^k$ in the point b equals to lift from point c of annihilator of all fields $\bar{X}^k_b$. Hence this annihilator is generated by forms $\{d\,q^a_\sigma - q^a_{\sigma+1}\,d\,x_i\}$, $|\sigma| \leq k$, because $d\,q^a_\sigma(\bar{X}^k_b) = \alpha_i\ q^a_{\sigma+i} = q^a_{\sigma+i}\,d\,x_i(\bar{X}^k_b)$.

4.5. The main geometric property of Cartan distribution is described by

Theorem 4.6. Every section of bundle π_{k+1} which is R-manifold actually represents the graph of some jet of section of bundle π .

< Let $\theta^{k+1} \in \Gamma(\pi_{k+1})$, $k \geq 0$, $b = \theta^{k+1}(x)$, $c = \pi_{k+1,k}(b)$ and

$\theta^{k+1}_x(T_x(M)) \subset \Delta_k|_b$, $\forall x \in M$. Let $\theta^{\ell} := \pi_{k+1,\ell} \cdot \theta^{k+1}$. The point b defines plane $\Pi_b \subset \Delta_{k-1}|_c$ which is tangent to graphs of jets of all sections $\gamma \varepsilon \Gamma(\pi)$ for which $j_{k+1}\gamma_x = b$. Differently speaking, Π_b consists of all vectors of form $\bar{X}^k_b$, $X \varepsilon D(M)$. By lemma 4.1, $I_k|_b = \pi^*_{k+1,k}|_b(\mathrm{Ann}\ \Pi_b \subset T^*_c(J^k\pi))$. By condition, $I_k|_b(\theta^{k+1}_x(T_xM)) = 0$ so $0 = I_k|_b(\theta^{k+1}_x(T_x(M))) = \pi^*_{k+1,k}|_b(\mathrm{Ann}\ \Pi_b \Theta^k_x(T_x(M))))$ hence $\Pi_b = \theta^k_x(T_x(M))$. When k=0 we get $\theta^1 = j_1(\theta^\circ)$, when $k=2 - \theta^2 = j_2(\theta^\circ)$, etc. until $\theta^{k+1} = j_{k+1}(\theta^\circ)$ >

4.6. Here we study symmetries of Cartan distribution, that is, such diffeomorphisms $A_{k+1}: J^{k+1}\pi \to J^{k+1}\pi$ which preserve Cartan distribution.

First of all, let us calculate dimensions of some R-planes, i.e. tangent spaces to R-manifolds. According to §1, the dimension of the fiber $J^k\pi \to J^{k-1}\pi$, $k > 0$, is the number of different symbols q^a_σ , $|\sigma| = k$. This is n(=dim E - dim M) times the number of solutions in nonnegative integers of the equation $t_1 + \ldots + t_n = |\sigma| = k \geq 1$, i.e. $n\binom{m+k-1}{k}$.

By the lemma 4.5, $\forall\, \eta \varepsilon \Delta_k|_b$, $b \varepsilon J^{k+1}\pi$, either $\pi_{k+1,k}(\eta) = 0$ or $\pi_{k+1}(\eta) \neq 0$. Let $\pi_{k+1}(\eta) \neq 0$. Consider all R-planes which contain η . For convenience we choose on M such coordinates that $\pi_{k+1}(\eta) = \partial/\partial x_1$. Let $\xi \varepsilon \Pi(\eta)$, then $\omega|_b(\xi) = d\omega|_b(\xi,\eta) = 0$, $\forall \omega \varepsilon I_k$. As $\eta \varepsilon \Delta_k|_b$, we have $\eta = \frac{\partial}{\partial x_1} + q^a_{\sigma+1} \frac{\partial}{\partial q^a_\sigma} + \beta^a_\tau \frac{\partial}{\partial q^a_\tau}$, $|\sigma| \leq k$, $|\tau| = k+1$, where β^a_τ are arbitrary, fixed numbers. By §4.4, local generators of I_k have the form $\mu^a_\sigma = dq^a_\sigma - q^a_{\sigma+i}\, dx_i$, $|\sigma| \leq k$. Hence the form $\eta \lrcorner d\mu^a_\sigma = (-\eta) \lrcorner (dq^a_{\sigma+i} \wedge dx_i)$ equals to $dq^a_{\sigma+1} - q^a_{\sigma+1+i}\, dx_i$ when $|\sigma| \leq k-1$ and to $dq^a_{\sigma+1} - \beta^a_{\sigma+i}\, dx_i$ when $|\sigma| = k$. Only forms $dq^a_{\sigma+1} - \beta^a_{\sigma+i}\, dx_i$, $|\sigma| = k$, are linearly independent with I_k. The amount of such forms is $n\binom{m+k-1}{k} = n\binom{m+k-1}{m-1} > 1\binom{m}{m-1} = m$ $(n > 1,\ k \geq 1)$. On the other hand, the fiber of $\pi_{k+1,k}$ in the point b is defined by $\{I_k|_b = 0$, $dx_1 = \ldots = dx_m = 0\}$, i.e. by m additional forms linearly independent with $I_k|_b$. This means that, when $n > 1$, $k > 0$, R-manifolds of maximal dimension are the fibers of projection $\pi_{k+1,k}$. Consider now case k=0, n>1 . Let Π be R-plane, $1 \leq \dim \pi_{k+1}(\Pi) = s \leq m$. Choose basis $(\eta^1,\ldots,\eta^s,\xi^i)$ in Π and coordinates in M such that

$\pi_1(\eta^i) = \partial/\partial x_i$, $\pi_1(\xi) = 0$. Let $\{\mu^a = dq^a - q^a_i dx_i\}$ be the basis in $I_0|_b$. Then $\eta^i \lrcorner d\mu^a = dq^a_i - \beta^{a,i}_j dx_j$, where $\eta^i = \frac{\partial}{\partial x_i} + q^a_i \frac{\partial}{\partial q^a} + \beta^{a,i}_j \frac{\partial^j}{\partial q^a}$. As such, the set (ξ^i) is defined by the system $\{dq^a_1 = \ldots = dq^a_s = 0\}$, the codimension of this system is ns and $\dim \Pi = s + (mn - ns) = mn - s(n-1) < mn = \dim \pi^{-1}_{1,0}|_b$. We thus proved

Lemma 4.7. If $n > 1$ then R-planes of maximal dimension are fibers of $\pi_{k+1,k}$.

Symmetries of Cartan distribution are also called classified transformations. We shall first consider some special class of of such transformations.

Let $A: E \to E$ be the diffeomorphis of the bundle π , i.e. there exists the diffeomorphism $\bar{A}: M \to M$ such that $\bar{A} \cdot \pi = \pi \cdot A$. Such A generates morphism of sections $\tilde{A}: \Gamma(\pi) \to \Gamma(\pi)$ and, hence, the diffeomorphism A^k transforms graphs of sections of the form $(j_k\gamma)(M)$ into **itself**. By theorem 4.4, A^k preserves Cartan distribution, i.e. A^k is classified transformation.

Theorem 4.8. Let $A_k: J^k\pi \to J^k\pi$ be classified transformation, $n > 1$. Then $A_k = B^k$ for some $B: E \to E$.

< A_k moves R-manifolds into R-manifolds, so R-manifolds of maximal dimension go to R-manifolds of maximal dimension. The above remarks show that diffeomorphism A_k is preserving fibers of $\pi_{k,k-1}$ and thus generates some diffeomorphism $A_{k-1}: J^{k-1}\pi \to J^{k-1}\pi$.

If $k - 1 > 0$ then $J^{k-1}\pi$ has its own Cartan distribution Δ_{k-2}. Let us see how it interacts with A_{k-1}. By theorem 4.4, Δ_{k-2} on $J^{k-1}\pi$ (corr. Δ_{k-1} on $J^k\pi$) is linearly generated by tangent spaces to the graphs of jets $j_{k-1}\gamma$ (corr. $j_k\gamma$). As $j_{k-1}\gamma = \pi_{k,k-1}(j_k\gamma)$, we see that Δ_{k-2} is generated by $\pi_{k,k-1}(\Delta_{k-1})$. But $A_k(\Delta_{k-1}) = \Delta_{k-1}$, $A_{k-1} \cdot \pi_{k,k-1} = \pi_{k,k-1} \cdot A_k$, hence $A_{k-1}(\Delta_{k-2}) = \Delta_{k-2}$, i.e. A_{k-1} is also a classified transformation. Continuing in the same manner we get the classified transformation $A_1: J^1\pi \to J^1\pi$ and diffeomorphism $A_0: E \to E$.

Let W,V be domains in $J^k\pi$, $A_k:W\to V$ is a classified diffeomorphism and $A_{k-1}:\overline{W} = \pi_{k,k-1}(W) \to \overline{V} = \pi_{k,k-1}(V)$ is the corresponding diffeomorphism constructed above. As A_k preserves fibers of $\pi_{k,k-1}$, it moves those planes which are transversal to fibers of $\pi_{k,k-1}$ into transversal ones. By lemma 4.5 and theorem 4.6, graphs of k-jets which belong to W move under A_k into the graphs. A_{k-1} produces the same effect on graphs of (k-1)-jets which belong to $\overline{W}$. Continuing, we meet the following possibilties:

1) $\pi_{k,1}(W) = J^1\pi = \pi_{1,0}^{-1}(\pi_{k,0}(W) = J^0\pi = E)$, as in conditions of theorem 4.8. In this case A_0 transforms $\Gamma(\pi)$ into $\Gamma(\pi)$, i.e. A_0 is diffeomorphis of the bundle π >

2) $\pi_{k,1}(W) \neq \pi_{1,0}^{-1}(\pi_{k,0}(W))$. In this case A_0 transforms some transversal (to the fibers of π) planes into nontransversal ones, i.e. some graphs of sections are turning into the "multivalued" sections. Hence A_0 is arbitrary diffeomorphism $\pi_{k,0}(W) \to \pi_{k,0}(V)$ and thus we proved

Theorem 4.9. Let $A_k:W\to V$ be a classified diffeomorphism of domains $W,V\in J^k\pi$. Then A_k preserves fibers of the projection $\pi_{k,0}$ and is naturally lifted from the diffeomorphism $A_0:\pi_{k,0}(W)\to\pi_{k,0}(V)$ of domains of E.

Remark. The case $n=1$ is well known (see [1]). The same arguments prove

Theorem 4.10. Let $n=1$ and $A_k:W\to V$ be a classified diffeomorphism of domains $W,V\in J^k\pi$. Then A_k preserves fibers of the projection $\pi_{k,1}$ and is naturally generated by some classified diffeomorphism $A_1:\pi_{k,1}(W)\to\pi_{k,1}(V)$.

It should be noted that the idea of the proofs of theorems 4.8 and 4.9 presented here is due to A.M. Vinogradov.

§5. Formal symmetries and evolution equations. Operator τ.

5.1. When one studies symmetries of any kind it's useful to consider infinitesimal symmetries.

Let $X \in D(J^k\pi)$ and $X \leftrightarrow \{A_t\}$ be a (local) one parameter group of translations along the trajectories of the field X. The field X is called an infinitesimal classified transformation, or infinitesimal symmetry, or simply symmetry if $\{A_t\}$ is group of classified transformations. Because of equaltiy $X = (d/dt)A_t^*|t = 0$ infinitesimal version of theorems 4.9 and 4.10 is

Theorem 5.1. Let $X \in D(J^k\pi)$ be the symmetry. If $n > 1$ then there exists such $Y \in D(E)$ that X is naturally lifted from Y and vise versa: $\forall Y \in D(E)$, $\exists ! \ X \in D(J^k\pi): X \cdot \pi^*_{k,0} = \pi^*_{k,0} \cdot Y$ and X is symmetry (if $n=1$, Y is a classified field on $J^1\pi$, see [1]).

§5.2. $A: J^{k+1}\pi \to J^{k+1}\pi$ is symmetry if it moves distribution Δ_k into Δ_k, or, equivalently, $A^*(I_k) \subset I_k$. The infinitesimal version of this condition, as it's well known, is $X(I_k) \subset I_k$. By theorem 5.1, when $n > 1$, $\exists\, Y \in D(E): X = Y^{k+1}$. Obviously, fields Y^p and Y^q are compatible with respect to the projection $\pi_{p,q}: Y^p \cdot \pi^*_{p,q} = \pi^*_{p,q} \cdot Y^q$. Thus the set of fields $\{Y = Y^0, Y^1, Y^2, \ldots\}$ can be considered as a "field" on $J^\infty\pi$, that is, the differentiation $\overline{Y}: K \to K$ of the ring $K = C^\infty(J^\infty\pi)$. Among others elements from $D(K)$, field of the form $\overline{Y}$ are distinguished by the condition: $\forall k \geq 0$, $\overline{Y}(I_k) \subset I_k$, where I_k are considered as subsets in $\Lambda'(K)$.

Thus $Y(I^1) \subset I^1$. Now it's helpful "to go up" in $J^\infty\pi$ and to put the problem of symmetries on the formal ground, i.e. to look for such $X \in D(K)$ - formal symmetries - that $X(I^1) \subset I^1$. We know that $I^1 = \mathrm{Ann}\, \overline{\Pi}(D(\pi_\infty))$. Let $\omega \in I^1$, $Y \in \overline{D(\pi_\infty)} \overset{def}{=} \overline{\Pi}(D(\pi_\infty))$. As $(X(\omega)(Y) = X(\omega(Y)) - \omega([X,Y])$, we see that the set $\mathrm{sym}\, I^1$ of formal symmetries is the normalizator of the Lie algebra $\overline{D(\pi_\infty)}$, i.e. $X \in \mathrm{sym}\, I^1 \leftrightarrow [X, \overline{D(\pi_\infty)}] \subset \overline{D(\pi_\infty)}$.

Corollary 5.2. Formal symmetries form Lie algebra.

Theorem 5.3. Every symmetry X from $\mathrm{sym}\, I^1$ is uniquely defined by its value $X \cdot \pi^*_{\infty,0} \in D(\pi_{\infty,0})$. Conversely, every field $Z \in D(\pi_{\infty,0})$ is uniquely lifted in $D(K)$ to be $\overline{Z} \in \mathrm{sym}\, I^1$.

< Denote $D^V(K) = \{X \in D(K) | X \cdot \pi^*_\infty = 0\}$, $D^V(\pi_{\infty,0}) = \{X \in D(\pi_{\infty,0}) | X \cdot \pi^* = 0\}$. $\forall X \in D(K)$, $X \cdot \pi^* \in D(\pi_\infty)$ and $X = (X - \overline{X \cdot \pi^*_\infty}) + \overline{X \cdot \pi^*_\infty}$, i.e. there is canonical decomposition of K-modules

$$D(K) = D^V(K) \oplus \overline{D(\pi_\infty)} . \tag{5.1}$$

Thus it's enough to look for formal symmetries in $D^V(K)$. $\overline{D(\pi_\infty)}$ is (locally) generated by $\Pi(D(M))$ over K, so the equality $[X,f\overline{Y}] = X(f)\overline{Y} + f[X,\overline{Y}]$, $f \in K$, $Y \in D(M)$, shows that $X \in \text{sym}\, I^1 \leftrightarrow [X, \Pi(D(M))] \subset \overline{D(\pi_\infty)}$. If, in addition, $X \in D^V(K)$, then $[X, \Pi(D(M))] \cdot \pi^*_\infty = 0$, hence $X \in \text{sym}\, I^1 \cap D_v(K) \leftrightarrow [X, (D(M))] = 0$. Let, locally, $X = A^a_\sigma \frac{\partial}{\partial q^a_\sigma}$. The equality $[X, \Pi(D(M))] = 0$ it's sufficient to check out for the basis of $D(M)$, e.g. $Y_i = \frac{\partial}{\partial x_i} \in D(M)$. We have $[X, \frac{\overline{\partial}}{\partial x_i}] = [A^a_\sigma \partial^a_\sigma, D_i] = -D_i(A^a_\sigma)\partial^a_\sigma + A^a_\sigma[\partial^a_\sigma, \partial_i + q^b_{\tau+i}\partial^b_\tau] = -D_i(A^a_\sigma)\partial^a_\sigma + A^a_{\sigma+i}\partial^a_\sigma$, where $D_i = \frac{\overline{\partial}}{\partial x_i}$, $\partial^a_\sigma = \frac{\partial}{\partial q^a_\sigma}$. Thus $[X,D_i] = 0$, $\forall_i = 1,\dots,m \leftrightarrow A^a_{\sigma+i} = D_i(A^a_\sigma)$ and induction on $|\sigma|$ shows that $A^a_\sigma = D_\sigma(A^a)$, where $D_\sigma = D_{i_1} \cdot \ldots \cdot D_{i_{|\sigma|}}$. At last, $X \cdot \pi^*_\infty = (X \cdot \pi^*_{\infty,0}) \cdot \pi^*$ >

<u>Corollary 5.4</u>. If $Y \in D(E)$, then $\overline{Y}|_{J^k\pi} = Y^k$.

< The two lifts of Y into $D(K) \cap \text{sym}\, I^1$ by the methods of theorems 5.1 and 5.3, must coincide due to theorem 5.3 as they are coincident on $F(E)$ >

<u>Theorem 5.5</u>. $\forall b \in J^k\pi$, vectors of the form $Y^k|_b$, $Y \in D(E)$, form the basis in $T_b(J^k\pi)$.

< If $Y \in D^V(E)$, $Y = A^a\partial^a$, then $Y^k = D_\sigma(A^a)\partial^a_\sigma$ as has been shown in the proof of theorem 5.3. Let $b = [\gamma]^k_{x_0}$, then $Y^k|_b = \frac{\partial^{|\sigma|}(\gamma^*(A^a))}{\partial x_\sigma}\Big|_{x=x_0} \partial^a_\sigma$. The set of numbers of the form $\frac{\partial^{|\sigma|}(\gamma^*(A^a))}{\partial x_\sigma}\Big|_{x=x_0}$ may be as arbitrary as is desired. Thus we have the basis in $T_b(J^k\pi) \cap (\pi_k|_b)^{-1}(0)$. Now let $Y \in D(E)$ and $Y = 0 \cdot \partial^a + \partial_i$. Then $Y = (-q^a_i\, \partial^a) + (\partial_i + q^a_i\, \partial^a)$, $\overline{Y} = D_\sigma(-q^a_i)\, \partial^a_\sigma + (\partial_i + q^a_{\sigma+i} q^a_\sigma)$ $= -q^a_{\sigma+i}\, \partial^a_\sigma + \partial_i + q^a_{\sigma+i}\, \partial^a_\sigma = \partial_i$. This provides the rest in $T_b(J^k\pi)$ >

5.3. Let us find geometric interpretation of the fields $\overline{X}$, $X \in D(\pi_{\infty,0})$. If X was a (vector) field on a manifold N, then X had trajectories, i.e. such maps $\gamma: \mathbb{R}^1 \to N$, that $\gamma^* \cdot X = \frac{\partial}{\partial t} \cdot \gamma^*$. Our situation is quite different: $\overline{X}$ is not a field on a manifold. If one defines trajectory of $\overline{X}$ as such a smooth (on t) one-parameter family $\theta : M \to J^\infty\pi$ that $\theta^* \cdot \bar{X} = \partial/\partial t \cdot \theta^*$, then θ may exist and not to be of the kind $j\gamma^t$ whatever a family $\gamma^t \in \Gamma(\pi)$ may be (it's easy to construct corresponding examples). Thus it's necessary to require additionally that θ be of the form $j\gamma^t$. Now, if $X \cdot \pi^* \neq 0$ then the action of the operator $[(j\gamma^{t*}) \cdot \bar{X} - \partial/\partial t \cdot (j\gamma^t)^*] = 0$ on $\pi_\infty^*(\mathcal{F}(M))$ provides that the trajectory of the field $\bar{X}$ has to be at the same time the trajectory of the field $\bar{X}^v = \overline{X - X \cdot \pi^*}$ and, additionally, the equality $(j\gamma^t)^* \cdot \overline{X \cdot \pi^*} = 0$ must hold. This is impossible in general (unless $X \cdot \pi^* \in \pi_\infty^* \cdot D(M)$). Hence we require that $X \in D^v(\pi_{\infty,0})$.

Trajectories of a field on a manifold are (locally) solutions of the corresponding system of (autonomous) ODE. Let us see what is analogous to this interpretation in our situation. Let, locally $X = A^a\partial^a$, $A^a \in K$, $\bar{X} = D_\sigma(A^a)\partial_\sigma^a$. Then $(j\gamma^t)^* \cdot \bar{X} - \partial/\partial t \cdot (j\gamma^t)^* = (j\gamma^t)^* D_\sigma(A^a)\partial_\sigma^a - \partial/\partial t \cdot (j\gamma^t)^*(q_\sigma^a)\partial_\sigma^a = \{\frac{\partial^{|\sigma|}}{\partial x_\sigma} [(\gamma^t)^*(A^a)] - \frac{\partial}{\partial t} \cdot \frac{\partial^{|\sigma|}}{\partial x_\sigma} [(\gamma^t)^*(q_\sigma^a)]\} \partial_\sigma^a$, i.e. $\frac{\partial^{|\sigma|}}{\partial x_\sigma} \{(\gamma^t)^*(A^a) - \frac{\partial}{\partial t} [(\gamma^t)^*(q^a)]\} = 0$, i.e.

$$\frac{\partial}{\partial t}(\gamma^t)^*(q^a) = (\gamma^t)^*(A^a) . \tag{5.2}$$

This is the system of evolution equations. Thus we proved

Theorem 5.6. Let $X \in D^v(\pi_{\infty,0})$. Then trajectories of the field $\bar{X}$ defined by the equation $(j\gamma^t)^* \cdot \bar{X} = \partial/\partial t \cdot (j\gamma^t)^*$ are (locally) solutions of the evolution system (5.2).

Remark. Trajections may not exist; Cauchy problem for them may have nonunique solution. This is the difference from the case of a field on a manifold.

5.4. Among all sections from $\Gamma(\pi_k)$ only those of the form $j_k\pi$, $\gamma\in\Gamma(\pi)$, are "meaningful". In most cases the behavior of any differential geometric objects (such as forms and tensors) in $J^k\pi$ is necessary to know only when these objects are restricted on graphs of jets of sections of $\Gamma(\pi)$. As such, we introduce and investigate the important operator τ which makes the restriction (another important operator τ^+ is considered in §2.II).

Let $\omega \in \Lambda^s(\pi_k)$. Define form $\tau\omega \in \Lambda_0^s(\pi_{k+1})$ by the universal property $(j_{k+1}\gamma)^*(\tau\omega) = (j_k\gamma)^*(\omega)$. As the value of $(j_k\gamma)_x^*(\omega)$ is uniquely defined by plane $(j_k\gamma)_x(T_xM)$ which in turn is uniquely defined by point $j_{k+1}\gamma_x \in J^{k+1}\pi$, τ is correctly defined. Let's list some properties of operator $\tau : \Lambda^s \to \Lambda_0^s$.

Lemma 5.7. τ is K-homomorphism of K-algebras $\Lambda^* \to \Lambda_0^*$. $\tau^2 = \tau$, i.e. τ is projector.

< $\forall \omega_1, \omega_2 \in \Lambda^*$, $(j\gamma)^*(\omega_1 \wedge \omega_2) = (j\gamma)^*(\omega_1) \wedge (j\gamma)^*(\omega_2)$. The rest is evidently >

Corollary 5.8. $\forall \omega \in \Lambda^*$, $\exists!$ decomposition $\omega = \psi + \mu$: $\tau\psi = \psi$, $\tau\mu = 0$. $\mathrm{Ker}\,\tau = I_m(1-\tau)$. $I := \mathrm{Ker}\,\tau$ is ideal in Λ^* .

Lemma 5.9. $\tau\Lambda^s = 0$ when $s > m$.

< $\Lambda^s(M) = 0$ when $s > m = \dim M$ >

Lemma 5.10. $(j\gamma)^*(1-\tau) = 0$, $\forall \gamma \in \Gamma(\pi)$.

< $(j\gamma)^* = (j\gamma)^* \cdot \tau$ >

Theorem 5.11. $\Lambda^1 \cap I = I^1$.

< Let $X \in D(M)$, $\omega \in \Lambda^1(J^k\pi)$, $b = [\gamma]_x^{k+1} \in J^{k+1}\pi$, $a = [\gamma]_x^k \in J^k\pi$. Then $[(\pi_{k+1,k}^* \omega)(\bar{X}^k)](b) = \omega|_a(\bar{X}_b^k) = [(j_k\gamma)_x^*(\omega)](X|_x) = [(j_{k+1}\gamma)_x^*(\tau\omega)](X|_x)$. Hence $\omega \in I^1 \leftrightarrow (\pi_{k+1,k}^* \omega)(\bar{X}^k) = 0$, $\forall X \in D(M) \leftrightarrow (\tau\omega)(\bar{X}) = 0$, $\forall X \in D(M) \leftrightarrow \tau\omega = 0$, because $\tau\omega \in \Lambda_0^1$ >

Corollary 5.12. Ideal generated by I^1 in Λ^* , lies in I .

< $\tau(I^1 \wedge \Lambda^*) = \tau I^1 \wedge \tau\Lambda^* = 0$ >

Theorem 5.13. Let $\omega \in \Lambda^s(\pi_k)$. $\tau\omega = 0 \leftrightarrow (\pi_{k+1,k}^* \omega)(\bar{X}_1^k, \ldots, \bar{X}_s^k) = 0$, $\forall X_i \in D(M)$

< As in the proof of theorem 5.11, $[(\pi_{k+1,k}^* \omega)(\bar{X}_1^k, \ldots, \bar{X}_s^k)](b) = \omega|_a(\bar{X}_1^k|_b, \ldots, \bar{X}_s^k|_b) = [(j_k\gamma)_x^*(\omega)](X_1|_x, \ldots, X_s|_x) = [(j_{k+1}\gamma)_x^*(\tau\omega)](X_1|_x, \ldots, X_s|_x)$ >

Corollary 5.14. $\tau\omega = 0 \leftrightarrow \tau(\bar{X} \lrcorner \omega) = 0$, $\forall X \in D(M)$.

< Make induction by $s = \deg\omega$ by beginning in theorem 5.11 and with the step of theorem 5.13 >

Note, that $\overline{fX} = f\overline{X}$, $\forall f\varepsilon K$, $\forall X\varepsilon D(M)$, and $\tau(\overline{fX}\lrcorner\omega) = f\tau(\overline{X}\lrcorner\omega)$. Hence theorem 5.13 and corollary 5.14 can be formulated as follows

Lemma 5.15. $\forall\omega\varepsilon\Lambda^s(K)$, $\tau\omega = 0 \leftrightarrow \tau(\overline{X}\lrcorner\omega) = 0$, $\forall X\varepsilon D(\pi_\infty) \leftrightarrow \omega(\overline{X}_1,\ldots,\overline{X}_s) = 0$, $\forall X_i \varepsilon D(\pi_\infty)$.

Let's find out the symmetries of ideal I, i.e. such fields $X \varepsilon D(K)$, that $X(I)\subset I$.

Theorem 5.16. $\mathrm{Sym}\, I = \mathrm{Sym}\, I^1$.

< $I^1 = I \cap \Lambda^1$ hence $\mathrm{Sym}\, I \subset \mathrm{Sym}\, I^1$. Now, if $X \varepsilon \mathrm{Sym}\, I^1$, $\omega \varepsilon I \cap \Lambda^s$, then for $\forall X_1,\ldots,X_s \varepsilon D(\pi_\infty)$, $X(\omega)(\overline{X}_1,\ldots,\overline{X}_s) = X(\omega(\overline{X}_1,\ldots,\overline{X}_s)) + \Sigma\omega(\overline{X}_1,\ldots,[\overline{X}_i,X],\ldots,\overline{X}_s) = 0$, because $[X, \overline{D(\pi_\infty)}] \subset \overline{D(\pi_\infty)}$ >

Lemma 5.17. Let $X \varepsilon D^v(\pi_{\infty,0})$. Then $\overline{X}(\mathrm{Im}\tau) \subset \mathrm{Im}\tau$.

< $\overline{X}(\tau\omega) = \overline{X}\lrcorner d\tau\omega$, $d\tau\omega \varepsilon \Lambda_1^{s+1}$, $(\overline{X}\lrcorner d\tau\omega) \varepsilon \Lambda_0^s = \tau\Lambda_0^s$ >

Theorem 5.18. Let $X \varepsilon D(\pi_{\infty,0})$. Then $\tau\overline{X} = \tau\overline{X}\tau$. If, in addition, $X \varepsilon D^v(\pi_{\infty,0})$, then $\tau\overline{X} = \overline{X}\tau$.

< $\tau\overline{X} - \tau\overline{X}\tau = \tau\overline{X}(1-\tau)$, $\mathrm{Im}(1-\tau) = \ker\tau = I$, $\overline{X}(I) \subset I$, $\tau I = 0$. If $X \varepsilon D^v(\pi_{\infty,0})$ then $\overline{X}(\mathrm{Im}\tau) \subset \mathrm{Im}\tau$ and $\tau\overline{X}\tau = \overline{X}\tau$ >

Lemma 5.19. $\tau(dq_\sigma^a) = q_{\sigma+i}^a\, dx_i$, $\tau(dx_j) = dx_j$.

< By definition of τ >

The natural properties of morphism τ are considered in §2.II .

Theorem 5.20. $dI\subset I$, i.e. ideal I is differentially closed.

< Let $\omega\varepsilon\Lambda^s \cap I$. $\forall X_i\varepsilon D(M)$, $d\omega(\overline{X}_1,\ldots,\overline{X}_{s+1}) = \pm\overline{X}_i(\omega(\ldots,\hat{\overline{X}}_i,\ldots)) \pm \omega([\overline{X}_i,\overline{X}_j],\ldots) = 0$ because $[\overline{X}_i,\overline{X}_j] = \overline{[X_i,X_j]} \varepsilon \overline{D(\pi_\infty)}$ >

Corollary 5.21. $\tau d\tau = \tau d$. Hence $(\tau d)^2 = 0$

< $\tau d - \tau d\tau = \tau d(1-\tau)$. $\mathrm{Im}(1-\tau) = I$, $dI \subset I$, $\tau I = 0$ >

Theorem 5.22. Ideal I coincides with ideal $\tilde{I}$ which generated by I^1 in Λ^* .

< Let's proceed by induction on $s = \deg\omega$, $\omega \varepsilon I$. For $s=1$ $I^1 = \Lambda^1 \cap I$

(theorem 5.11). Let the theorem be true for all $s \leq N$, $\omega \in \Lambda^{N+1} \cap I$. Let's find representation $\omega = \omega^i \wedge \nu^i$, $\omega^i \in \Lambda^1$, $\nu^i \in \Lambda^N$. Then $\omega = \omega^i \wedge (1-\tau)\nu^i + \omega^i \wedge \tau\omega^i$. By assumption, $(1-\tau)\nu^i \in \tilde{I}$, hence $\omega^i \wedge (1-\tau)\nu^i \in I$. Then $0 = \tau\omega$ $\tau\omega^i \wedge \tau\nu^i$, hence $\omega^i \wedge \tau\nu^i = (1-\tau)\omega^i \wedge \tau\nu^i$, but $(1-\tau)\omega^i \in I^1 >$

CHAPTER II

THE STRUCTURE OF LAGRANGIAN FORMALISM

§1. Formulae for the first variation

Until the end of this chapter Ω denotes an arbitrary element from $\Lambda_0^m = \Lambda_0^m(\pi_\infty)$.

Definition. Formula for the first variation (of form Ω) is the identity

$$\tau\bar{X}(\Omega) = D(\bar{X}\lrcorner\delta\Omega) + \tau(X\lrcorner\partial\Omega) \quad , \quad D \overset{\text{def}}{=} \tau d \quad , \tag{1.1}$$

$\forall X\in D(\pi_{\infty,0})$ and some forms $S\Omega\in\Lambda_1^m(\pi_\infty)$, $\partial\Omega\in\Lambda_1^{m+1}(\pi_\infty,\pi_{\infty,0})$ not depending on X . Operators $S:\Omega\to S\Omega$, $\partial = \Omega\to\partial\Omega$ (a priori they might not exist) are called Legendre transformation and Euler-Lagrange operator correspondingly.

Existence of forms $S\Omega$ and $\partial\Omega$ will be proved in §3. Note that the difficulty for their construction is due to the globality property. Indeed, it is easy to construct them locally as it usually is done in textbooks on the calculus of variations. However, it is not enough as the difference $\tau\bar{X}(\Omega) - \tau(X\lrcorner\partial\Omega)$ has to be D-exact globally.

Consider now the problem of uniqueness of forms $S\Omega$ and $\partial\Omega$. Let $S_1\Omega$ and $\partial_1\Omega$ be another pair of forms satisfying (1.1) , i.e. $\forall X\in D(\pi_{\infty,0})$

$$\tau\bar{X}(\Omega) = D(\bar{X}\lrcorner S_1\Omega) + \tau(X\lrcorner\partial_1\Omega) \quad . \tag{1.2}$$

Denote $S_2\Omega = S_1\Omega - S\Omega$, $\partial_2\Omega = \partial\Omega - \partial_1\Omega$ and substruct (1.2) from (1.1); one gets

$$D(\bar{X}\lrcorner S_2\Omega) = \tau(X\lrcorner\partial_2\Omega) \tag{1.3}$$

Let us prove that $\partial_2\Omega = 0$. Let $\theta\in J^\infty\pi$, $x = \pi_\infty(\theta)$. Let V be a small neighborhood of x and U be the boundary of V . Let $\gamma\in\Gamma(\pi|_{\pi^{-1}(V)})$,

$j\gamma = \theta$ and $X\varepsilon D^v(E)$, $X|_{\pi^{-1}(U)} = 0$. Then, by Stoks theorem, $\int_V (j\gamma)^*(X\lrcorner\partial_2\Omega) = 0$. Hence, by Duboua-Reimond lemma, $\partial_2\Omega|_\theta = 0$, i.e. $\partial_2\Omega \equiv 0$.

Remark. The arguments above also work if we don't require $S\Omega$ to be in Λ_1^m : it can be in Λ^m as well.

From (1.3) we conclude that $\partial_1\Omega = \partial\Omega$, $S_1\Omega = S\Omega + S_2\Omega$, where $S_2\Omega$ must satisfy $D(\overline{X}\lrcorner S_2\Omega) \equiv 0$.

When $m>1$, there are plenty of such forms $S_2\Omega$:

Lemma 1.1. Let $m>1$, $w\varepsilon\Lambda_0^{m-2}$, $\mu = dDw$. Then $D(\overline{X}\lrcorner\mu) = 0$, $\forall X\varepsilon D(\pi_{\infty,0})$.

< $D(\overline{X}\lrcorner dDw) = \tau d(\overline{X}\lrcorner dDw) = \tau[\overline{X}(dDw) - \overline{X}\lrcorner d(dDw)] = \tau\overline{X}(dDw) = \tau\overline{X}\tau(dDw) = \tau\overline{X}D^2w = 0$ >

If $m = 1$ ("classical case") the form $S\Omega$ is unique.

Lemma 1.2. $\tau S\Omega = \Omega$.

<Let $Y\varepsilon D(M)$, $X = Y^\circ$. Then $\overline{X} = \overline{Y}$. We have $\tau\overline{X}(\Omega) = \tau(\overline{Y}\lrcorner d\Omega) + \tau d(\overline{Y}\lrcorner\Omega) = \tau d(\overline{Y}\lrcorner\Omega) = \tau d(\overline{X}\lrcorner S\Omega) + \tau(X\lrcorner\partial\Omega) = \tau d(\overline{X}\lrcorner S\Omega)$, because $\tau(\overline{Y}\lrcorner\Lambda^{m+1}) = 0$ (see §5.I). Then $\tau d(\overline{X}\lrcorner S\Omega) = \tau\overline{X}(S\Omega) - \tau(\overline{X}\lrcorner dS\Omega) = \tau\overline{X}(S\Omega) = \tau\overline{X}(\tau S\Omega) = \tau d(\overline{Y}\lrcorner\tau S\Omega)$. Thus, $\tau d(\overline{Y}\lrcorner\Omega) = \tau d(\overline{Y}\lrcorner\tau S\Omega)$, i.e. $\tau d(\overline{Y}\lrcorner\mu) = 0$, where $\mu = (\Omega - \tau S\Omega)\varepsilon\Lambda_0^m$. Let $f\varepsilon F(M)$ then $\overline{fY} = f\overline{Y}$ (see §3.I), so $0 = \tau d(\overline{fY}\lrcorner\mu) = \tau d(f\overline{Y}\lrcorner\mu) = f\tau d(\overline{Y}\lrcorner\mu) + Df\wedge\tau(\overline{Y}\lrcorner\mu) = Df\wedge(\overline{Y}\lrcorner\mu)$. f is arbitrary, hence $\overline{Y}\lrcorner\mu = 0$; Y is arbitrary too, so $\mu = 0$, hence $\tau S\Omega = \Omega$ >

Remark 1.3. On the way we have proven that for every $X = Y^0$, $Y\varepsilon D(M)$, the arbitrary pair of forms $S\Omega\varepsilon\Lambda^m$, $\partial\Omega\varepsilon\Lambda_1^{m+1}$ $[\pi_{\infty,0}]$ which satisfies $\tau S\Omega = \Omega$, also satisfies (1.1). The same is true, of course, if $Y\varepsilon D(\pi_\infty)$. Due to (5.1.I) we need to check (1.1) now only for $X\varepsilon D^v(\pi_{\infty,0})$.

Lemma 1.4. Let $m = 1$, $\tilde{S}\varepsilon\Lambda_1^1$ $[\pi_{\infty,k}]$ and $D(\overline{X}\lrcorner\tilde{S}) = 0$, $\forall X\varepsilon D(\pi_{\infty,0})$.

Then $\tilde{S} = 0$.

< As $\dim M = 1$, equality $D(\bar{X}\lrcorner\tilde{S}) = 0$ means that $\bar{X}\lrcorner\tilde{S} = c(X) = \text{const}$. But operator $X\mapsto(\bar{X}\lrcorner\tilde{S})$ is differential operator of order $\leq k$, hence $\tilde{S} = 0$>

1.2. In view of remark 1.3, we can restrict ourself to look for such forms $S\Omega$ and $\partial\Omega$ that $\tau S\Omega = \Omega$ and (1.1) is valid for vertical fields $X\epsilon D^{v}(\pi_{\infty,0})$. We can proceed farther.

Theorem 1.5. If (1.1) is valid for every $X\epsilon D^{v}(E)$ thena it is valid $\forall X\epsilon D^{v}(\pi_{\infty,0})$.

< Let $X\epsilon D^{v}(E)$. Then $\tau\bar{X}(\Omega) = \bar{X}(\Omega)$, $\tau(X\lrcorner\partial\Omega) = X\lrcorner\partial\Omega$, hence $\bar{X}(\Omega) = \bar{X}\lrcorner d\Omega = D(\bar{X}\lrcorner S\Omega) + X\lrcorner\partial\Omega = \tau\bar{X}(S\Omega) - \tau(\bar{X}\lrcorner dS\Omega) + X\lrcorner\partial\Omega$. By lemma 1.2, $\tau\bar{X}(S\Omega) = \tau X(\tau S\Omega) = \tau\bar{X}(\Omega) = \bar{X}(\Omega)$, and we get

$$X\lrcorner\partial\Omega = \tau(\bar{X}\lrcorner dS\Omega). \tag{1.4}$$

$S\Omega\epsilon\Lambda_1^m$, hence $dS\Omega\epsilon\Lambda_2^{m+1}$. Let us define form $\tau^+ dS\Omega\epsilon\Lambda_1^{m+1}$ by the property $Y\lrcorner\tau^+ dS\Omega = \tau(Y\lrcorner dS\Omega)$, $\forall Y\epsilon D^{v}(K)$. Because a form from Λ_1^{m+1} is uniquely defined by the values of contractions $(Y\lrcorner\cdot)$ for all $Y\epsilon D^{v}(K)$, the above definition is correct. More details about operator τ^+ see in §2. Now rewrite (1.4) as

$$X\lrcorner\partial\Omega = \bar{X}\lrcorner\tau^+ dS\Omega. \tag{1.5}$$

This means that

$$\partial\Omega = \tau^+ dS\Omega. \tag{1.6}$$

Because of coincidence of both forms, formulas (1.5), (1.4) and (1.1) are valid for $\forall X\epsilon D^{v}(\pi_{\infty,0})$ too >

Thus, the main problem of Lagrangian formalism, that is, the construction of formula (1.1), is reduced to construction of the operator $S:\Lambda_0^m \to \Lambda_1^m$ which satisfies the conditions

$$\tau S = \mathrm{id}, \quad \mathrm{Im}(\tau^+ dS) \subset \Lambda_1^{m+1}[\pi_{\infty,0}]. \tag{1.7}$$

Nevertheless, this reformulation doesn't help to construct S (even if one requires so called "Hamilton-Cartan principle", see § 4) and it needs to be reformulated farther. It will be done in § 3. At the moment let us remark that operator of Euler-Lagrange (uniquely defined) is derived from the Legendre transformation (which is defined nonuniquely).

1.3. Here we consider the case $\Omega \in \Lambda_0^m(\pi_1)$, i.e. the first nontrivial case (if $\Omega \in \Lambda_0^m(\pi)$, one can put $S\Omega = \Omega$, $\partial\Omega = d\Omega$).

Suppose formula (1.1) is constructed and we handle some $S\Omega \in \Lambda_1^m$, $\partial\Omega \in \Lambda_1^{m+1}[\pi_{\infty,0}]$. Let $f \in K$, $X \in D^v(\pi_{\infty,0})$. We have $(f\cdot\overline{X})(\Omega) = f.\overline{X}(\Omega) + df \wedge (\overline{X} \lrcorner \Omega) = f\cdot\overline{X}(\Omega)$, hence, denoting $\overline{X}_f = \overline{fX} - f\overline{X}$, we get

$$\overline{X}_f(\Omega) = D(\overline{fX} \lrcorner S\Omega) - fD(\overline{X} \lrcorner S\Omega). \tag{1.8}$$

Now $\overline{fX} = \overline{X}_f + f\overline{X}$, so $D(\overline{fX} \lrcorner S\Omega) = D(\overline{X}_f \lrcorner S\Omega) + D(f\overline{X} \lrcorner S\Omega) = D(\overline{X}_f \lrcorner S\Omega) + Df \wedge (\overline{X} \lrcorner S\Omega) + fD(\overline{X} \lrcorner S\Omega)$, and (1.8) can be rewritten as

$$\overline{X}_f(\Omega) = D(\overline{X}_f \lrcorner S\Omega) + Df \wedge (\overline{X} \lrcorner S\Omega). \tag{1.9}$$

Observe now that if we find $S\Omega \in \Lambda_1^m$ such that $\tau S\Omega = \Omega$ and (1.9) is valid, this guarantees (1.1).

Let us require that $S\Omega \in \Lambda_1^m[\pi_{\infty,0}]$, i.e. $S\Omega$ is the horizontal over E. $\overline{X}_f$ is vertical over E, i.e. $\overline{X}_f \cdot \pi_{\infty,0}^* = 0$, so (1.9) turns into

$$\overline{X}_f(\Omega) = Df \wedge (\overline{X} \lrcorner S\Omega) \tag{1.10}$$

$\Omega\varepsilon\Lambda_0^m(\pi_1)$, so $\overline{X}_f(\Omega)$ is the differentiation of K into $\Lambda_0^m(\pi_\infty)$ (with respect to f). Besides, both sides of (1.10) are K-linear with respect to X. Hence (1.10) uniquely defines $S\Omega\varepsilon\Lambda_1^m(\pi_1;\pi_{1,0})$. Now (1.6) shows that $\partial\Omega\varepsilon\Lambda_1^{m+1}(\pi_2;\pi_{2,0})$, and we get

Lemma 1.6. Let $\Omega\varepsilon\Lambda_0^m(\pi_k)$, k=1. Then $\exists!$ the pair of forms $S\Omega\,\varepsilon\,\Lambda_1^m(\pi_1;\pi_{1,0})$, $\partial\Omega\varepsilon\Lambda_1^{m+1}(\pi_2;\pi_{2,0})$ for (1.1)

§2. Operator τ^+ and its geometry

2.1 When deriving (1.6) we have seen how operator τ^+ arises. Here we study its properties in more detail. Note that account is parallel to that of §5.I.

2.2 τ^+ is K-homomorphism from Λ^{m+k} in Λ_k^{m+k}, $k>0$, Forms from Λ_k^{m+k} are "k times vertical" so it's enough to define their value on sets of k vertical (with respect to π_∞) fields $X_i\varepsilon D^v(K)$.

Definition. $X_1\lrcorner\ldots\lrcorner X_k\lrcorner\tau^+\omega = \tau(X_1\lrcorner\ldots\lrcorner X_k\lrcorner\omega)$, $\forall\omega\varepsilon\Lambda^{m+k}, \forall X_i\varepsilon D^v(K)$.

Lemma 2.1. $(\tau^+)^2 = \tau^+$, i.e. τ^+ is the projector.

< $\tau^2 = \tau$ >

Lemma 2.2. It's sufficient to define the form $\tau^+\omega$ only on evolution fields.

< By theorem 5.5.I, fields of the form X^r, $X\varepsilon D^v(E)$, constitute the basis in vertical subspace for all points of $J^r\pi$ >

Lemma 2.3. If $\omega\varepsilon\Lambda^{m+k}(J^r\pi)$, then $\tau^+\omega\varepsilon\Lambda_k^{m+k}(\pi_{r+1};\pi_{r+1,r})$, i.e. $\tau^+\omega$ is horizontal over $J^r\pi$.

< If $Y\varepsilon D(J^{r+1}\pi)$ and $Y\cdot\pi_{r+1,r} = 0$, then $Y\lrcorner\pi^*_{r+1,r}\omega = 0$ >

Lemma 2.4. Let $X\varepsilon D^v(\pi_{\infty,0})$, $\omega\varepsilon\Lambda_k^{m+k}$. Then $\overline{X}(\omega)\varepsilon\Lambda_k^{m+k}$, i.e. $\overline{X}(\mathrm{Im}\tau^+)\subset\mathrm{Im}\,\tau^+$.

< Let $d^m x$ be a local volume element on M, then, locally, $\omega = \theta \wedge d^m x$, $\theta \in \Lambda^k$. From $\bar{X}(d^m x) = 0 \Rightarrow \bar{X}(\omega) = \bar{X}(\theta) \wedge d^m x \in \Lambda^{m+k}_k$ >

Lemma 2.5. $\forall \omega \in \Lambda^{m+k}$, $\exists!$ decomposition $\omega = \psi + \mu$, $\tau^+ \psi = \psi$, $\tau^+ \mu = 0$.

< By lemma 2.1, $\psi = \tau^+ \omega$, $\mu = (1-\tau^+)\omega$ >

Lemma 2.6. $\mathrm{Ker}\, \tau^+ = \mathrm{Im}(1-\tau^+)$.

< $(\tau^+)^2 = \tau^+$ >

Lemma 2.7. Let $X \in D(\pi_\infty)$, $\omega \in \Lambda^{m+k}$. Then $(\bar{X} \lrcorner \omega) \in \mathrm{Ker}\, \tau^+$ when $k > 1$ and $(\bar{X} \lrcorner \omega) \in \mathrm{Ker}\, \tau$ when $k = 1$.

< $X_1 \lrcorner \ldots \lrcorner X_{k-1} \lrcorner \tau^+(\bar{X} \lrcorner \omega) = \tau(X_1 \lrcorner \ldots \lrcorner X_{k-1} \lrcorner \bar{X} \lrcorner \omega) = (-1)^{k-1} \tau(\bar{X} \lrcorner \mu)$, where $\mu = (X_1 \lrcorner \ldots \lrcorner X_{k-1} \lrcorner \omega) \subset \Lambda^{m+1} \subset I$, I is Cartan ideal (see §5.I). Hence $(\bar{X} \lrcorner \mu) \in I$ and $\tau(\bar{X} \lrcorner \mu) = 0$ >

Lemma 2.8. Let $X \in D(\pi_\infty)$, $\omega \in \Lambda^{m+k}$. Then $\tau^+ \bar{X}(\omega) = \tau^+ d(\bar{X} \lrcorner \omega)$ when $k > 0$ and $\tau \bar{X}(\omega) = \tau d(\bar{X} \lrcorner \omega)$ when $k = 0$.

< $\bar{X}(\omega) = \bar{X} \lrcorner d\omega + d(\bar{X} \lrcorner \omega)$. By Lemma 2.7, $\tau^+(\bar{X} \lrcorner d\omega) = 0$ when $k > 0$ and $\tau(\bar{X} \lrcorner d\omega) = 0$ when $k = 0$ >

Let us denote $I^+ = \mathrm{Ker}\, \tau^+$.

Lemma 2.9. I^+ is K-submodule in Λ^*.

< Obviously >

Lemma 2.10. $dI^+ \not\subset I^+$.

<Using local coordinates introduced in §1.I, let us pick up $\omega = dq^a_\sigma \wedge dq^b_\nu \wedge (\partial_i \lrcorner d^m x)$. Let $\mu = (1-\tau^+)\omega \in I^+$. We have $\mu = (1-\tau^+)\omega = dq^a_\sigma \wedge dq^b_\nu \wedge (\partial_i \lrcorner d^m x) - (q^b_{\nu+i} dq^a_\sigma - q^a_{\sigma+i} dq^b_\nu) \wedge d^m x$. As $d\omega = 0$, so $d\mu = -d\tau^+ \omega = (dq^a_{\sigma+i} \wedge dq^b_\nu - dq^b_{\nu+i} \wedge dq^a_\sigma) \wedge d^m x \neq 0$. But $d\mu = \tau^+ d\mu$ >

2.3. Thus we have in Λ^* two objects: the differentially closed ideal $I, dI \subset I$, and the differentially unclosed submodule $I^+, dI^+ \not\subset I^+$. Symmetries of I were calculated in §5.I. Let us calculate now symmetries sym I^+ of I^+, i.e. we look for such $X\in D(K)$, that $X(I^+)\subset I^+$.

Theorem 2.11. Sym I^+ = symI.

< Let $\omega\in\Lambda^{m+k}$, $k>0$. $\tau^+\omega = 0 \Longleftrightarrow \overline{Y}^{(m)}\lrcorner\,\overline{X}^{(k)}\lrcorner\,\omega \overset{\text{def}}{=} \overline{Y}_1\lrcorner\ldots\lrcorner\overline{Y}_m\lrcorner\overline{X}_1\lrcorner\ldots\lrcorner\overline{X}_k\lrcorner\,\omega = 0$, $\forall Y_i\in D(\pi_\infty)$, $\forall X_i\in D^v(\pi_{\infty,0})$. Equivalently, $\tau^+\omega=0 \Longleftrightarrow \overline{Y}^{(m)}\lrcorner\,\overline{Z}^{(k)}\lrcorner\,\omega = 0$, $\forall Y_i\in D(\pi_\infty)$, $\forall Z_i\in D^v(K)$. Let $X\in D(K)$, then $X(\omega)\,(\overline{Y}^{(m)},\overline{X}^{(k)}) = X(\omega(\overline{Y}^{(m)},\overline{X}^{(k)})) + \omega([\overline{Y}^{(m)},X],\overline{X}^{(k)}) + \omega(\overline{Y}^{(m)},[\overline{X}^{(k)},X]$, where $[\overline{Y}^{(m)},X] \overset{\text{def}}{=} ([\overline{Y}_1,X],\ldots,[\overline{Y}_m,X])$ and the same for $[\overline{X}^{(k)},X]$. For $\omega\in I^+$ we thus get

$$X(\omega)(\overline{Y}^{(m)},\overline{X}^{(k)}) = \omega([\overline{Y}^{m},X],\overline{X}^{(k)}) + \omega(\overline{Y}^{(m)},[\overline{X}^{(k)},X]). \qquad (2.2)$$

If $X = \overline{Y}$, $Y\in D(\pi_\infty)$, then $[\overline{Y}^{(m)},X]\in\overline{D(\pi_\infty)}^{(m)}$, $[\overline{X}^{(k)},X]\in\overline{D(\pi_\infty)}^{(k)}$, and r.h.s. of (2.2) vanishes. Hence $\overline{D(\pi_\infty)}\subset$ sym I^+. But $D(K) = D^v(K)\oplus\overline{D(\pi_\infty)}$ (formula (5.1.I)), so let now $X\in D^v(K)$. Replacing $\overline{X}^{(k)}$ by $Z^{(k)}$ in (2.2) and observing that $[Z_i,X]\in D^v(K)$, we rewrite (2.2) as

$$X(\omega)\,(\overline{Y}^{(m)},Z^{(k)}) = \omega([\overline{Y}^{(m)},X],Z^{(k)}). \qquad (2.3)$$

Thus $X\in\text{Sym}\,I^+\cap D^v(K) \Longleftrightarrow \omega([\overline{Y}^{(m)},X],Z^{(k)}) = 0$, $\forall Y_i\in D(\pi_\infty)$, $\forall Z_i\in D^v(K)$, $\forall\omega\in\Lambda^{m+k}\cap I^+$. Consider $\omega^{(k)} \overset{\text{def}}{=} (Z^{(k)}\lrcorner\,\omega)\in\Lambda^m$. As $\omega^{(k)}$ may take an arbitrary value in $I\cap\Lambda^m$, we get that $X\in\text{Sym}\,I^+\cap D^v(K) \Longleftrightarrow X(I\cap\Lambda^m)\subset I\cap\Lambda^m$. Let us take $\omega^{(k)}\in(I^1\wedge\Lambda_0^{m-1})\subset\Lambda^m\cap I$. Then for X to belong to Sym I^+ it's necessary $X(I^1)\subset I^1$. By theorem 5.3.I, $X\in\overline{D^v(\pi_{\infty,0})}$. On the other hand, (2.2) garantees that if $X\in\overline{D^v(\pi_{\infty,0})}$ then $X\in\text{Sym}\,I^+$>

Lemma 2.12. If $X\in D(\pi_{\infty,0})$ then $\tau^+\overline{X} = \tau^+\overline{X}\tau^+$. If in addition, $X\in D^v(\pi_{\infty,0})$ then $\tau^+\overline{X} = \overline{X}\,\tau^+$.

< $\tau^+\overline{X} - \tau^+\overline{X}\,\tau^+ = \tau^+\,\overline{X}(1-\tau^+)$. $\text{Im}(1-\tau^+) = \text{Ker}\,\tau^+ = I^+$ (lemma 2.6), $\overline{X}(I^+)\subset I^+$, $\tau^+I^+ = 0$, so $\tau^+\overline{X}(1-\tau^+) = 0$. In the case $X\in D^v(\pi_{\infty,0})$ we have

$\tau^+ \overline{X}\tau^+ = \overline{X}\,\tau^+$ because $\overline{X}(\mathrm{Im}\,\tau^+) \subset \mathrm{Im}\,\tau^+$ (lemma 2.4) >

2.4. Let us consider now natural properties of operators τ, τ^+, of ideal I and of submodule I^+.

Let $\Delta : \Gamma(\pi) \to \Gamma(\nu)$ be a differential operator of order $\leq k$ and $\phi_\Delta : J^k\pi \to F$, $\phi^s_\Delta : J^{k+s}\pi \to J^s\nu$, $\overline{\phi}_\Delta : J^\infty\pi \to J^\infty\nu$ be corresponding maps (see §1.I). Let $Y \in D(M)$, $\overline{Y}^s_\pi \in D(\pi_{s+1,s})$, $\overline{Y}^s_\nu \in D(\nu_{s+1,s})$ be corresponding lifts of Y into the bundles π and ν.

Lemma 2.13. $\overline{Y}^{k+s}_\pi \cdot \phi^{s*}_\Delta = \phi^{s+1*}_\Delta \cdot \overline{Y}^s_\nu$. Hence $\overline{Y}_\pi \cdot \overline{\phi}^*_\Delta = \overline{\phi}^*_\Delta \cdot \overline{Y}_\nu$.

< $\forall \gamma \in \Gamma(\pi)$ we have $j_r(\tilde{\gamma}) = \phi^r_\Delta \cdot j_{r+K}(\gamma)$, where $\tilde{\gamma} = \Delta \cdot \gamma$. Hence $(j_{k+s+1}\gamma)^* \cdot \overline{Y}^{k+s}_\pi \cdot \phi^{s*}_\Delta = Y \cdot (j_{k+s}\gamma)^* \cdot \phi^{s*}_\Delta = Y \cdot (\phi^s_\Delta \cdot j_{k+s}\gamma)^* = Y \cdot (j_s\tilde{\gamma})^* = (j_{s+1}\tilde{\gamma})^* \cdot \overline{Y}^s_\nu = (j_{k+s+1}\gamma)^* \cdot \phi^{s+1*}_\Delta \cdot \overline{Y}^s_\nu$ >

Lemma 2.14. $\overline{\phi}^*_\Delta\, I^1_\nu \subset I^1_\pi$. Hence $\overline{\phi}^*_\Delta\, I_\nu \subset I_\pi$.

< $I^1_\nu = \mathrm{Ann}_\nu \overline{D(M)}$ (see §4.I). Let $\omega \in I^1_\nu$. Then $(\overline{\phi}^*_\Delta\omega)(\overline{Y}_\pi) = \overline{\phi}^*_\Delta(\omega(\overline{Y}_\nu)) = 0$ >

Lemma 2.15. $\overline{\phi}^*_\Delta \tau_\nu = \tau_\pi \overline{\phi}^*_\Delta$.

< Let $\omega \in \Lambda^*(\nu_\infty)$, $\omega = \tau_\nu\omega + (1-\tau_\nu)\omega$. Then $\tau_\pi \overline{\phi}^*_\Delta(\omega) = \tau_\pi \overline{\phi}^*_\Delta[\tau_\nu\omega + (1-\tau_\nu)\omega] = \tau_\pi \overline{\phi}^*_\Delta \tau_\nu\omega + \tau_\pi \overline{\phi}^*_\Delta(1-\tau_\nu)\omega = \tau_\pi \overline{\phi}^*_\Delta \tau_\nu\omega$, because $(1-\tau_\nu)\omega \in I_\nu$, $\overline{\phi}^*_\Delta I_\nu \subset I_\pi$, $\tau_\pi I_\pi = 0$. Now $\tau_\nu\omega \in \Lambda^*_0(\pi_\infty)$, hence $\overline{\phi}^*_\Delta\tau_\nu\omega \in \Lambda^*_0(\pi_\infty)$, hence $\tau_\pi \overline{\phi}^*_\Delta \tau_\nu\omega = \overline{\phi}^*_\Delta \tau_\nu\omega$ >

Lemma 2.16. $\overline{\phi}^*_\Delta\, I^+_\nu \subset I^+_\pi$.

< Let $\omega \in \Lambda^{m+s}(\nu_\ell)$, $\omega \in I^+_\nu$, i.e. $\omega(\overline{Y}^{(m)}_\nu, X^{(s)}) = 0$, $\forall Y_i \in D(M)$, $\forall X_i \in D^v(K_\nu)$, where $K_\nu \overset{\mathrm{def}}{=} \lim\,\mathrm{ind}\, F(J^r\nu)$. Let $Z_1,\ldots,Z_s$ be vertical tangent vectors from $T_\theta(J^{k+\ell}\pi)$, $\theta \in J^{k+\ell}\pi$. Let $X_i\big|_{\phi^\ell_\Delta(\theta)} = \phi^\ell_{\Delta*}(Z_i)$, then $(\overline{\phi}^*_\Delta\omega)(\overline{Y}^{(m)}_\pi, Z^{(s)})\big|_\theta = \omega(\overline{Y}^{(m)}_\nu, X^{(s)})\big|_{\phi^\ell_\Delta(\theta)} = 0$, i.e. $(\phi^*_\Delta\omega) \in I^+_\pi$ >

Lemma 2.17. $\tau^+_\pi \overline{\phi}^*_\Delta = \overline{\phi}^*_\Delta \tau^+_\nu$.

< This stands for lemma 2.16 as lemma 2.15 stands for lemma 2.14 >

Lemma 2.18. $D_\pi \bar{\phi}^*_\Delta = \bar{\phi}^*_\Delta D_\nu$.

< $D_\pi \bar{\phi}^*_\Delta = \tau_\pi d\, \bar{\phi}^*_\Delta = \tau_\pi \bar{\phi}^*_\Delta d = \bar{\phi}^*_\Delta \tau_\nu d = \bar{\phi}^*_\Delta D_\nu$ >

§3. Construction of operators S and ∂.

3.1. Let $\Omega \in \Lambda^m_0(\pi_k)$. We shall construct forms $S\Omega$ and $\partial\Omega$ by induction on k. The case $k = 1$ was considered in §1.4. Suppose now that we have constructed forms $S\Omega$ and $\partial\Omega$ for all $k \leq N$. Let us consider a transition $N \to N + 1$.

We begin with the differential operator $j_\ell = j_\ell(\pi)$: $\Gamma(\pi) \to \Gamma(\pi_\ell)$ which corresponds to the morphism 1_ℓ: $J^\ell\pi \to J^\ell\pi$ (see § 2.I.). Let us consider $\forall_k > 0$, the commutative diagram (see 1.I)

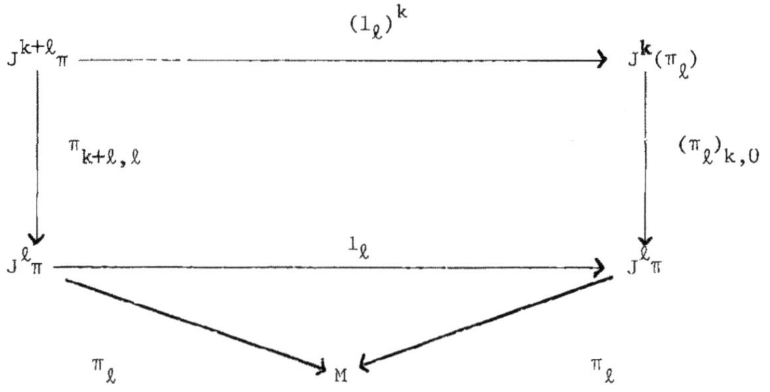

Let $X \in D^v(E)$ and $X^\ell \in D(J^\ell\pi)$, $X^{k+\ell} \in D(J^{k+\ell}\pi)$, $X^{\ell,k} \in D(J^k(\pi_\ell))$ be corresponding lifts of X (see §5.I.). Let $X \Longleftrightarrow \{A_t\}$, $\{A^\ell_t\} \Longleftrightarrow X^\ell$, $\{A^{k+\ell}_t\} \Longleftrightarrow X^{k+\ell}$, $\{A^{\ell,k}_t\} \Longleftrightarrow X^{\ell,k}$.

Lemma 3.1. $(1_\ell)^k \cdot A^{k+\ell}_t = A^{\ell,k}_t \cdot (1_\ell)^k$.

< If $\gamma \in \Gamma(\pi)$ then $A^{k+\ell}_t(j_{k+\ell}\gamma_x) = j_{k+\ell}(\gamma(t)_x)$, where $\gamma(t) = A_t \cdot \gamma$. Hence $(1_\ell)^k \cdot A^{k+\ell}_t(j_{k+\ell}\gamma_x) = (1_\ell)^k (j_{k+\ell}\gamma(t)_x) = j_k(\pi_\ell)(j_\ell\gamma(t)_x) = j_k(\pi_\ell)(A_t(j_\ell\gamma_x)) = A^{\ell,k}_t (j_k(\pi_\ell)(j_\ell\gamma_x)) = A^{\ell,k}_t \cdot (1_\ell)^k \cdot j_{k+\ell}\gamma_x$ >

Remark. Obviously, lemma 3.1 remains valid if X is an infinitesimal automorphism of the bundle π, $X \in \mathrm{Aut}(\pi) \subset D(E)$, and $\{A_t\} \Longleftrightarrow X$.

Corollary 3.2. Let $X \in \mathrm{Aut}(\pi)$. Then $X^{k+\ell} \cdot (1_\ell)^{k*} = (1_\ell)^{k*} \cdot X^{\ell,k}$.

< From lemma 3.1 we have $A_t^{k+\ell} \cdot (1_\ell)^{k*} = (1_\ell)^{k*} \cdot A_t^{\ell,k*}$. Applying $d/dt|t=0$ one gets the desired formula >

Let $\ell = 1$, $k = N$, $\Omega \in \Lambda_0^m(\pi_{N+1})$. The map $(1_1)^N : J^{N+1}\pi \to J^N(\pi_1)$ is an immersion, hence in $J^N(\pi_1)$ there exists a tube neighborhood U^N of the image $(1_1)^N(J^{N+1}\pi)$ together with a projection $v^N : U^N \to (1_1)^N(J^{N+1}\pi)$. Translating Ω from $J^{N+1}\pi$ on $(1_1)^N(J^{N+1}\pi)$ we can then carry this form over the whole U^N by virtue of $v^N{}^*$. Finally one can extend this form smoothly over $J^N(\pi_1)$. The resulting form will be denoted $\tilde{\Omega}$.

By induction assumption, we can construct such forms $S\tilde{\Omega} \in \Lambda_1^m((\pi_1)_\infty)$ and $\partial\tilde{\Omega} \in \Lambda_1^{m+1}((\pi_1)_\infty; (\pi_1)_{\infty,0})$ that

$$\tau_{\pi_1} S\tilde{\Omega} = \tilde{\Omega} \quad \text{and} \quad X^{1,N}(\tilde{\Omega}) = \tau_{\pi_1} d(\overline{X^1} \lrcorner\, S\tilde{\Omega}) + X^1 \lrcorner\, \partial\tilde{\Omega}, \quad \forall X \in D^v(E) . \tag{3.2}$$

We require now, in addition, that for $\Omega \in \Lambda_0^m(\pi_k)$, $S\Omega$ be in $\Lambda_1^m(\pi_{2k-1}; \pi_{2k-1,k-1})$. This is true for $k=1$ (lemma 1.6).

Then from (1.6) and lemma 2.3 one gets that $\partial\Omega \in \Lambda_1^{m+1}(\pi_{2k}; \pi_{2k,0})$, and we can rewrite (3.2) as

$$X^{1,N}(\tilde{\Omega}) = \tau_{\pi_1} d(X^{1,N-1} \lrcorner\, S\tilde{\Omega}) + X^1 \lrcorner\, \partial\tilde{\Omega} . \tag{3.3}$$

Applying to (3.3) from the left the operator $(1_1)^{2N*}$, using lemma 2.18, corollary 3.2 and definition of $\tilde{\Omega}$, we get

$$X^{N+1}(\Omega) = \tau_\pi d(X^N \lrcorner\, (1_1)^{2N-1*} S\tilde{\Omega}) + X^1 \lrcorner\, (1_1)^{2N*} \partial\tilde{\Omega}. \tag{3.4}$$

By inductive assumption, $S\tilde{\Omega} \in \Lambda_1^m((\pi_1)_{2N-1}; (\pi_1)_{2N-1,N-1})$, $\partial\tilde{\Omega} \in \Lambda_1^{m+1}((\pi_1)_{2N}; (\pi_1)_{2N,0})$, hence $\alpha \overset{\mathrm{def}}{=} (1_1)^{2N-1*} S\tilde{\Omega} \in \Lambda_1^m(\pi_{2N}; \pi_{2N,N})$, $\beta \overset{\mathrm{def}}{=} (1_1)^{2N*} \partial\tilde{\Omega} \in \Lambda_1^{m+1}(\pi_{2N+1}; \pi_{2N+1,1})$. Besides, from $\tau_{\pi_1} S\tilde{\Omega} = \tilde{\Omega}$ and lemma 2.15 we conclude that $\tau\alpha = \Omega$.

Thus

$$X^{N+1}(\Omega) = D(X^N \lrcorner\, \alpha) + X^1 \lrcorner\, \beta \ . \tag{3.5}$$

3.2. The formula (3.5) is not exactly of the type we sought: β is horizontal over $J^1\pi$ and not over E, so operator $X \to \bar{X} \lrcorner\, \beta$ is a differential operator of the first order instead of a desirable homomorphism. Thus we face the need to find decompositions of the form

$$\tau(\bar{X} \lrcorner\, \beta) = D(\bar{X} \lrcorner\, \hat{S}\beta) + \tau(X \lrcorner\, \hat{\partial}\beta), \tag{3.6}$$

$\forall X \in D(\pi_{\infty,0})$, $\forall\ \beta \in \Lambda^{m+1}_m$, with some forms $\hat{S}\,\beta \in \Lambda^m_1$, $\hat{\partial}\beta \in \Lambda^{m+1}_1[\pi_{\infty,0}]$.

This is a more general statement of a problem than the formula for the first variation (1.1). This last one is a very particular case of (3.6) when $\beta = d\,\Omega : \tau\bar{X}(\Omega) = \tau(\bar{X} \lrcorner\, d\Omega) + D(\bar{X} \lrcorner\, \Omega)$. In particular, $\partial = \hat{\partial}\, d$. Construction of the formula (3.6) may be called extended Lagrangian formalism. Here we won't solve this problem in full generality, i.e. to construct operators $\hat{S}$ and $\hat{\partial}$ for arbitrary β, because for our purposes we need only β's which are horizontal over $J^1\pi$, see (3.5). It might be remarked that in complete parallelism of results of §1, one can easily find that: $\hat{\partial}$ is unique; $\hat{S}$ in nonunique module such $\mu \in \Lambda^m_1$ that $D(X \lrcorner\, \mu) \equiv 0$, $\forall X \in D(\pi_{\infty,0})$; $\hat{\partial} = \tau^+(1 + d\hat{S})$; $\tau\hat{S} = 0$ (lemma 3.3).

<u>Lemma 3.3</u>. Formula (3.6) holds $\forall X = Y^\circ$, $Y \in D(\pi_\infty) \Longleftrightarrow \tau\hat{S}\,\beta = 0$.

< $\tau(\bar{Y} \lrcorner\, \beta) = \tau(\bar{Y} \lrcorner\, \hat{\partial}\,\beta) = 0$, because $\beta, \hat{\partial}\,\beta \in \Lambda^{m+1} \subset I$. Then $0 = D(\bar{Y} \lrcorner\, \hat{S}\beta) = \tau[\bar{Y}(\hat{S}\beta) - \bar{Y} \lrcorner\, d\hat{S}\beta] = \tau\bar{Y}(\hat{S}\beta) = \tau\bar{Y}(\tau\hat{S}\beta)$, $\forall Y \in D(\pi_\infty) \Longleftrightarrow \tau\hat{S}\beta = 0$, because $\tau\hat{S}\,\beta$ is horizontal >

Thus, repeating remark 1.3, it's sufficient to check out (3.6) only for $X \in D^v(\pi_{\infty,0})$ provided $\tau\hat{S}\,\beta = 0$. In addition, from (3.5) it's clear that it's enough to consider only $X \in D^v(E)$ (this is also a corollary of the formula $\hat{\partial} = \tau^+(1 + d\hat{S})$).

Now we construct (3.6) for $\beta \in \Lambda^{m+1}_1[\pi_{\infty,1}]$.

Acting as in 1.4 we get, $\forall X \in D^V(E)$,

$$\overline{X}_f \lrcorner \beta = D(\overline{X}_f \lrcorner \hat{S}\beta) + Df \wedge (\overline{X} \lrcorner \hat{S}\beta). \tag{3.7}$$

We now require that $\hat{S}\beta$ be horizontal over E. Then $\overline{X}_f \lrcorner \hat{S}\beta = 0$ and (3.7) turns into

$$\overline{X}_f \lrcorner \beta = Df \wedge (X \lrcorner \hat{S}\beta) . \tag{3.8}$$

Both sides of (3.8) are: differentiation with respect to $f \in K$ and homomorphism with respect to X ($\hat{S}\beta$ is horizontal over E). Hence (3.8) uniquely defined $\hat{S}\beta$, and we see that if $\beta \in \Lambda_1^{m+1}(\pi_r; \pi_{r,1})$ then $\hat{S}\beta \in \Lambda_1^m(\pi_r;\pi_{r,1})$ and $\hat{\partial}\beta \in \Lambda_1^{m+1}(\pi_{r+1};\pi_{r+1,0})$. Thus we have proven

Theorem 3.4. $\forall \beta \in \Lambda_1^{m+1}(\pi_r;\pi_{r,1})$ $\exists\,!$ pair of forms $\hat{S}\beta \in \Lambda_1^m(\pi_r;\pi_{r,0})$, $\hat{\partial}\beta \in \Lambda_1^{m+1}(\pi_{r+1};\pi_{r+1,0})$ such that (3.6) holds $\forall X \in D(\pi_{\infty,0})$.

Referring back to (3.5), we apply theorem 3.4 to our β with $r = 2N + 1$ and find some forms $\hat{S}\beta \in \Lambda_1^m(\pi_{2N+1};\pi_{2N+1,0})$, $\tau\hat{S}\beta = 0$, and $\hat{\partial}\beta \in \Lambda_1^{m+1}(\pi_{2N+2};\pi_{2N+2,0})$. Inserting (3.6) into (3.5) we get

$$X^{N+1}(\Omega) = D(X^N \lrcorner (\alpha + \hat{S}\beta)) + X \lrcorner \hat{\partial}\beta . \tag{3.9}$$

Let $S\Omega = \alpha + \hat{S}\beta$, $\partial\Omega = \hat{\partial}\beta$. Then $S\Omega \in \Lambda_1^m(\pi_{2N+1};\pi_{2N+1,N})$, $\partial\Omega \in \Lambda_1^{m+1}$. Thus we have completed the induction step. Formula (1.1) is constructed.

Corollary 3.5. $\partial D = 0$.

< Let $\mu \in \Lambda^{m-1}$. Then $\tau\overline{X}(D\mu) = \tau\overline{X}(\tau d\mu) = \tau\overline{X}(\tau d\tau\mu) = \tau\overline{X}(d\tau\mu) = \tau d\overline{X}(\tau\mu) = \tau d[\overline{X} \lrcorner d\tau\mu + d(\overline{X} \lrcorner \tau\mu)] = D(\overline{X} \lrcorner d\tau\mu)$. On the other hand, (1.1) gives us $\tau\overline{X}(D\mu) = D(\overline{X} \lrcorner SD\mu) + \tau(X \lrcorner \partial D\mu)$. But ∂ is unique so $\partial D\mu = 0$ >

§4. Hamilton-Cartan principle

4.1. Thus far, the basic objects of Lagrangian formalism were horizontal forms $\Omega \in \Lambda_0^m$. Nevertheless, it's sometimes useful to consider nonhorizontal forms in Λ^m, namely those in Λ_1^m. For example, in mechanics ($k = m = 1$) instead of $\Omega = L(\mathbf{q}, \dot{\mathbf{q}}, t)dt$ it's helpful to study $S\Omega = Ldt + L_{\dot{\mathbf{q}}}(d\mathbf{q} - \dot{\mathbf{q}}dt)$ on $J^1\pi$. One of the reasons for such substitution is the coincidence of extremals of both forms Ω and $S\Omega$, if the form Ω is, in a sense, nondegenerate. This fact is called Hamilton-Cartan principle in mechanics. It allows us to go from Lagranian formalism on $T(M) \times \mathbb{R} \approx J^1(\pi: M \times \mathbb{R} \to \mathbb{R})$ to Hamiltonian formalism on $T^*(M) \times \mathbb{R}$ (when $\partial L/\partial t \equiv 0$, one can get rid of $\mathbb{R}$). The "nondegenerationness" is the condition of analytical type, and we are trying to avoid all analysis of the calculus of variations. As such, the Hamiltonian-Cartan principle stated below is some statement about relations of extremals of forms Ω and $S\Omega$.

4.2. Let $\omega \in \Lambda_1^m(\pi)$, $\tau\omega \in \Lambda_0^m(\pi_1)$ We showed in §3 that $\partial\Lambda_0^m(\pi_1) \subset \Lambda_1^{m+1}(\pi_2;\pi_{2,0})$.

<u>Lemma 4.1.</u> $\partial\tau\omega \in \Lambda_1^{m+1}(\pi_1;\pi_{1,0})$.

< By (1.1), $\tau\bar{X}(\tau\omega) = D(X \lrcorner S\tau\omega) + \tau(X \lrcorner \partial\tau\omega)$. On the other hand, $\tau\bar{X}(\tau\omega) = \tau\bar{X}(\omega) = \tau d(X \lrcorner \omega) + \tau(X \lrcorner d\omega) = D(X \lrcorner \omega) + +\tau(X \lrcorner \tau^+ d\omega)$. By the uniqueness of ∂, $\partial\tau\omega = \tau^+ d\omega$. But $d\omega \in \Lambda_2^{m+1}(\pi)$, and by lemma 2.3, $\tau^+ d\omega \in \Lambda_1^{m+1}(\pi_1)$ >

<u>Definition.</u> Let $\Omega \in \Lambda_0^m(\pi_k)$, $\partial\Omega \in \Lambda_1^{m+1}(\pi_{2k};\pi_{2k,0})$. Equation of extremals $\Gamma_\Omega \subset J^{2k}\pi$ is the set (possibly, empty) of those points in $J^{2k}\pi$ where $\partial\Omega$ vanishes. Clearly, Γ_Ω is Euler-Lagrange equation. Extremals of the form Ω are solutions of this equation, i.e. such $\lambda \in \Gamma(\pi)$ that $j_{2k}\lambda(M) \subset \Gamma_\Omega$.

It's important to stress that extremals can not exist. In §5 we mean, without mentioning it explicitly, that equation Γ_Ω can be prolonged as many times as needed. This means, essentially, that $\forall x \in M$, every jet $j_{2k}\lambda_x \in \Gamma_\Omega$ of a solution of Γ_Ω can be prolonged to jet $j_{2k+r}\lambda_x$ of the same solution, $\forall r>0$.

Let again $\Omega \in \Lambda_0^m(\pi_k)$, $\partial\Omega \in \Lambda_1^{m+1}(\pi_{2k};\pi_{2k,0})$, $S\Omega \in \Lambda_1^m(\pi_{2k-1};\pi_{2k-1,k-1})$. Let us consider form $S\Omega$ on the bundle $\pi_{2k-1}: J^{2k-1}\pi \to M$. By lemma, 4.1, $\partial\tau_{\pi_{2k-1}} S\Omega \in \Lambda_1^{m+1}((\pi_{2k-1})_1; (\pi_{2k-1})_{1,0})$. Let us go to the diagramm

$$
\begin{array}{ccc}
J^{2k}\pi & \xrightarrow{(1_{2k-1})^1} & J^1(\pi_{2k-1}) \\
\downarrow{\scriptstyle \pi_{2k,2k-1}} & & \downarrow{\scriptstyle (\pi_{2k-1})_{1,0}} \\
J^{2k-1}\pi & \xrightarrow{1_{2k-1}} & J^{2k-1}\pi
\end{array}
\qquad (4.1)
$$

<u>Theorem 4.2.</u> $(1_{2k-1})^{1*}\,\partial\tau_{\pi_{2k-1}} S\Omega = \partial\Omega$. Hence $(1_{2k-1})^1(\Gamma_\Omega) \subset \Gamma_{\tau_{\pi_{2k-1}} S\Omega}$

(Hamilton-Cartan principle).

< When proving lemma 4.1, we found that $\partial\tau\omega=\tau^{+}d\omega$. Let us apply this fact to $\omega=S\Omega$. We have $(1_{2k-1})^{1*}\partial\tau_{\pi_{2k-1}}S\Omega=(1_{2k-1})^{1*}\tau^{+}_{\pi_{2k-1}}dS\Omega=$ (by lemma 2.17) $\tau^{+}_{\pi}(1_{2k-1})^{*}dS\Omega=\tau^{+}_{\pi}dS\Omega=\partial\Omega$ by formula (1.6) >

Corollary 4.3. As $(1_{2k-1})^{1}$ is monomorphism, so $(1_{2k-1})^{1}(\Gamma_{\Omega})=\Gamma_{\tau_{\pi_{2k-1}}S\Omega}\cap(1_{2k-1})^{1}(J^{2k}\pi)$.

Remarks. 1) The above statements don't depend upon the nonuniqueness of S. 2) As $(1_{2k-1})^{1}(j_{2k}(\pi)\gamma)=j_{1}(\pi_{2k-1})(1_{2k-1}(j_{2k-1})(\pi)\gamma))$, $\forall\gamma\in\Gamma(\pi)$, we see that extremals of Ω go to extremals of $\tau_{\pi_{2k-1}}S\Omega$.

§5. Symmetries and conservation laws

5.1. A field $X\in D(\pi_{\infty,0})$ such that $(j\gamma)^{*}\bar{X}(\Omega)=0$, for every extremal γ (of form Ω) is called a symmetry of Ω.

A form $\theta\in\Lambda^{m-1}_{0}$ such that $d(j\gamma)^{*}\theta=0$, $\forall$ extremal γ is called the conservation law (of Ω).

A field $Y\in D(\pi_{\infty})$ such that $d(j\gamma)^{*}(Y\lrcorner\Omega)=0$, $\forall$ extremal γ is called a current (of Ω).

Clearly, a current Y defines the conservation law $Y\lrcorner\Omega$. For the converse to be true it's sufficient that Ω not vanishes. From (1.1) we get

Theorem 5.1 (Formal Noether Theorem). If X is symmetry then $\tau(\bar{X}\lrcorner S\Omega)$ is a conservation law. Conversely, if Y is current then $X=Y^{0}$ is symmetry.

< The first statement is a corollary of (1.1). Then $\tau\bar{Y}(\Omega)=\tau[d(Y\lrcorner\Omega)+\bar{Y}\lrcorner d\Omega]=D(Y\lrcorner\Omega)$, $(j\gamma)^{*}\bar{Y}(\Omega)=d(j\gamma)^{*}(Y\lrcorner\Omega)=0$ >

Thus, Noether theorem is a trivial consequence of the formula for the first variation.

Usually by symmetry one understands such infinitesimal automorphism X of the bundle π that $\bar{X}(\Omega)\equiv 0$. Such symmetries evidently form Lie algebra and corresponding translations constitute Lie group (local). The reason for extending the usual notion of symmetry (and to loose the Lie property) is that conservation laws which one gets from "usual" symmetries, $\tau(\bar{X}\lrcorner S\Omega)$, belong essentially in $\Lambda^{m-1}_{0}(\pi_{2k-1})$. Note that arbitrariness in the choice of $S\Omega$ doesn't kill this situation (i.e. it's impossible to get nontrivial conservation laws in $\Lambda_0^{m-1}(\pi_r)$ for $r>2k-1$) because one can change $S\Omega$ by such $S_2\Omega$ that $D(\overline{X}\lrcorner S_2\Omega)\equiv 0$, $\forall X\in D(\pi_{\infty,0})$. Thus addition is a trivial conservation law. On the other hand, as, for example, the popular Korteveg-De Vries equation shows, conservation laws may lie in arbitrary high jets.

Sometimes one can meet in literature also the next definition: a field $Z\in D(\pi_{\infty,0})$ is called symmetry if $\exists\mu\in\Lambda_0^{m-1}$ such that $(j\gamma)^*\overline{Z}(\Omega)=d(j\gamma)^*(\mu)$, $\forall$ extremal γ. Clearly, if $\overline{Z}$ is symmetry in the last sense then $(\tau(Z\lrcorner S\Omega)-\mu)$ is a conservation law.

§6. First complex for operator ∂.

6.1. For Langrangian formalism on the bundle $\pi:E\to M$ the bundle $\tilde{\pi} \stackrel{\text{def}}{=} \pi\times 1_{\mathbb{R}}$: $E\times\mathbb{R}\to M\times\mathbb{R}$ plays a very important role. In particular, operator $\partial_{\tilde{\pi}}$ allows very simply to include operator ∂_π in a complex. It's shown below that equality $\partial_{\tilde{\pi}}\circ\partial_\pi=0$ is corollary of (1.6) which, in turn, is corollary of (1.1). This disadvantage of this method, that is, moving out of the basic bundle, will be improved in §7 by complication of a prolonging operator which acts not on forms but on tensors.

6.2. Let ϕ denote each of projections: $M\times\mathbb{R}\to M$, $E\times\mathbb{R}\to E$, $J^k\tilde{\pi}\to J^k\pi$; from the context it will be clear where ϕ really acts. Let χ denote projection $M\times\mathbb{R}\to\mathbb{R}$. Let us describe the map ϕ on $J^k\tilde{\pi}$. Let t be coordinate on $\mathbb{R}$, then $\tilde{\gamma}\in\Gamma(\tilde{\pi})$ can be regarded as one-parameter family $\gamma(t)\in\Gamma(\pi)$. Let $\tilde{\theta}\in J^k\tilde{\pi}$, $t_0=\chi\circ\tilde{\pi}_k(\tilde{\theta})$, $x=\phi\circ\tilde{\pi}_k(\tilde{\theta})$. Let $\tilde{\gamma}\in\Gamma(\tilde{\pi})$ be such that $j_k(\tilde{\pi})\tilde{\gamma}_{(x,t_0)}=\tilde{\theta}$. Let us consider $\tilde{\gamma}^{t_0}\in\Gamma(\tilde{\pi})$: $\tilde{\gamma}^{t_0}(x,t)=\tilde{\gamma}(x,t_0)$. Then $\exists!\ \gamma\in\Gamma(\pi):\phi^*(\gamma)=\gamma^{t_0}$. The k-jet of $\tilde{\gamma}$ uniquely defines the k-jet of $\tilde{\gamma}^{t_0}$, so ϕ is correctly defined.

We choose local coordinates in $J^k\tilde{\pi}$ such functions $(x_i,\ t,\ q^a_{\sigma\oplus n},\ |\sigma|+n\leq k)$ that $(j_k\tilde{\gamma})^*(q^a_{\sigma\oplus n})=\dfrac{\partial^{|\sigma|}}{\partial x_\sigma}\dfrac{\partial^n}{\partial t^n}\tilde{\gamma}^*(q^a)$. Then $\phi|_{J^k\pi}:(x_i,t,q^a_{\sigma\oplus n})\mapsto(x_i,t,q^a_\sigma=q^a_{\sigma\oplus 0})$.

Lemma 6.1. $\tau_{\tilde{\pi}}\phi^*=\tau_{\tilde{\pi}}\phi^*\tau^+_\pi$ on $\Lambda_2^{m+1}(\pi_\infty)$.

< Let $\omega\in\Lambda_2^{m+1}(\pi_\infty)$ then locally $\omega=\tilde{\omega}+\tilde{\mu}$, where $\tilde{\omega}=B^{a,\sigma}dq^a_\sigma\wedge d^mx$, $\tilde{\mu}=B^{a,\sigma,i}_{b,\nu}$ $dq^a_\sigma\wedge dq^b_\sigma\wedge(\partial_i\lrcorner d^mx)$. As $\tau^+_\pi\tilde{\omega}=\tilde{\omega}$, it's enough to check that $\tau_{\tilde{\pi}}\phi^*(1-\tau^+_\pi)\mu=0$ where $\mu=dq^a_\sigma\wedge dq^b_\sigma\wedge(\partial_i\lrcorner d^mx)$. We have $\tau^+_\pi\mu=(q^b_{\nu+i}\,dq^a_\sigma-q^a_{\sigma+i}\,dq^b_\nu)\wedge d^mx$, $\phi^*\tau^+_\pi\mu=(q^b_{\nu+i\oplus 0}\,dq^a_{\sigma\oplus 0}-q^a_{\sigma+i\oplus 0}dq^b_{\nu\oplus 0})\wedge d^mx$, $\tau_{\tilde{\pi}}\phi^*\tau^+_\pi\mu=(q^b_{\nu+i\oplus 0}\,q^a_{\sigma\oplus 1}-q^a_{\sigma+i\oplus 0}\,q^b_{\nu\oplus 1})$ $dt\wedge d^mx$. On the other hand, $\tau_{\tilde{\pi}}\phi^*\mu=\tau_{\tilde{\pi}}(dq^a_{\sigma\oplus 0}\wedge dq^b_{\nu\oplus 0}\wedge(\partial_i\lrcorner d^mx))=$

$(q^a_{\sigma+j\oplus 0}\, dx_j + q^a_{\sigma\oplus 1}\, dt) \wedge (q^b_{\nu+k\oplus 0}\, dx_k + q^b_{\nu\oplus 1}\, dt) \wedge (\partial_i \lrcorner\, d^m x) = (q^a_{\sigma\oplus 1}\, q^b_{\nu+i\oplus 0} - q^a_{\sigma+i\oplus 0}\, q^b_{\nu\oplus 1})$ $dt \wedge d^m x$ >

Theorem 6.2. $\partial_{\tilde\pi}\, \tau_{\tilde\pi}\, \phi^* \partial_\pi = 0$.

< Let $\Omega \in \Lambda^m_0(\pi_\infty)$. By (1.6), $\partial_\pi \Omega = \tau^+_\pi\, dS\Omega$. As $S\Omega \in \Lambda^m_1(\pi_\infty)$, so $dS\Omega \in \Lambda^{m+1}_2(\pi_\infty)$, and with the lemma 6.1 we get $\tau_{\tilde\pi}\phi^*\partial_\pi\Omega = \tau_{\tilde\pi}\phi^*\tau^+_\pi\, dS\Omega = \tau_{\tilde\pi} d\phi^* S\Omega = D_{\tilde\pi}\, \phi^* S\, \Omega$. By corollary 3.5, $\partial_{\tilde\pi}\, D_{\tilde\pi} = 0$ >

Thus we get an infinite complex of differential forms prolonging operator ∂_π:

$$\Lambda^0(\pi_\infty) \xrightarrow{D_\pi} \cdots \xrightarrow{D_\pi} \Lambda^{m-1}_0(\pi_\infty) \xrightarrow{D_\pi} \Lambda^m_0(\pi_\infty)$$
$$\xrightarrow{\partial_\pi} \Lambda^{m+1}_1(\pi_\infty;\pi_{\infty,0}) \xrightarrow{\partial_{\tilde\pi}\tau_{\tilde\pi}\phi^*} \Lambda^{m+2}_1(\tilde\pi_\infty;\tilde\pi_{\infty,0}) \longrightarrow \cdots \qquad (6.1)$$

7. "Higher" Lagrangian Formalism. Second Complex for the Euler-Lagrange Operator .

§7.1. Equation of extremals $\Gamma_\Omega = \{\partial\Omega = 0\}$ depends, in fact, on $\partial\Omega$ and not upon Ω itself. Hence, say, for the study of infinitesimal symmetries of the equation Γ_Ω, that is, classified fields which are tangent to "prolongation" $\overline{\Gamma_\Omega}$ of Γ_Ω , it's natural to consider the action of $D(\pi_{\infty,0})$ on $\partial\Omega$. More generally, we will consider the action of $D(\pi_{\infty,0})$ on forms $\omega \in \Lambda^{m+1}_1(\pi_\infty)$. The study of such action in the meaning of formula (7.1) will be called the (first) higher Lagrangian formalism.

7.2. Let us define the "formula for the first variation" as equality

$$\tau^+\bar{X}(\omega) = D^+(S^1\omega(\bar{X})) + \partial^1\omega(X) \ , \ D^+ \overset{\text{def}}{=\!=} \tau^+ d \ , \qquad (7.1)$$

$\forall X \in D(\pi_{\infty,0})$, with some forms (more precisely, tensors) $S^1\omega \in \Lambda^1 \otimes \Lambda^m_1$, $\partial^1\omega \in \Lambda^1[\pi_{\infty,0}] \otimes \Lambda^{m+1}_1$, satisfying restrictions stated below. The proof of existence of $S^1\omega$, $\partial^1\omega$ and the uniqueness of $\partial^1\omega$ is analogous to that in § 1, § 3 for S and ∂. So we only point out the sequence of steps.

1) $\tau^+\, \bar{X}(\omega) = \tau^+(\bar{X} \lrcorner\, d\omega) + D^+(\bar{X} \lrcorner\, \omega)$, hence it's enough to construct the formula

$$\tau^{+}(\bar{X} \lrcorner d\omega) = D^{+}(S^2\omega(\bar{X})) + \partial^1\omega(X), \tag{7.2}$$

where $S^2\omega(\bar{X}) = S^1\omega(\bar{X}) - \bar{X} \lrcorner \omega$.

2) Let us suppose that we found forms $\widetilde{S^2\omega}$ and $\widetilde{\partial^1\omega}$ satisfying (7.2) on $D^V(\pi_{\infty,0})$. As $\tau^{+}(\bar{X} \lrcorner d\omega) = 0$, $\forall X \in D(\pi_\infty)$, equalities $\partial^1\omega(X) = \widetilde{\partial^1\omega}(X)$, $\partial^1\omega(Y) = 0$, $S^2\omega(\bar{X}) = \widetilde{S^2\omega}(\bar{X})$, $S^2\omega(\bar{Y}) = 0$, $\forall X \in D^V(\pi_{\infty,0})$, $\forall Y \in D(\pi_\infty)$, uniquely define forms $S^2\omega$ and $\partial^1\omega$. Hence, we need to look to (7.2) only for vertical fields.

3) Uniqueness. Let $\overline{S^2\omega}$ and $\overline{\partial^1\omega}$ be another pair of forms satisfying (7.2). We act as in §1: let $\omega' = \partial^1\omega - \overline{\partial^1\omega}$; subtracting one equality (7.2) from another, applying to the difference from the left operator $\tau_{\tilde{\pi}}\phi^*$ (see §6), we get $\tau_{\tilde{\pi}}\phi^*\omega'(X) \equiv 0$, $\forall X \in D(\pi_{\infty,0})$. As $\operatorname{Ker} \tau_{\tilde{\pi}}\phi^*\big|_{\Lambda_1^{m+1}(\pi_\infty)} = \{0\}$, so $\omega'(X) \equiv 0$, i.e. $\omega' = 0$. Thus operator ∂^1 is unique . Note that the uniqueness of operator ∂^1 doesn't depend upon future restrictions on $\partial^1\omega$ and $S^2\omega$.

4) This is a key point: here we prove the theorem about the reduction of $D^V(\pi_{\infty,0})$ to $D^V(E)$ (analog of theorem 1.5). We need two technical facts.

<u>Lemma 7.1</u>. Let $\omega \in \Lambda_0^{m-1}$, θ , $\mu \in \Lambda^1$. Then

$$\tau^{+}(\theta \wedge \mu \wedge \omega) = \tau\theta \wedge \mu \wedge \omega + \theta \wedge \tau\mu \wedge \omega \tag{7.3}$$

< If $X \in D^V(K)$ then $X \lrcorner \tau^{+}(\theta \wedge \mu \wedge \omega) = \tau(X \lrcorner (\theta \wedge \mu \wedge \omega)) = \tau[\theta(X)\mu \wedge \omega - \mu(X)\theta \wedge \omega]$ $\theta(X)\tau\mu \wedge \omega - \mu(X)\tau\theta \wedge \omega = X \lrcorner (\theta \wedge \tau\mu \wedge \omega + \tau\theta \wedge \mu \wedge \omega)$ >

<u>Lemma 7.2</u>. Let $S \in \Lambda_1^m$, $\theta \in \Lambda^1$. Then

$$\tau^{+}(\theta \wedge S) = \theta \wedge \tau S + \tau\theta \wedge S \ . \tag{7.4}$$

< $S = \tau S + (1-\tau)S$, $\tau^{+}(\theta \wedge \tau S) = \theta \wedge \tau S$, so it's enough to check out that $\tau^{+}(\theta \wedge (1-\tau)S) = \tau\theta \wedge S$. $\forall X \in D^V(K)$, we have $X \lrcorner \tau^{+}(\theta \wedge (1-\tau)S) = \tau[X \lrcorner (\theta \wedge (1-\tau)S)] =$ $\tau[\theta(X)(1-\tau)S - \theta \wedge (X \lrcorner (1-\tau)S] = -\tau\theta \wedge (X \lrcorner S) = X \lrcorner (\tau\theta \wedge S)$ >

<u>Corollary 7.3</u>. Let $S \in \Lambda_1^m$, $f \in K$. Then

$$\tau^+(df\wedge S) = df\wedge\tau S + Df\wedge S\,, \tag{7.5}$$

To get an analog of formula (1.6) we must be able to express $D^+(S^2\omega(\bar{X}))$ as $\psi(\bar{X})$ with some $\psi\in\Lambda^1\otimes\Lambda_1^{m+1}$. Let, locally, $S^2\omega = (dq^a_\sigma - q^a_{\sigma+j}\,dx_j)\otimes S^{a,\sigma}$, where $S^{a,\sigma}\in\Lambda_1^m$, $X = A^a\,\partial^a$, $\bar{X} = D_\sigma(A^a)\partial^a_\sigma$ (see §5.I). Then $\tau^+dS^2(\bar{X}) = \tau^+d[D_\sigma(A^a)S^{a,\sigma}] = \tau^+[dD_\sigma(A^a)\wedge S^{a,\sigma} + D_\sigma(A^a)dS^{a,\sigma}] = (7.5)\ dD_\sigma(A^a)\wedge\tau S^{a,\sigma} + DD_\sigma(A^a)\wedge S^{a,\sigma} + D_\sigma(A^a)D^+S^{a,\sigma}$. As $DD_\sigma(A^a) = D_{\sigma+i}(A^a)dx_i$, so for $D^+(S^2\omega(\bar{X}))$ to be of the form $\psi(\bar{X})$, i.e. to be linear on $D_\sigma(A^a)$, it's necessary and sufficient that $\tau S^{a,\sigma} = 0$ holds. This means that $\tau S^2\omega(X) = 0$, $\forall X\in D(\pi_{\infty,0})$. If we let $(\hat{D}^+(\psi))(\bar{X}) \overset{\text{def}}{=\!=} D^+(\psi(\bar{X}))$, $\forall\psi\in I^1\otimes(\Lambda_1^m\cap I)$, we thus get the differential operator of the first order $\hat{D}^+:[I\cap\Lambda^1(\pi_r)]\otimes(\Lambda_1^m\cap I)\to[I\cap\Lambda^1(\pi_{r+1})]\otimes\Lambda_1^{m+1}$, $\forall r\geq 0$.

Taking into account that $D^+(S^2\omega(\bar{X})) = 0$, $\forall X\in D(\pi_\infty)$, the above calculations shows that

$$\hat{D}^+[(dq^a_\sigma - q^a_{\sigma+j}dx_j)\otimes S^{a,\sigma}] = (dq^a_\sigma - q^a_{\sigma+j}dx_j)\otimes D^+S^{a,\sigma} +$$

$$(dq^a_{\sigma+i} - q^a_{\sigma+i+j}\,dx_j)\otimes(dx_i\wedge S^{a,\sigma})\,,\quad \tau S^{a,\sigma} = 0\,. \tag{7.6}$$

<u>Corollary 7.4</u>. Let $\eta\in I^1\otimes(\Lambda_1^m\cap I)$, $\theta\in\Lambda_0^1$, $f\in K$. Define $(f\eta)(\bar{X}) \overset{\text{def}}{=\!=} f\cdot\eta(\bar{X})$; $(\theta\wedge\eta)(\bar{X}) \overset{\text{def}}{=\!=} \theta\wedge\eta(\bar{X}) : (\theta\wedge\eta)\in I^1\otimes\Lambda_1^{m+1}$. Then

$$\hat{D}^+(f\eta) = f\hat{D}^+(\eta) + Df\wedge\eta\,. \tag{7.7}$$

Thus, (7.2) can be rewritten as $\tau^+(\bar{X}\lrcorner d\omega) = (\hat{D}^+S^s\omega + \partial^1\omega)(\bar{X})$. Note that there exists natural embedding $\Lambda_2^{m+2}\hookrightarrow\Lambda^1\otimes\Lambda_1^{m+1}$, so we can define $\hat{d}^+\omega\in I^1\otimes\Lambda_1^{m+1}$ by equality $(\hat{d}^+\omega)(\bar{X}) = \tau^+(\bar{X}\lrcorner d\omega)$ and we get $(\hat{d}^+\omega)(\bar{X}) = (\hat{D}^+S^2\omega + \partial^1\omega)(\bar{X})$, or

$$d^+\omega = \hat{D}^+S^2\omega + \partial^1\omega\,. \tag{7.8}$$

This is an analog of (1.6). Thus we see that (7.2) has to be checked only on $D^v(E)$, because (7.8) being equivalent to (7.2), is true on $D^v(\pi_{\infty,0})$ if it is true on $D^v(E)$.

5) The basis of induction is the following: let $\omega \in \Lambda_1^{m+1}(\pi_k)$, k=1. We require that $S^2\omega \in \Lambda^1[\pi_{\infty,0}] \otimes \Lambda_1^m$. Then, variation of (7.2) by some $f \in K$ gives us

$$\overline{X}_f \lrcorner\, d\omega = \tau^+(df \wedge S^2\omega(X)) = df \wedge \tau S^2\omega(X) + Df \wedge S^2\omega(X) \ . \tag{7.9}$$

The form $\overline{X}_f \lrcorner\, d\omega$, when X is fixed, depends on f through Df. In order for (7.9) to be solvable it's necessary that term with df in r.h.s. disappears, i.e. $\tau S^2\omega(X) = 0$. We already have seen that this condition when required equality $D^+(S^2\omega(\overline{X})) = \psi(\overline{X})$ is possible. Thus

$$\overline{X}_f \lrcorner\, d\omega = Df \wedge S^2\omega(X) \ , \ \tau S^2\omega(X) = 0 \tag{7.10}$$

Clearly, this equation is uniquely solvable and automatically $S^2\omega(Y) = 0$, $\forall Y \in \overline{D(\pi_\infty)}$. Also it's clear that $S^2\omega(X) \in \Lambda_1^m(\pi_2)$, i.e. $S^2\omega \in \Lambda^1(\pi) \otimes \Lambda_1^m(\pi_2)$. In view of $\hat{D}^+ [\Lambda^1(\pi_r) \otimes (\Lambda_1^m(\pi_\ell) \cap I)] \subset \Lambda^1(\pi_{r+1}) \otimes \Lambda_1^{m+1}(\pi_{\ell+1})$, we get, using (7.8), that $\partial^1 \omega \in \Lambda^1(\pi) \otimes \Lambda_1^m(\pi_3)$.

6) Induction step: let S^2 and ∂^1 are constructed for all $k \leq N$ and $\omega \in \Lambda_1^{m+1}(\pi_{N+1})$. We require that $S^2(\Lambda_1^{m+1}(\pi_k)) \subset \Lambda^1(\pi_{k-1}) \otimes \Lambda_1^m(\pi_{3k-1})$, $\partial^1(\Lambda_1^{m+1}(\pi_k)) \subset \Lambda^1(\pi) \otimes \Lambda_1^{m+1}(\pi_{3k})$, which holds for k=1. Exactly as in §3 we can find $\tilde{\omega} \in \Lambda_1^{m+1}((\pi_1)_N)$ such that $(1_1)^{N*}\tilde{\omega} = \omega$. By induction assumption we can construct forms $S^2\tilde{\omega} \in \Lambda^1((\pi_1)_{N-1}) \otimes \Lambda_1^m((\pi_1)_{3N-1})$ and $\partial^1\tilde{\omega} \in \Lambda^1(\pi_1) \otimes \Lambda_1^{m+1}((\pi_1)_{3N})$ such that

$$X^{1,N} \lrcorner\, d\tilde{\omega} = \tau_{\pi_1}^+ \ d(S^2\tilde{\omega}(X^{1,N-1})) + \partial^1\tilde{\omega}(X^1) \ . \tag{7.11}$$

Applying to (7.11) from the left operator $(1_1)^{3N*}$ and taking into account lemma 2.17, we get

$$X^{N+1} \lrcorner\, d\omega = \tau_\pi^+ d(A(X^N)) + B(X^1) \tag{7.12}$$

where $A(X^N) \overset{\text{def}}{=} (1_1)^{3N-1*} S^2\tilde{\omega}(X^{1,N-1})$, $B \overset{\text{def}}{=} (1_1)^{3N*} \partial^1 \tilde{\omega}$, i.e. $A \in \Lambda^1(\pi_N) \otimes \Lambda_1^m(\pi_{3N})$, $B \in \Lambda^1(\pi_1) \otimes \Lambda_1^{m+1}(\pi_{3N+1})$. Note that from $\tau_{\pi_1} S^2\tilde{\omega}(X^{1,N-1}) = 0$ follows $\tau_\pi A(X^N) = 0$.

Now, as in §3, we are looking for the decomposition

$$B(X^1) = \tau^+ d(\hat{S}^1 B(X)) + \hat{\partial}^1 B(X) \ , \tag{7.13}$$

where $\hat{S}^1 B \in \Lambda^1(\pi) \otimes \Lambda_1^m(\pi_{3N+2})$, $\hat{\partial}^1 B \in \Lambda^1(\pi) \otimes \Lambda_1^{m+1}(\pi_{3N+3})$, and we impose the condition $\tau\hat{S}^1 B(X) = 0$. By already standard procedure, (7.13) is equivalent to

$$B(\bar{X}_f) = Df \wedge \hat{S}^1 B(X) \ , \tag{7.14}$$

which is uniquely solvable. It results in $S^2\omega = A + \hat{S}^1 B$, $\partial^1\omega = \hat{\partial}^1 B$. Besides, from (7.14), (7.13) it follows that $S^2\omega \in \Lambda^1(\pi_N) \otimes (\Lambda_1^m(\pi_{3N+2}) \cap I)$, $\partial^1\omega \in \Lambda^1(\pi) \otimes \Lambda_1^{m+1}(\pi_{3N+3})$. This ends the induction step and proof of existence of (7.2).

Corollary 7.5. $\partial^1 \tau^+ d = 0$ on Λ_1^m .

< Let $\omega \in \Lambda_1^m$, then $\forall X \in D^V(\pi_{\infty,0})$, $\tau^+\bar{X}(\tau^+ d\omega) = \tau^+\bar{X}(d\omega) = D^+(\bar{X}(\omega)) = D^+((S^1\tau^+ d\omega)(\bar{X})) + (\partial^1\tau^+ d\omega)(X)$. In view of the uniqueness of ∂^1 , $\partial^1\tau^+ d\omega = 0$ >

Corollary 7.6. $\partial^1\partial = 0$.

< $\partial = \tau^+ dS$ + corollary 7.5>

The uniqueness problem for S^2 is more complicated than for S . Here I only remark that, when $m = 1$, S^2 is unique. This follows from (7.6) by simple computation.

7.3. We now face two operators prolonging ∂ : $\partial_{\tilde{\pi}}\tau_{\tilde{\pi}}\phi^*$ and ∂^1 . Let us clarify their relationship. For this we rewrite (7.8) as

$$\partial^1\omega = \hat{d}^+\omega - \hat{D}^+ S^2\omega \ , \quad \omega \in \Lambda_1^{m+1}(\pi_\infty) \ . \tag{7.15}$$

Applying from the left operator alt $\cdot\, \tau_{\tilde{\pi}} \cdot \phi^*$, where $\tau_{\tilde{\pi}}$ acts (as above) on the second tensor multiplier and "alt" means alternation of tensor into the differential form, we get

$$\mathrm{alt} \cdot \tau_{\tilde{\pi}} \cdot \phi^* \partial^1\omega = \mathrm{alt} \cdot \tau_{\tilde{\pi}} \cdot \phi^* \hat{d}^+\omega - \mathrm{alt} \cdot \tau_{\tilde{\pi}} \cdot \phi^* \hat{D}^+ S^2\omega \ . \tag{7.16}$$

Straightforward computation with the help of (7.6) shows that

$$\mathrm{alt} \cdot \tau_{\tilde{\pi}} \cdot \phi^* \cdot \hat{d}^+ = \tau_{\tilde{\pi}}^+ \cdot d \cdot \phi^* \quad \text{on } \Lambda_1^{m+1}(\pi_\infty) \ , \tag{7.17}$$

$$\mathrm{alt}\ \ \tau_{\tilde{\pi}} \cdot \phi^* \cdot \hat{D}^+ = -\tau_{\tilde{\pi}}^+ \cdot d \cdot \mathrm{alt} \cdot \tau_{\tilde{\pi}} \cdot \phi^* \ , \text{ on } I_\pi^1 \otimes (\Lambda_1^m(\pi_\infty) \cap I_\pi) \ . \tag{7.18}$$

Hence we can rewrite (7.16) as

$$\mathrm{alt}\cdot\tau_{\tilde{\pi}}\cdot\phi^{*}\cdot\partial^{1}\omega = D^{+}_{\tilde{\pi}}(\phi^{*}\omega + \mathrm{alt}\cdot\tau_{\tilde{\pi}}\cdot\phi^{*}\, S^{2}\omega) \ . \tag{7.19}$$

It's also not hard to see that

$$\tau_{\tilde{\pi}}\cdot\mathrm{alt}\cdot\tau_{\tilde{\pi}}\cdot\phi^{*} = 0 \quad \text{on} \quad I^{1}_{\pi}\otimes(\Lambda^{m}_{1}(\pi_{\infty})\cap I_{\pi}) \ . \tag{7.20}$$

Let $\tilde{\omega} = (1-\tau_{\tilde{\pi}})\phi^{*}\omega + \mathrm{alt}\cdot\tau_{\tilde{\pi}}\cdot\phi^{*}S^{2}\omega$. Then from (7.19) we get

$$\mathrm{alt}\cdot\tau_{\tilde{\pi}}\cdot\phi^{*}\partial^{1}\omega = D^{+}[\tau_{\tilde{\pi}}\,\phi^{*}\omega + \tilde{\omega}] \ , \quad \Lambda^{m+1}_{1}(\tilde{\pi}_{\infty}) \ . \tag{7.21}$$

As $S^{2}\omega\in I^{1}_{\pi}\otimes(\Lambda^{m}_{1}(\pi_{\infty})\cap I_{\pi})$, so $\tau_{\tilde{\pi}}\tilde{\omega} = 0$ by (7.20). Besides, $\partial^{1}\omega\in\Lambda^{1}[\pi_{\infty,0}]\otimes \Lambda^{m+1}_{1}(\pi_{\infty})$, hence $\mathrm{alt}\cdot\tau_{\tilde{\pi}}\cdot\phi^{*}\partial^{1}\omega\in\Lambda^{m+2}_{1}(\tilde{\pi}_{\infty};\tilde{\pi}_{\infty,0})$.

On the other hand, we can write the formula for the first variation for $\tau_{\tilde{\pi}}\ \phi^{*}\omega\in\Lambda^{m+1}_{0}(\tilde{\pi}_{\infty})$. We write it in the form (1.6):

$$\partial_{\tilde{\pi}}\ \tau_{\tilde{\pi}}\ \phi^{*}\omega = D^{+}_{\tilde{\pi}}\ S\ \tau_{\tilde{\pi}}\ \phi^{*}\omega \ . \tag{7.22}$$

By setting $\bar{S} = S - 1$ we write (7.22) as

$$\partial_{\tilde{\pi}}\ \tau_{\tilde{\pi}}\ \phi^{*}\omega = D^{+}_{\tilde{\pi}}\ [\tau_{\tilde{\pi}}\ \phi^{*}\omega + \bar{S}\,\tau_{\tilde{\pi}}\phi^{*}\ ] \ . \tag{7.23}$$

As $\tau_{\tilde{\pi}}\ S = \mathrm{id}$, so $\tau_{\tilde{\pi}}\bar{S} = 0$, i.e. $\tau_{\tilde{\pi}}\ \bar{S}\tau_{\tilde{\pi}}\ \phi^{*}\omega = 0$.

The uniqueness property of ∂ can be reformulated as: $\forall\,\omega\in\Lambda^{m}_{0}(\pi_{\infty})$, $\exists\,!$ $\partial\,\omega\in\Lambda^{m+1}_{1}[\pi_{\infty,0}] : \partial\omega = D^{+}_{\pi}(\omega + \bar{S}\omega)$ with some $\bar{S}\,\omega\in\Lambda^{m}_{1}(\pi_{\infty})\cap I_{\pi}$. Applying this reformulation to (7.21) and (7.23), we get

<u>Theorem 7.7</u>. $\mathrm{alt}\cdot\tau_{\tilde{\pi}}\cdot\ \phi^{*}\ \cdot\ \partial^{1} = \partial_{\tilde{\pi}}\cdot\tau_{\tilde{\pi}}\cdot\phi^{*}$ on $\Lambda^{m+1}_{1}(\pi_{\infty})$.

<u>Corollary 7.8</u>. $\mathrm{Ker}\ \partial^{1} = \mathrm{Ker}(\partial_{\tilde{\pi}}\cdot\tau_{\tilde{\pi}}\cdot\phi^{*})$.
< $\mathrm{alt}\cdot\tau_{\tilde{\pi}}\cdot\ \phi^{*}$ is a monomorphism on $\{(\Lambda^{1}[\pi_{\infty,0}]\cap I^{1}_{\pi})\otimes\ \Lambda^{m+1}_{1}(\pi_{\infty})\}\supset \mathrm{Im}\ \partial^{1}$ >

7.4. Here we consider natural properties of Lagrangian formalism. Knowledge of such properties is important for Hamiltonian formalism in PDE. We restrict ourselves to the operator ∂ .

Let $\bar{\phi}_\Delta : J^\infty\pi \to J^\infty\nu$ be the map associated with a differential operator $\Delta: \Gamma(\pi) \to \Gamma(\nu)$ (see §1.I). Let $\Omega' \in \Lambda_0^m(\nu_\infty)$, $\Omega = \bar{\phi}_\Delta^*(\Omega') \in \Lambda_0^m(\pi_\infty)$. If Δ is differential operator of order s then $\bar{\phi}_\Delta^*(\partial_\nu \Omega') \in \Lambda_1^{m+1}[\pi_{\infty,s}]$ while $\partial_\pi \Omega \in \Lambda_1^{m+1}[\pi_{\infty,0}]$.

In what follows we use the operator $\hat{\partial} = \hat{\partial}_\pi : \Lambda_1^{m+1}(\pi_\infty) \to \Lambda_1^{m+1}[\pi_{\infty,0}]$ (see §3).

Lemma 7.9. $\hat{\partial}D^+ = \partial\tau$ on $\Lambda_1^m(\pi_\infty)$.

< $\forall \omega \in \Lambda_1^m(\pi_\infty)$, $\forall X \in D(\pi_{\infty,0})$, we have $\tau(\bar{X} \lrcorner\ \tau^+ d\omega) = \tau(\bar{X} \lrcorner\, d\omega) = \tau[\bar{X}(\omega) - d(\bar{X} \lrcorner\, \omega)] = \tau\bar{X}(\tau\omega) - D(\bar{X} \lrcorner\, \omega) = D(\bar{X} \lrcorner\, (S\tau\omega - \omega)) + \tau(X \lrcorner\, \partial\tau\omega)$. On the other hand, $\tau(\bar{X} \lrcorner\ \tau^+ d\omega) = \tau(\bar{X} \lrcorner\, D^+\omega) = D(\bar{X} \lrcorner\, \hat{S}D^+\omega) + \tau(X \lrcorner\, \hat{\partial} D^+\omega)$, hence $\hat{\partial}D^+\omega = \partial\tau\omega$ >

Theorem 7.10. $\partial_\pi \bar{\phi}_\Delta^* = \hat{\partial}_\pi \bar{\phi}_\Delta^* \partial_\nu$.

< $\bar{\phi}_\Delta^* \partial_\nu \Omega' = \bar{\phi}_\Delta^* \tau_\nu^+ dS_\nu \Omega' = \tau_\pi^+ \bar{\phi}_\Delta^* dS_\nu \Omega' = \tau_\pi^+ d\bar{\phi}_\Delta^* S_\nu \Omega' = D_\pi^+ \bar{\Omega}$, where $\bar{\Omega} = \bar{\phi}_\Delta^* S_\nu \Omega' \in \Lambda_1^m(\pi_\infty)$. As $\tau_\pi \bar{\phi}_\Delta^* = \bar{\phi}_\Delta^* \tau_\nu$, so $\tau_\pi \bar{\Omega} = \tau_\pi \bar{\phi}_\Delta^* S_\nu \Omega' = \bar{\phi}_\Delta^* \tau_\nu S_\nu \Omega' = \bar{\phi}_\Delta^* \Omega' = \Omega$. By lemma 7.9, $\hat{\partial}_\pi \bar{\phi}_\Delta^* \partial_\nu \Omega' = \hat{\partial}_\pi D_\pi^+ \bar{\Omega} = \partial_\pi \tau_\pi \bar{\Omega} = \partial_\pi \Omega = \partial_\pi \bar{\phi}_\Delta^* \Omega'$ >

§8. Local structure of the Kernel and Image of operator ∂

8.1. Formulas for the first variation (1.1) and (7.1) allow us to construct homotopy formulas for pairs of operators (D,∂) and (D^+,∂^1) correspondingly.

Let $U \subset E$ be a domain in E with local coordinates (x,q), $\bar{U} = \pi_{\infty,0}^{-1}(U) \subset J^\infty\pi$. Let us consider the family of maps $A_t : U \to U$ by the rule $A_t : (x_i, q^a) \mapsto [(1-t)x_i, \alpha(t)q^a]$, where $\alpha(t) = \exp[1-(1-t)^{-2}]$. Clearly, $A_0 = id|_U$, $A_1 : U \to \{0\}$, and A_t are diffeomorphisms of the bundle π when $0 \le t < 1$. As is well known (see, e.g. [3]), for the family $\{A_t\}$ there exists a one-parameter family of fields X_t such that A_t is the translation operator for $\{X_t\}$. As we know, both $\{A_t\}$ and $\{X_t\}$ can be lifted to the family of classified transformations $\bar{A}_t^\ell$ and classified fields $\{\bar{X}_t^\ell\}$ on $\bar{U}^\ell = \bar{U} \cap J^\ell\pi$, $\forall \ell > 0$; correspondingly. Besides, $\bar{A}_1 : \bar{U} \to \{0\}$.

Let us write formula (1.1) for $\bar{X}_t$. As $\bar{X}_t$ is automorphism of the bundle $\pi_\infty|_{\bar{U}}$, so $\tau\bar{X}_t(\Omega) = \bar{X}_t(\Omega)$. Applying from the left operator $\bar{A}_t^*$ we get

$\frac{d}{dt}\bar{A}^*_t(\Omega) = \bar{A}^*_t\,\bar{X}_t(\Omega) = \bar{A}^*_t\,\tau d(\bar{X}_t \lrcorner S\Omega) + \bar{A}^*_t\,\tau(X_t \lrcorner \partial\Omega) = D\bar{A}^*_t(\bar{X}_t \lrcorner S\Omega) + \tau\bar{A}^*_t(X_t \lrcorner \partial\Omega)$, because $\tau\bar{A}^*_t = \bar{A}^*_t\tau$ by lemma 2.15. Note that despite the fact that field $\bar{X}_1$ is not defined, nevertheless $\exists \lim_{t\to 1} \bar{A}^*_t(X_t \lrcorner \omega)$, $\forall \omega \in \Lambda^*$. Hence we can integrate the previous equality: $\int_0^1 dt \frac{d}{dt} \bar{A}^*_t(\Omega) = \bar{A}^*_1(\Omega) - \bar{A}^*_0(\Omega) = -\Omega = D\int_0^1 dt\, \bar{A}^*_t(\bar{X}_t \lrcorner S\Omega)$ $+ \tau\int_0^1 dt\, \bar{A}^*_t(X_t \lrcorner \partial\Omega) = D\tau\int_0^1 dt\, \bar{A}^*_t(\bar{X}_t \lrcorner S\Omega) + \tau\int_0^1 dt\, \bar{A}^*_t(X_t \lrcorner \partial\Omega)$, because $D = D\tau$ by corollary 5.21.I. Let us denote $\psi_1 : \Lambda^m_{0_1} \to \Lambda^{m-1}_0$, $\psi_2 : \Lambda^{m+1}_0[\pi_{\infty,0}] \to \Lambda^m_0$, $\psi_1(\Omega) = -\tau\int_0^1 dt\, \bar{A}^*_t(\bar{X}_t \lrcorner S\Omega)$, $\psi_2(\mu) = -\tau\int_0^1 dt\, \bar{A}^*_t(X_t \lrcorner \mu)$, $\forall \mu \in \Lambda^{m+1}_1[\pi_{\infty,0}]$. Thus we have on $\Lambda^m_0(\pi_\infty|_{\bar{U}})$

$$1 = D\,\psi_1 + \psi_2\,\partial \tag{8.1}$$

Analogous reasoning for the formula (7.1) together with the formula $\bar{A}^*_t\tau^+ = \tau^+\bar{A}^*_t$ (lemma 2.17), allows us to construct operators $\psi_3 : I^1 \otimes \Lambda^m_1 \to \Lambda^m_1$, $\psi_4 : (\Lambda^1[\pi_{\infty,0}] \cap I^1) \otimes \Lambda^{m+1}_1 \to \Lambda^{m+1}_1$ by formulas $\psi_3(\theta) = -\int_0^1 dt\, \bar{A}^*_t\,\theta(\bar{X}_t)$, $\psi_4(\tilde{\theta}) = -\int_0^1 dt\, \bar{A}^*_t\,\tilde{\theta}(X_t)$, such that

$$1 = D^+\psi_3 + \psi_4\,\partial^1 \quad \text{on } \Lambda^{m+1}_1(\pi_\infty|_{\bar{U}}) \tag{8.2}$$

<u>Theorem 8.1.</u> Let U be a domain in E . Then $\operatorname{Ker}\partial = \operatorname{Im}(D|_{\Lambda^{m-1}_0})$, $\operatorname{Ker}\partial^1 = \operatorname{Im}(D^+|_{\Lambda^m_1})$ in $\pi^{-1}_{\infty,0}(U)$.

< If $\partial\Omega = 0$, then $\Omega = D\,\psi_1\,\Omega$; if $\partial^1\omega = 0$, then $\omega = D^+\psi_3\omega$ >

8.2. By corollary 7.6, $\operatorname{Im}\partial \subset \operatorname{Ker}\partial^1$.

<u>Theorem 8.2.</u> $\operatorname{Im}\partial = \operatorname{Ker}(\partial^1|_{\Lambda^{m+1}_1[\pi_{\infty,0}]})$ in $\bar{U}$.

< Let $\omega \in \Lambda^{m+1}_1[\pi_{\infty,0}]|_{\bar{U}}$ and $\partial^1\omega = 0$. By theorem 8.1, $\omega = \tau^+ d\,\psi_3\omega$. Let $\Omega = \tau\psi_3\omega$. Then we can choose $\psi_3\omega$ as $S\Omega$, because $\tau(\psi_3\omega) = \Omega$ and $\tau^+d(\psi_3\Omega) = \omega \in \Lambda^{m+1}_1[\pi_{\infty,0}]|_{\bar{U}}$. Thus $\omega = \partial\Omega$ by (1.7)>

<u>Corollary 8.3.</u> Complex $\Lambda^0 \xrightarrow{D} \dots \xrightarrow{D} \Lambda^m_0 \xrightarrow{\partial} \Lambda^{m+1}_1[\pi_{\infty,0}] \xrightarrow{\partial^1} \dots$ is locally exact in members Λ^m_0 and $\Lambda^{m+1}_1[\pi_{\infty,0}]$.

By corollary 7.8, complex (6.1) is also locally exact in members $\Lambda_0^m(\pi_\infty)$ and $\Lambda_1^{m+1}[\pi_{\infty,0}]$.

8.3. Summing up, note that theorem 8.2 gives us the practical tool for answering question: Is the (locally) given form $\omega\varepsilon\Lambda_1^{m+1}[\pi_{\infty,0}]$ "functional derivative" of some lagrangian (speaking in classical manner)? This is the case iff $\partial^1\omega = 0$, or , equivalently, $\partial_{\tilde{\pi}}\, \tau_{\tilde{\pi}}\, \phi^*\omega = 0$, and lagrangian itself is simply $\tau\, \psi_3\, \omega$. It's self-understood, of course, that in practice the real question is more complicated: given an equation $\Gamma \subset J^\infty\pi$, is it of the form Γ_Ω for some Ω ? Nevertheless, in this situation it's also possible to look for such $\omega \varepsilon \Lambda_1^{m+1}[\pi_{\infty,0}]$ that $\partial^1\omega = 0$ and $\omega|_\Gamma = 0$.

Note, finally, that the first half of theorem 8.1 states that if "functional derivative of a lagrangian is zero, then lagrangian itself is (locally) divergence".

Chapter III.

HAMILTONIAN FORMALISM IN FIELD THEORY

§1. Cotangent bundle to bundle. Universal form ρ. Poisson bracket.

1.1 Let $\bar{\bar{\pi}}: T^*\pi \to E$ be the vector bundle whose sections represent forms from $\Lambda_1^{m+1}(\pi)$, that is, $\Lambda_1^{m+1}(\pi) \ni \omega \leftrightarrow S_\omega \in \Gamma(\bar{\bar{\pi}})$. Let $\bar{\pi} = \pi \circ \bar{\bar{\pi}}: T^*\pi \to M$. The bundle $\bar{\pi}_\infty: J^\infty \bar{\pi} \to M$ is called the cotangent bundle to the bundle π. By analogy with mechanics, $\exists!$ universal form $\rho \in \Lambda_1^{m+1}(\bar{\pi}) \cap \Lambda^{m+1}(\bar{\bar{\pi}})$ which is uniquely defined by its universal property: $S_\omega^*(\rho) = \omega$, $\forall \omega \in \Lambda_1^{m+1}(\pi)$.

1.2 Let us introduce special local coordinates in $T^*\pi$. If $\omega \in \Lambda_1^{m+1}(\pi)$, then locally $\omega = \alpha^i dq^i \wedge d^m x$, where $d^m x \overset{\text{def}}{=\!=} dx_1 \wedge \ldots \wedge dx_m$, $\alpha^i \in F(E)$. We choose in $T^*\pi$ local coordinates (p^i, q^i, x) in such manner that $\alpha^i = S_\omega^*(p^i)$. Then $\dim T^*\pi - \dim E = n = \dim E - \dim M$, $\rho = p dq \wedge d^m x \overset{\text{def}}{=\!=} p^i dq^i \wedge d^m x$, $d\rho = dp \wedge dq \wedge d^m x$, $d\rho \in \Lambda_2^{m+2}(\bar{\pi})$.

1.3 The form $d\rho$ defines the isomorphism $\Gamma: D^V(\bar{\pi}_{\infty,0}) \to \Lambda_1^{m+1}(\bar{\pi}_\infty; \bar{\pi}_{\infty,0})$, $\Gamma(X) = -(X \lrcorner d\rho)$. Γ is called Hamiltonian map.

Elements of $\Lambda_0^m(\bar{\pi}_\infty)$ are called hamiltonians. Hamiltonian vector field X_H corresponding to hamiltonian H is defined as $X_H = \Gamma^{-1}(\partial H)$. In other words, $X_H \lrcorner d\rho = -\partial H$. Thus, trivial hamiltonians belong to $\operatorname{Ker} \partial$ (constants in mechanics). Poisson bracket $\{H, F\}$ of hamiltonians H and F is defined as $(X_F \lrcorner X_H \lrcorner d\rho) \in \Lambda_0^m(\bar{\pi}_\infty)$. By definition, $\{H, F\} = -\{F, H\}$

We consider the form $\Omega = d\rho$ as 2-form on fibers $(\bar{\pi}_\infty)^{-1}$ with values in $\bar{\pi}_\infty^*(\Lambda^m M)$. Then $\{H, F\} = \Omega(X_H, X_F)$.

Lemma 1.1 $\{H, F\} = X_H \lrcorner \partial F = -X_F \lrcorner \partial H$.

◂ By definition ▸

Lemma 1.2 $\{H, F\} = \bar{X}_H(F) - D(\bar{X}_H \lrcorner SF) = -\bar{X}_F(H) + D(\bar{X}_F \lrcorner SH)$.

◂ Corollary of (1.1.II) ▸

1.4 Let, locally, $H = Hd^m x$, $H \in \bar{K} \stackrel{\text{def}}{=} \lim \operatorname{ind} F(J^k \bar{\pi})$. Then $X_H = \frac{\delta H}{\delta p^i}\frac{\partial}{\partial q^i} - \frac{\delta H}{\delta q^i}\frac{\partial}{\partial p^i}$. Equations of trajectories of the field X_H take the form $\frac{\partial q^i}{\partial t} = \frac{\delta H}{\delta p^i}$, $\frac{\partial p^i}{\partial t} = -\frac{\delta H}{\delta q^i}$. If $F = Fd^m x$, then $\{H,F\} = (\frac{\delta H}{\delta p^i}\frac{\delta F}{\delta q^i} - \frac{\delta H}{\delta q^i}\frac{\delta F}{\delta p^i})d^m x$.

Note: The more mathematical definition of Poisson bracket $\{H,F\}$ should be $\bar{X}_H(F)$. The reason why we choose $\Omega(X_H, X_F)$ is to be as close to classical mechanics as possible. On the other hand, $\bar{X}_H(F) = \Omega(X_H, X_F) + D(\bar{X}_H \lrcorner SF)$ (lemma 1.2), so no difference mod Im D.

§2. Jacobi identity for Poisson bracket.

2.1 In mechanics, for every triple of hamiltonians H,F,G we have Jacobi identity $J(H,G,F) \stackrel{\text{def}}{=} \{H,\{G,F\}\} + \text{c.p.} = 0$, where "c.p." denotes "cyclic permutation". This fact may be (but not necessarily) deduced from the important formulae $X_{\{H,F\}} = [X_H, X_F]$. In field theory we have the following analogs:

Theorem 2.1.

$$\bar{X}_{\{H,F\}} = [\bar{X}_H, \bar{X}_F]. \tag{2.1}$$

$$J(H,G,F) = D[\bar{X}_{\{H,G\}} \lrcorner SF - \bar{X}_F \lrcorner S\{H,G\} - \bar{X}_H(\bar{X}_G \lrcorner SF - \bar{X}_F \lrcorner SG) + \text{c.p.}]. \tag{2.2}$$

◀ We write $A \sim B$ if $(A-B) \in \operatorname{Im}(D|_{\Lambda_0^{m-1}(\bar{\pi}_\infty)})$. We have $\{H,\{G,F\}\} = \bar{X}_H\{G,F\} - D(\bar{X}_H \lrcorner S\{G,F\})$, so $J(H,G,F) = \bar{X}_H\{G,F\} + \text{c.p.} - D(\bar{X}_H \lrcorner S\{G,F\} + \text{c.p.})$. (2.3)
Then $\{H,\{G,F\}\} \sim \bar{X}_H\{G,F\} \sim \bar{X}_H(\bar{X}_G F)$ because $\bar{X}D = D\bar{X}$ by theorem 5.18.I. Similarly, $\{G,\{F,H\}\} \sim -\bar{X}_G\{H,F\} \sim -\bar{X}_G(\bar{X}_H F)$, $\{F,\{H,G\}\} \sim -\bar{X}_{\{H,G\}}(F)$, so

$$J(H,G,F) \sim ([\bar{X}_H, \bar{X}_G] - \bar{X}_{\{H,G\}})(F). \tag{2.4}$$

$d\Omega = 0$, so $0 = d\Omega(\bar{X}_H, \bar{X}_G, \bar{X}_F) = \bar{X}_H(\Omega(X_G, X_F)) + \text{c.p.} - (\Omega([\bar{X}_H, \bar{X}_G], \bar{X}_F) + \text{c.p.}) = \bar{X}_H\{G,F\} + \text{c.p.} - ([\bar{X}_H, \bar{X}_G] \lrcorner \partial F + \text{c.p.})$, hence $\bar{X}_H\{G,F\} + \text{c.p.} = [\bar{X}_H, \bar{X}_G] \lrcorner \partial F + \text{c.p.} = [\bar{X}_H, \bar{X}_G](F) - D([\bar{X}_H, \bar{X}_G] \lrcorner SF) + \text{c.p.}$ Next, $[\bar{X}_H, \bar{X}_F](F) + \text{c.p.} = \bar{X}_H(\bar{X}_G F) - \bar{X}_G(\bar{X}_H F) + \text{c.p.} = \bar{X}_H(\{G,F\} + D(\bar{X}_G \lrcorner SF)) + \bar{X}_G(\{F,H\} - D(\bar{X}_H \lrcorner SF)) + \text{c.p.} = 2\bar{X}_H\{G,F\} + D[\bar{X}_H(\bar{X}_G \lrcorner SF) - \bar{X}_G(\bar{X}_H \lrcorner SF)] + \text{c.p.}$ Hence,

$$\bar{X}_H\{G,F\} + \text{c.p.} = D\{[\bar{X}_H, \bar{X}_G] \lrcorner SF + \bar{X}_G[\bar{X}_H \lrcorner SF - X_F \lrcorner SH] + \text{c.p.}\}. \tag{2.5}$$

From (2.3), (2.5) we conclude that $J(H,G,F) \sim 0$. Let $\bar{Y} = [\bar{X}_H, \bar{X}_G] - \bar{X}_{\{H,G\}}$. Then (2.4) says that $\bar{Y}(F) \sim 0$, $\forall F$. From the lemma 2.2 (below) we see that $Y = 0$, i.e. (2.1). From (2.3), (2.5) and (2.1) after c.p. one gets (2.2) ▶

Lemma 2.2 . Let $Y \in D^{v}(\pi_{\infty,0})$ and $\bar{Y}(F) \sim 0$, $\forall F \in \Lambda^{m}(\pi_{\infty})$. Then $Y = 0$.

◀ This is a corollary of lemma 2.3, because $\partial D = 0$ ▶

Lemma 2.3 . Let $Y \in D^{v}(\pi_{\infty,0})$ and $\partial\bar{Y}(F) = 0$, $\forall F \in \Lambda^{m}_{0}$. Then $Y = 0$.

◀ Let $Y \in D^{v}(\pi_{k,0})$ and, locally, $Y = B^{a}\partial^{a}$, $B^{a} \in F(J^{k}\pi)$.
Let $y \in M$, $y = (x^{0}_{1},\ldots, x^{0}_{m}) = (x^{0})$, $s = \pi_{\infty}^{-1}(y)$. Let $F = q^{c}(x - x^{0})^{\sigma} d^{m}x$, $|\sigma| = k$.
Then $\bar{Y}(F) = B^{c}(x - x^{0})^{\sigma} d^{m}x$, $0 = \partial\bar{Y}(F)\big|_{s} = \frac{\delta[B^{c}(x - x^{0})^{\sigma}]}{\delta q^{a}}\Big|_{s} dq^{a}\wedge d^{m}x =$
$(-1)^{|\sigma'|} D_{\sigma'}\left(\frac{\partial B^{c}}{\partial q^{a}_{\sigma'}}(x - x^{0})^{\sigma}\right)\Big|_{x=x^{0}} dq^{a}\wedge d^{m}x = (-1)^{|\sigma|} \frac{\partial B^{c}}{\partial q^{a}_{\sigma}}\Big|_{x=x^{0}} dq^{a}\wedge d^{m}x$, $|\sigma| = k$,
hence (induction on k) $B^{a} = B^{a}(x)$, $Y = B^{a}(x)\partial^{a}$. Let now $F = (q^{c})^{2} d^{m}x$.
Then $Y(F) = 2B^{c}(x)q^{c}d^{m}x$, $\partial Y(F) = 2B^{c}dq^{c}\wedge d^{m}x = 0$ (no summation over c),
i.d. $Y = 0$ ▶

§3. Hamiltonians which are linear in momenta .

3.1. In mechanics, hamiltonians which are linear on momenta represent symbols of vector fields on configuration space. The situation in field theory is analogous.

$\bar{\bar{\pi}}: T^{*}\pi \to E$ is vector bundle, hence $\bar{\bar{\pi}}^{k}: J^{k}\bar{\pi} \to J^{k}\pi$ is a vector bundle too. As such, the notion of hamiltonian $H \in \Lambda^{m}_{0}(\bar{\pi}_{k})$ which is linear on momenta (i.e. on fibers of $\bar{\bar{\pi}}^{k}$) is correctly defined. Besides, we have "standard hamiltonians" $\in \Lambda^{m}_{0}(\pi_{k}\cdot\bar{\bar{\pi}}(\pi_{k,0}))$ which are linear on fibers of vector bundle $\bar{\bar{\pi}}(\pi_{k,0})$: $\bar{\bar{\pi}}*(J^{k}\pi) \to J^{k}\pi$. These hamiltonians evidently are in one-to-one correspondence with fields from $D^{v}(\pi_{k,0})$ due to the map $D^{v}(\pi_{k,0}) \ni X \mapsto (X \lrcorner \rho) \in \Lambda^{m}_{0}(\pi_{k}\cdot\bar{\bar{\pi}}(\pi_{k,0}))$. By natural projection $J^{k}\bar{\pi} \to \bar{\bar{\pi}}*(J^{k}\pi)$, standard hamiltonians can be embedded in all linear hamiltonians. Obviously, $\partial_{\bar{\pi}}(X \lrcorner \rho) = 0 \Leftrightarrow X = 0$, $X \in D^{v}(\pi_{\infty,0})$.

Theorem 3.1 . Let $H \in \Lambda^{m}_{0}(\bar{\pi}_{\infty})$ be a hamiltonian which is linear on momenta.
$\exists! \ X \in D^{v}(\pi_{\infty,0})$: $\partial_{\bar{\pi}}(H - X \lrcorner \rho) = 0$.

◀ The uniqueness of such X is evident from the note above. To prove its existence, let $H \in \Lambda^{m}_{0}(\pi_{\infty}\cdot\bar{\bar{\pi}}^{k}(\pi_{\infty,k}))$, i.e. locally $H = A^{a}_{\sigma}\, p^{a}_{\sigma}\, d^{m}x$, where $A^{a}_{\sigma} \in K$, $|\sigma| \le k$. We have $A^{a}_{\sigma}\, p^{a}_{\sigma}\, d^{m}x = D(A^{a}_{\sigma}\, p^{a}_{\sigma-i}\, \partial_{i} \lrcorner d^{m}x) - D_{i}(A^{a}_{\sigma})p^{a}_{\sigma-i}\, d^{m}x \sim -D_{i}(A^{a}_{\sigma})p^{a}_{\sigma-i}\, d^{m}x$: we made the step $k \mapsto k - 1$. And so forth until $k = 0$. ▶

3.2. The field $X \in D^v(\pi_\infty, \mathcal{O})$ is lifted to the hamiltonian field $\hat{X} \in D^v(\bar{\pi}_{\infty,0})$ with hamiltonian $X \lrcorner \rho = \hat{X} \lrcorner \rho$. Obviously, fields X and $\hat{X}$ are compatible with respect to projection $(\overline{\bar{\pi}})$, i.e. $\hat{X} \circ (\overline{\bar{\pi}})^* = (\overline{\bar{\pi}}) \circ \bar{X}$. In mechanics we have the formulae: $\widehat{[X,Y]} = [\hat{X},\hat{Y}]$ and $\{\rho(X),\rho(Y)\} = \rho([X,Y])$. In field theory we have the following analogs:

Theorem 3.2.

$$\overline{\widehat{[X,Y]}} = [\bar{\hat{X}},\bar{\hat{Y}}] \tag{3.1}$$

$$\{X \lrcorner \rho,\ Y \lrcorner \rho\} = [X,Y] \lrcorner \rho + D[\bar{\hat{Y}} \lrcorner S(X \lrcorner \rho) - \bar{\hat{X}} \lrcorner S(Y \lrcorner \rho)]\ . \tag{3.2}$$

< $\bar{\hat{X}}(Y \lrcorner \rho) = D(\bar{\hat{X}} \lrcorner S(Y \lrcorner \rho)) + X \lrcorner \partial(Y \lrcorner \rho) = D(\bar{\hat{X}} \lrcorner S(Y \lrcorner \rho)) + \{X \lrcorner \rho, Y \lrcorner \rho\}$, by lemma 1.1. Next, $\{X \lrcorner \rho, Y \lrcorner \rho\} = d\rho(\hat{X},\hat{Y}) = \bar{\hat{X}}(Y \lrcorner \rho) - \bar{\hat{Y}}(X \lrcorner \rho) - [\bar{\hat{X}},\bar{\hat{Y}}] \lrcorner \rho = D(\bar{\hat{X}} \lrcorner S(Y \lrcorner \rho)) + \{X \lrcorner \rho, Y \lrcorner \rho\} - D(\bar{\hat{Y}} \lrcorner S(X \lrcorner \rho)) - \{Y \lrcorner \rho, X \lrcorner \rho\} - [\bar{\hat{X}},\bar{\hat{Y}}] \lrcorner \rho$. Pairs $(\bar{\hat{X}},\bar{X})$ and $(\bar{\hat{Y}},\bar{Y})$ are compatible with respect to $(\overline{\bar{\pi}})$; ρ is horizontal over $J^\infty\pi$. Hence $[\bar{\hat{X}},\bar{\hat{Y}}] \lrcorner \rho = [X,Y] \lrcorner \rho$, and from the above calculations we obtain (3.2). If we write (3.2) as $\{X \lrcorner \rho, Y \lrcorner \rho\} \sim [X,Y] \lrcorner \rho$ and apply (2.1) we get (3.1) >

4. Theory of Symmetries

4.1. Consider now the problem of symmetries of Hamiltonian structure in field theory. Continuing our analogy with mechanics (see [2],[3]), we say that a field $X \in D^v(\bar{\pi}_{\infty,0})$ is a (infinitesimal) canonical field if $\partial^1\bar{X}(\rho) = 0$. If $X = X_H$ is a hamiltonian field, then $\partial^1\bar{X}_H(\rho) = \partial^1(X_H \lrcorner d\rho + d(X_H \lrcorner \rho)) = \partial^1(X_H \lrcorner d\rho) = \partial^1(-\partial H) = 0$ by corollaries 7.5, 7.6. Therefore, hamiltonian fields are canonical. A field $X \in D^v(\bar{\pi}_{\infty,0})$ is called locally-hamiltonian if $\partial^1(X \lrcorner d\rho) = 0$.

Theorem 4.1. Canonical and locally-hamiltonian fields coincide.

< $\partial^1\bar{X}(\rho) = \partial^1(X \lrcorner d\rho + d(X \lrcorner d\rho)) = \partial^1(X \lrcorner d\rho)$ >

Theorem 4.2. Let X,Y be locally-hamiltonian fields. Then $[X,Y]$ is a hamiltonian field with hamiltonian $Y \lrcorner X \lrcorner d\rho$

< We have to check that

$$\partial(X \lrcorner Y \lrcorner \Omega) = [X,Y] \lrcorner \Omega\ , \tag{4.1}$$

if $\partial^1(X \lrcorner \Omega) = \partial^1(Y \lrcorner \Omega) = 0$. Let us use the classical formula of differential geometry: if $U,V \in D(M)$, $\omega \in \Lambda^2(M)$, $d\omega = 0$, then

$$[U,V] \lrcorner \omega = d(\omega(U,V)) + U(V \lrcorner \omega) - V(U \lrcorner \omega) \quad . \tag{4.2}$$

We apply this formula to our situation under the same agreement as we made proving Jacobi identity: formula (4.2) is true if $\omega \in \Lambda^{m+2}_2(\overline{\pi}_\infty)$, $d\omega = 0$, $U,V \in D^V(\overline{K})$. This correspondence can be made by choosing (local) volume-form μ on M and representing $\omega = \tilde{\omega} \wedge \mu$, where $d\tilde{\omega} = 0$ and $\tilde{\omega}$ is considered as 2-form on fibers $\overline{\pi}_\infty^{-1}$; at the same time, $U(\mu) = V(\mu) = 0$. Thus,

$$[X,Y] \lrcorner \Omega = d(Y \lrcorner X \lrcorner \Omega) + \overline{X}(Y \lrcorner \Omega) - \overline{Y}(X \lrcorner \Omega) \quad . \tag{4.3}$$

We have $\overline{X}(Y \lrcorner \Omega) = d(X \lrcorner Y \lrcorner \Omega) + \overline{X} \lrcorner d(Y \lrcorner \Omega) = d(X \lrcorner Y \lrcorner \Omega) + D^+(S^2(Y \lrcorner \Omega)(\overline{X}))$, because $\partial^1(Y \lrcorner \Omega) = 0$. Analogously, $\overline{Y}(X \lrcorner \Omega) = d(Y \lrcorner X \lrcorner \Omega) + D^+(S^2(X \lrcorner \Omega)(\overline{Y}))$. Denote $\alpha = S^2(Y \lrcorner \Omega)(\overline{X}) - S^2(X \lrcorner \Omega)(\overline{Y})$. Because $\tau\alpha = 0$ the above calculation substituted into (4.3) results in

$$[X,Y] \lrcorner \Omega = d(X \lrcorner Y \lrcorner Y) + \tau^+ d\alpha \, , \quad \tau\alpha = 0 \tag{4.4}$$

$\forall Z \in D^V(\overline{\pi}_{\infty,0})$, we have $\overline{Z}(X \lrcorner Y \lrcorner \Omega) = \overline{Z} \lrcorner d(X \lrcorner Y \lrcorner \Omega) = \overline{Z} \lrcorner \{[X,Y] \lrcorner \Omega - \tau^+ d\alpha\} =$ $Z \lrcorner [X,Y] \lrcorner \Omega - \overline{Z} \lrcorner \tau^+ d\alpha$. Next, $\overline{Z} \lrcorner \tau^+ d\alpha = \tau(\overline{Z} \lrcorner d\alpha) = \tau\{\overline{Z}(\alpha) - d(\overline{Z} \lrcorner \alpha)\} =$ $-\tau d(\overline{Z} \lrcorner \alpha) = -D(\overline{Z} \lrcorner \alpha)$. On the other hand, $\overline{Z}(X \lrcorner Y \lrcorner \Omega) \sim Z \lrcorner \partial(X \lrcorner Y \lrcorner \Omega)$. By the uniqueness of ∂, this implies (4.1) >

4.2. The role of locally-hamiltonian fields in mechanics is the preservation of Hamiltonian structure, i.e. the form $d\rho$. These fields interact naturally with the Hamiltonian map Γ : if X is locally-hamiltonian then the following formula due to A.M. Vinogradov holds:

$$X \cdot \Gamma + \Gamma \cdot X = 0 \tag{4.5}$$

In field theory we have analogs:

<u>Theorem 4.3.</u> $\forall X \in D^V(\overline{\pi}_{\infty,0})$,

$$\Gamma \cdot \overline{X} + \overline{X} \cdot \Gamma = d\,\Gamma(X) \tag{4.6}$$

(this formula is true also in mechanics). If X locally-hamiltonian, then

$$\mathrm{Im}(\Gamma\cdot\overline{X}+\overline{X}\cdot\Gamma)\subset \mathrm{Im}D^{+}\subset \mathrm{Ker}\,\partial^{1} \qquad (4.7)$$

< $\forall Y\in D^{v}(\pi_{\infty,0})$ we have, using (4.3), $(\Gamma\cdot\overline{X}+\overline{X}\cdot\Gamma)(Y)=\Gamma([Y,X])-\overline{X}(Y\lrcorner\,\Omega)=$ $[X,Y]\lrcorner\,\Omega-\overline{X}(Y\lrcorner\,\Omega)=d(Y\lrcorner\,X\lrcorner\,\Omega)-\overline{Y}(X\lrcorner\,\Omega)=-\overline{Y}\lrcorner\,d(X\lrcorner\,\Omega)=\overline{Y}\lrcorner\,d\Gamma(X)$. This proves (4.6). If X is locally-hamiltonian, then $\overline{Y}\lrcorner\,d(X\lrcorner\,\Omega)=D^{+}(S^{2}(X\lrcorner\,\Omega)(\overline{Y})$ >

Chapter IV. HAMILTONIAN FORMALISM IN GENERAL

1. Since the Korteveg-DeVries boom began, it became clear that the classical basis for Hamiltonian formalism, i.e. nondegenerate closed 2-form, need to be generalized. Here we briefly present some features of Hamiltonian formalism which are common for all its very different faces. One rather exotic example is shown at the end of this chapter.

2. Definition. Let V be a k-module, k is commutative algebra. A map $\Gamma \in \mathrm{Hom}_k(V, \mathrm{End}_k V)$ is called Hamiltonian, if

$$\Gamma(\{v_1,v_2\}) = [\Gamma(v_1), \Gamma(v_2)] , \quad \forall v_1,v_2 \in V \tag{1}$$

where $\{v_1,v_2\} \overset{\text{def}}{=\!=} \Gamma(v_1)(v_2)$ is called Poisson bracket.

Γ provides V with the structure of algebra:

$$v_1 * v_2 \overset{\text{def}}{=\!=} \Gamma(v_1)(v_2) = \{v_1,v_2\} \tag{2}$$

Equality (1) can be rewritten as

$$(v_1*v_2)*v_3 = v_1*(v_2*v_3) - v_2*(v_1*v_3), \quad \forall v_i \in V , \tag{3}$$

or

$$(v_1*v_2)*v_3 - v_1*(v_2*v_3) = -v_2*(v_1*v_3) \tag{3'}$$

3. Denote $\mathrm{Ker}\,\Gamma$ by $\hat{V}$, $\hat{V} \subset V$. $\hat{V}$ is an abelian subalgebra.

Lemma 1. $(\{v_1,v_2 + v_2,v_1\}) \subset \hat{V}$, $\forall v_1,v_2 \; V$

$< \Gamma(\{v_1,v_2\} + \{v_2,v_1\}) = [\Gamma(v_1), \Gamma(v_2)] + [\Gamma(v_2), \Gamma(v_1)] = 0 >$

Lemma 2. $V * \hat{V} \subset \hat{V}$, $\hat{V} * V \subset \{0\}$

< see (1) >

Lemma 3. (Jacoby's identity). $\forall v_1,v_2,v_3 \in V$,

$$\{\{v_1,v_2\} , v_3\} + \{\{v_2,v_3\} , v_1\} + \{\{ v_3, v_1\} , v_2\} \subset \hat{V} . \tag{4}$$

< Applying Γ to (4), we obtain zero due to (1) >

4. Denote $L = V/\hat{V}$. From lemmas 2,3 we get

<u>Corollary 4</u>. L is Lie algebra. The action of L on $\hat{V}$ is a representation. The center of L is $\{0\}$.

5. Thus, Hamiltonian structure provides us with Lie algebra L together with representation L on $\hat{V}$. Let us consider inverse procedure.

Let L be a Lie algebra and ρ be its representation on a vector space $\hat{V}$. Let $V := L \oplus \hat{V}$. We introduce the following structure of algebra into V : $(g_1 \oplus v_1) * (g_2 \oplus v_2) := [g_1, g_2] \oplus \rho(g_1)v_2$. In other words, we constructed the linear map $\Gamma : V \to \mathcal{E}nd\ V$.

Lemma 5. Γ is Hamiltonian structure.

$<$ $[(g_1 \oplus v_1) * (g_2 \oplus v_2)] * (g_3 \oplus v_3) = ([g_1,g_2] \oplus \rho(g_1)v_2) * (g_3 \oplus v_3) = [[g_1,g_2],g_3] \oplus \rho[g_1,g_2](v_3)$. On the other hand, $(g_1 \oplus v_1) * ((g_2 \oplus v_2) * (g_3 \oplus v_3)) - (g_2 \oplus v_2) * ((g_1 \oplus v_1) * (g_3 \oplus v_3)) = (g_1 \oplus v_1) * ([g_2,g_3] \oplus \rho(g_2)v_3) - (g_2 \oplus v_2) * ([g_1,g_3] \oplus \rho(g_1)v_3) = [g_1,[g_2,g_3]] \oplus \rho(g_1)\,\rho(g_2)v_3 - [g_2,[g_1,g_3]] \oplus \rho(g_2)\,\rho(g_1)v_3$ $>$

6. Above we considered L as a fixed subspace in V . Clearly, we can consider L as another subspace, say, $\tilde{L}$. These two choices differ by some morphism $A : L \to \hat{V}$, $(\ell, v) \sim (\ell, v+A\ )$. The pair $(L,\hat{V})$ produces the same Hamiltonian structure as $(\tilde{L},\hat{V})$ if $\rho(\ell_1) \cdot A = A \cdot \mathrm{ad}_{\ell_1}$, $\forall \ell_1 \in L$.

7. In mechanics, every hamiltonian vector field is an infinitesimal symmetry of Hamiltonian structure. Let us consider how the analog of this fact takes place in a general framework.

Let $\Gamma_i : V_i \to \mathcal{E}nd\ V_i$, i=1,2 be two Hamiltonian structures, $\phi : V_1 \to V_2$ be an isomorphism, $\phi_{\mathcal{E}} : \mathcal{E}nd\ V_1 \to \mathcal{E}nd\ V_2$, $X \to \phi \cdot X \cdot \phi^{-1}$ be corresponding isomorphism in the $\mathcal{E}$nd-spaces. We call ϕ canonical if the following diagram

$$\begin{array}{ccc} V_1 & \xrightarrow{\ \phi\ } & V_2 \\ {\scriptstyle \Gamma_1}\downarrow & & \downarrow{\scriptstyle \Gamma_2} \\ \mathcal{E}nd\ V_1 & \xrightarrow{\ \phi_{\mathcal{E}}\ } & \mathcal{E}nd\ V_2 \end{array} \qquad (5)$$

is commutative, i.e., $\Gamma_2 \cdot \phi(v_1) = \phi \cdot \Gamma_1(v_1) \cdot \phi^{-1}$, $\forall v_1 \in V_1$. Consider formally the case when we have a one-parameter family of such canonical isomorphisms, $\phi = \phi(t) : \Gamma_2 \cdot \phi(t)(v_1) = \phi(t) \cdot \Gamma_1(v_1) \cdot \phi(t)^{-1}$. Differentiating this equality with respect to t when $t = 0$ and denoting $\phi_{t|t=0}$ as A, we get

$$\Gamma_2 \cdot A(v_1) = A \cdot \Gamma_1(v_1) - \Gamma_1(v_1) \cdot A, \quad \forall v_1 \in V_1 . \tag{6}$$

Every A satisfying this condition is called infinitesimal canonical transformation.

At last, consider the case when $V_1 = V_2 = V$, $\Gamma_1 = \Gamma_2 = \Gamma$. Equation (6) for $A \in \mathrm{End}\, V$ can be rewritten as

$$\Gamma \cdot A(v) = A \cdot \Gamma(v) - \Gamma(v) \cdot A, \quad \forall v \in V \tag{7}$$

Let $v_1 \in V$ and $A = \Gamma(v_1)$. Then (7) becomes

$$\Gamma(\Gamma(v_1)(v_1)) = [\Gamma(v_1), \Gamma(v)]. \tag{8}$$

Thus, we proved

<u>Theorem 6</u>. $\Gamma : V \to \mathrm{End}\, V$ is Hamiltonian iff $\mathrm{Im}\,\Gamma \subset$ {Infinitesimal symmetries (=canonical isomorphisms) of Γ} .

Let us denote this last space of infinitesimal symmetries of Γ as $\mathrm{Inf}\ s(\Gamma)$.

<u>Lemma 7</u>. $\mathrm{Inf}\ s(\Gamma)$ is a Lie algebra.

< Let $A_i \in \mathrm{Inf}\ s(\Gamma)$, $i=1,2$. Using (7) we have $\Gamma[A_1,A_2](v) = \Gamma(A_1A_2 - A_2A_1)(v) = \Gamma A_1(A_2v) - \Gamma A_2(A_1v) = A_1\Gamma A_2v - \Gamma(A_2v)A_1 - A_2\Gamma A_1v + \Gamma(A_1v)A_2 = A_1[A_2\Gamma(v) - \Gamma(v)A_2] - [A_2\Gamma v - \Gamma(v)A_2]A_1 - A_2[A_1\Gamma v - \Gamma(v)A_1] + [A_1\Gamma v - \Gamma(v)A_1]A_2 = [A_1,A_2]\Gamma v - \Gamma(v)[A_1,A_2]$ >

<u>Lemma 8.</u> $\mathrm{Im}\,\Gamma$ is ideal in $\mathrm{Inf}\ s(\Gamma)$.

< Formula (7) >

<u>Definition</u>. $H^1(\Gamma) = \mathrm{Inf}\ s(\Gamma)/\mathrm{Im}\,\Gamma$.

8. In practice, the role of End V plays special Lie algebra of differentiations ("evolution fields", see, e.g. Ch. I). Examples of nontrivial Hamiltonian structures can be found in [15], [16]. We apply freely notions and notations from these papers.

Consider the set of pairs of functions $\{u(x,y), h(x)\}$, $-\infty < x < \infty$, $0 \leq y \leq h$. Let us define moments $A_n \overset{\text{def}}{=} \int_0^h u^n \, dy$. The space $V \overset{\text{def}}{=} \lim\limits_{N \to \infty} \text{ind}\ C^\infty(A_1,\dots,A_{nj};\ A_1^{(1)},\dots,A_n^{(1)};\dots A_1^{(m)},\dots,A_n^{(m)} \mid n+m = N\}$, where $A_i^{(j)} = \frac{d^j A_i}{d x^j}$; $k = \mathbb{R}$. We consider V also as a commutative differential ring. Evolution fields are such $X \in D(V)$ that $[X,\partial] = 0$, $\partial \overset{\text{def}}{=} d/dx$. Hence it's enough to define action X on $\{A_k^{(m)}\}$ only for $m = 0$.

If $H \in V$, then X_H is defined by $X_H(A_n) = B_{nm}\, \delta H/\delta A_m$ where $\delta H/\delta A_m$ is functional derivative, and $B_{nm} = (n+m) A_{n+m-1}\, \partial + m A_{n+m-1}^{(1)}$. In [15] it was proven that this map $\Gamma : H \to X_H$ is Hamiltonian.

Equations of trajectories of the field X_H are $\dot{A}_n = B_{nm} H_{(m)}$, where $H_{(m)} = \delta H/\delta A_m$. Consider the system

$$\begin{cases} u_t = (u^j H_{(j)})_x - u_y \int_0^y d\,y (j\, u^{j-1} H_{(j)})_x\,, \\ h_t = (j\, A_{j-1}\, H_{(j)})_x\,, \quad A_j = \int_0^h u^j\, d\,y\,. \end{cases} \tag{9}$$

In (15) it was also proven that equations $\dot{A}_n = B_{nm}\, H_{(m)}$ are direct implications of (9).

Moreover,

Theorem 9. The map $\Gamma : H \to \{\text{system } (9)\}$ is a Hamiltonian map.

The proof will be published elsewhere.

Bibliography

[1] A.M. Vinogradov, Many-valued solutions, and a principle for the classification of nonlinear differential equations, Soviet Math. Dokl. 14(1973), 661-665.

[2] A.M. Vinogradov, I.S. Krasil'shchik, What is the Hamiltonian formalism? Russian Math. Surveys, 30:1(1975), 177-202.

[3] A.M. Vinogradov, B.A. Kupershmidt, The structures of Hamiltonian mechanics, Russian Math. Surveys, 32:4(1977), 177-243.

[4] B.A. Kupershmidt, On geometry of jet manifolds, Uspekhi Math. Nauk 30:5(1975), 211-212.

[5] B.A. Kupershmidt, The Lagrangian formalism in the calculus of variations, Func. Anal. Appl. 10(1976), 147-149.

[6] V.V. Lychagin, Local classification of nonlinear first order partial differential equations, Russian Math. Surveys 30:1(1975), 105-175.

[7] Whatever about singularities of smooth maps.

[8] H. Goldschmidt, Existence theorems for analytic linear partial differential equations, Ann. Math. (2), 86(1967), 246-270.

[9] H. Goldschmidt, Integrability criteria for systems of non-linear partial differential equations, J. Diff. Geometry 1(1967), 269-307.

[10] R. Hermann, Currents in classical field theories, J. Math. Phys. 13:1(1972), 97-99.

[11] Krupka, Lagrange theory in fibered manifolds, Reports Math. Phys. 2(1971) 121-133.

[12] J. Sniatycki, On the geometric structure of classical field theory in Lagrangian formulation, Proc. Cambr. Phil. Soc. 68:2(1970), 475-484.

[13] H. Goldschmidt. S. Sternberg, The Hamilton-Cartan formalism in the calculus of variations, Ann. Inst. Fourier, Grenoble 23(1973), 203-267.

[14] L.J.F. Broer, J.A. Kobussen, Canonical transformations and generating functionals, Phisica 62(1972), 275-288.

[15] B.A. Kupershmidt, Yu. I. Manin, Equations of long waves with a free surface. II, Hamiltonian structure and higher equations, Func. Anal. Appl. 12(1978), 20-29.

[16] Yu. I. Manin, Algebraic aspects of nonlinear differential equations, J. Sov. Math, (1979), 1-122.

INVOLUTION THEOREMS

Tudor Ratiu

§1. Introduction

This paper deals with general theorems on the involution of constants of the motion of a Hamiltonian system. In the proof of complete integrability of such a system the involution part is usually easier than the proof of independence and the finding of action-angle variables, but it is the first step in which one guesses the necessary number of integrals. The basic idea of all theorems presented here is that the constants of the motion came from hidden symmetries of the system. Such a situation is typical in the case of reduced Hamiltonian systems and all one has to do in order to find commuting integrals is to observe that a given system is a reduction of a system with obvious integrals. This is the hard part and the theorems in this review give some guidelines of how such a guess is made; it is based on the special form of the Hamiltonian vector field and throughout this paper these special expressions are emphasized as a second part of an involution theorem.

A brief review of the general reduction procedure of Marsden and Weinstein is given in §2. The examples of the Calogero and Moser-Sutherland systems due to Kazhdan, Kostant, Sternberg [7] follow in a general Lie algebra setting and the section closes with an involution theorem on reduced manifolds generalizing the Kostant-Symes theorem which together with its corollaries is treated in §3. §4 deals with Hamiltonian structures and recursion formulas of Lenard type, whereas §5 treats involution theorems obtained by translation of the argument of invariants of the structure (Kostant [8], Mishchenko, Fomenko [13]). Even though the theorems as stated apply only in the finite dimensional case, they can be used in the context of Katz-Moody Lie algebras or Lie algebras of pseudo-differential operators (Adler [2], Adler, van Moerbeke [3], Adler, Moser [4]).

I want to thank M. Adler for introducing me to this subject a year ago when he sketched for me the proof of the Kostant-Symes theorem not available at that time in the literature. Many thanks to J. Marsden for lots of discussions and suggestions regarding this paper. Conversations with B. Kupershmidt and W. Symes are also gratefully acknowledged.

§2. Involution Theorems Obtained by Reduction

We begin by recalling briefly the reduction procedure for Hamiltonian systems with symmetry as it is presented in Sections 4.2 and 4.3 of Abraham-Marsden [1]; for proofs see this reference.

Let G be a Lie group with Lie algebra $\mathfrak{g}$, exponential map $\exp:\mathfrak{g} \to G$, P a smooth manifold and $\phi: G \times P \to P$ a smooth action of G on P. $\xi_P(p) = \frac{d}{dt}\Big|_{t=0} \phi(\exp t\xi, p)$, $p \in P$, $\xi \in \mathfrak{g}$, $t \in \mathbb{R}$ will denote the infinitesimal generators of this action. If $G\cdot p = \{\phi(g,p) | g \in G\}$ denotes the G-orbit through $p \in P$, its tangent space at p is $T_p(G\cdot p) = \{\xi_P(p) | \xi \in \mathfrak{g}\}$. Later on three actions will be important.

- The action of G on itself by left-multiplication $L: G \times G \to G$, $L(g,h) = gh$; its infinitesimal generator is $\xi_G(g) = T_e R_g(\xi)$, where $R_g(k) = kg$ denotes right multiplication in G by g;

- The adjoint action $\mathrm{Ad}: G \times \mathfrak{g} \to \mathfrak{g}$ given by $\mathrm{Ad}_g = T_e(R_{g^{-1}} \circ L_g)$; its infinitesimal generator is $\xi_{\mathfrak{g}} = \mathrm{ad}\xi$, where $(\mathrm{ad}\xi)\eta = [\xi,\eta]$, $[\ ,\ ]$ denoting the Lie bracket in $\mathfrak{g}$;

- The co-adjoint action of G on $\mathfrak{g}^*$ is the dual of the adjoint action and is given by $g \mapsto \mathrm{Ad}^*_{g^{-1}}$; its infinitesimal generator is $\xi_{\mathfrak{g}^*} = -(\mathrm{ad}\xi)^*$.

Let (P,ω) be a symplectic manifold and $\phi: G \times P \to P$ a symplectic action, i.e. $\phi_g^*\omega = \omega$ for all $g \in G$. The map $J: P \to \mathfrak{g}^*$ is a momentum mapping for this action if

$$T_pJ(v_p)\cdot\xi = \omega_p(\xi_P(p), v_p)$$

for every $\xi \in \mathfrak{g}$, $p \in P$, $v_p \in T_pP$. Denoting by $\hat{J}(\xi):P \to \mathbb{R}$ the map defined by $\hat{J}(\xi)(p) = J(p)\cdot\xi$, the definition above says that ξ_P is a Hamiltonian vector field on P with Hamiltonian $\hat{J}(\xi)$, i.e. $X_{\hat{J}(\xi)} = \xi_P$ for all $\xi \in \mathfrak{g}$. We shall call (P,ω,ϕ, J) a <u>Hamiltonian G-space</u>. Since not every locally Hamiltonian vector field is globally Hamiltonian, not every action admits a momentum map. However, if a momentum map exists it is uniquely determined up to constants in $\mathfrak{g}^*$.

Momentum maps are important since they give conserved quantities. More precisely, if $H:P \to \mathbb{R}$ is a G-invariant Hamiltonian of (P,ω,ϕ,J), i.e. $H\circ\phi_g = H$ for all $g \in G$, then J is constant on the flow of the Hamiltonian vector field X_H.

The momentum map $J:P \to \mathfrak{g}^*$ is said to be <u>Ad*-equivariant</u> if $J(\phi_g p) = Ad^*_{g^{-1}}J(p)$ for all $p \in P$. In this case it is shown that

$$\{\hat{J}(\xi), \hat{J}(\eta)\} = \hat{J}[\xi,\eta]$$

for all $\xi, \eta \in \mathfrak{g}$, where $\{,\}$ denotes the Poisson bracket in P.

The following criterion gives a formula for the momentum mapping for exact symplectic manifolds. Assume $\omega = -d\theta$ and that $\phi_g^*\theta = \theta$ for all g. Then $J:P \to \mathfrak{g}^*$

$$J(p)\cdot\xi = (i_{\xi_P}\theta)(p)$$

in an Ad^*-equivariant momentum map for the action ϕ. Here i_{ξ_P} denotes the interior product of a form with the vector field ξ_P. Two special cases of this thoerem will be important later on.

– $P = T^*Q$ with the canonical symplectic structure when G acts on Q. The lift $\phi^{T^*}:G \times T^*Q \to T^*Q$, $\phi_g^{T^*} = T^*\phi_{g^{-1}}$ has a momentum mapping given by $J:T^*Q \to \mathfrak{g}^*$,

$$J(\alpha_q)\cdot\xi = \alpha_q\cdot\xi_Q(q) \ ,$$

for all $q \in Q$, $\alpha_q \in T_q^*Q, \xi \in \mathfrak{g}$.

– $P = TQ$ endowed with the symplectic structure induced from T^*Q via a pseudo-Riemannian metric $\langle,\rangle$ on Q. Let G act on Q

by isometries and lift this action to TQ by $\phi^T : G \times TQ \to TQ$, $\phi^T_g = T\phi_g$. ϕ^T has a momentum mapping $J : TQ \to \mathfrak{g}^*$ given by

$$J(v_q)\cdot\xi = \langle v_q, \xi_Q(q)\rangle_q$$

for all $q \in Q$, $v_q \in T_qQ$, $\xi \in \mathfrak{g}$.

Particular examples are:

a) $Q = \mathbb{R}^n$, $G = (\mathbb{R}^n, +)$, $(s,q) \in G \times Q \overset{\phi}{\longmapsto} s+q \in Q$, $\xi_Q(q) = \xi$ for $\xi \in \mathfrak{g} = \mathbb{R}^n$; the momentum map of ϕ^{T^*} is $J(q,p) = p$;

b) $Q = G$, $\phi = L$, $\xi_G(g) = T_eR_g(\xi)$; the action ϕ^{T^*} has a momentum map $J(\alpha_g) = (T_e^*R_g)(\alpha_g)$ for $g \in G$, $\alpha_g \in T_g^*G$.

Let (P,ω,ϕ,J) be a Hamiltonian G-space with Ad^*-equivariant momentum map. Denote by $G_\mu = \{g \in G \mid \mathrm{Ad}^*_{g^{-1}}\mu = \mu\}$ the isotropy subgroup of the co-adjoint action at $\mu \in \mathfrak{g}^*$. Assume that μ is a regular value for J so that $J^{-1}(\mu)$ is a (dim P - dim G)-dimensional submanifold of P. By Ad^*-equivariance G_μ acts on $J^{-1}(\mu)$. Assume that this action is proper and free so that $P_\mu = J^{-1}(\mu)/G_\mu$, the G_μ-orbit space of $J^{-1}(\mu)$, is a smooth (dim P - dim G - dim G_μ) - dimensional manifold with the canonical projection $\pi_\mu : J^{-1}(\mu) \to P_\mu$ a surjective submersion. The theorem of Marsden and Weinstein [12] states then that P_μ has a unique symplectic structure ω_μ satisfying $\pi_\mu^*\omega_\mu = i_\mu^*\omega$, where $i_\mu : J^{-1}(\mu) \hookrightarrow P$ is the canonical inclusion. P_μ is called the <u>reduced phase space</u>. The key to the proof of this theorem consists of the following two statements which shall be also used later on:

$T_p(G_\mu\cdot p) = T_p(G\cdot p) \cap T_p(J^{-1}(\mu))$

$T_p(J^{-1}(\mu))$ is the ω-orthogonal complement of $T_p(G\cdot p)$.

Under all the hypotheses above, let $H : P \to \mathbb{R}$ be a G-invariant Hamiltonian. Then the flow F_t of X_H leaves $J^{-1}(\mu)$ invariant (since J is a conserved quantity) and commutes with the G_μ-action on $J^{-1}(\mu)$ (since $\phi_g^*X_H = X_H$) so it induces canonically a flow $H_t : P_\mu \to P_\mu$ defined by $\pi_\mu \circ F_t = H_t \circ \pi_\mu$. Then the theorem of Marsden and Weinstein [12] asserts that H_t is a Hamiltonian flow on P_μ with the Hamiltonian H_μ induced by H i.e. $H_\mu \circ \pi_\mu = H \circ i_\mu$ and Hamiltonian vector field X_{H_μ} on P_μ which is π_μ-related to $X_H | J^{-1}(\mu)$, i.e. $T\pi_\mu \circ X_H = X_{H_\mu} \circ \pi_\mu$. H_μ is called the <u>reduced Hamiltonian</u> and X_{H_μ} the <u>reduced Hamiltonian vector field</u>.

The above framework of reduction enables us to state the first involution theorem:

<u>Proposition 1.1</u>. Let $\{,\}^\mu$ denote the Poisson bracket on P_μ. If $f,g : P \to \mathbb{R}$ are G-invariant, then $\{f,g\}$ is G-invariant and $\{f,g\}_\mu = \{f_\mu, g_\mu\}^\mu$. In particular, if f,g Poisson commute in P then they Poisson commute in P_μ.

<u>Proof</u>. Since ϕ_h is a symplectic diffeomorphism $h \in G$, $\phi_h^*\{f,g\} = \{\phi_h^* f, \phi_h^* g\} = \{f,g\}$ and thus $\{f,g\}$ is G-invariant. Let $[p]$ denote the class of $p \in J^{-1}(\mu)$ in P_μ. Since X_f, X_g are tangent to $J^{-1}(\mu)$, we have

$$
\begin{aligned}
\{f,g\}_\mu([p]) &= (\{f,g\}_\mu \circ \pi_\mu)(p) = (\{f,g\} \circ i_\mu)(p) = \{f,g\}(p) = \\
&= \omega_p(X_f(p), X_g(p)) = \omega(i_\mu(p))(T_p i(X_f(p)), T_p i(X_g(p)) \\
&= (i_\mu^* \omega)(p)(X_f(p), X_g(p)) \\
&= (\pi_\mu^* \omega_\mu)(p)(X_f(p), X_g(p)) \\
&= \omega_\mu[p]((T\pi_\mu \circ X_f)(p), (T\pi_\mu \circ X_g)(p)) \\
&= \omega_\mu[p](X_{f_\mu}[p], X_{g_\mu}[p]) = \{f_\mu, g_\mu\}^\mu[p]. \quad \blacksquare
\end{aligned}
$$

This simple proposition has remarkable applications as was shown by Kazhdan, Kostant and Sternberg [7].

<u>Example 1</u>. Let G be a semisimple Lie group with Lie algebra $\mathfrak{g}$ and Killing form κ. Let G act on $\mathfrak{g} \times \mathfrak{g}$ by $\phi(g,(\xi,\eta)) = (\mathrm{Ad}_g \xi, \mathrm{Ad}_g \eta)$, where Ad is the adjoint representation of G on $\mathfrak{g}$. If $\zeta \in \mathfrak{g}$, its infinitesimal generator is $\zeta_{\mathfrak{g} \times \mathfrak{g}} = (\mathrm{ad}\zeta, \mathrm{ad}\zeta)$, where

$(\mathrm{ad}\zeta)(\Xi) = [\zeta,\Xi]$. $\mathfrak{g}\times\mathfrak{g}$ is exact sympletic, the canonical one-form being given by $\Theta(\eta,\xi)\cdot(\eta,\xi,\zeta_1,\zeta_2) = \kappa(\xi,\zeta_2)$ and symplectic form $\Omega = -d\Theta$. The momentum mapping exists for this action and is given by $J(\eta,\xi)\cdot\zeta = i_{\zeta_{\mathfrak{g}\times\mathfrak{g}}}\Theta(\eta,\xi) = \kappa([\eta,\xi],\zeta)$, since Θ is G-invariant. Its tangent map is $T_{(\eta,\xi)}J(\zeta_1,\zeta_2) = \kappa([\zeta_1,\xi],\cdot) + \kappa([\eta,\zeta_2],\cdot)$. For a generic $\varepsilon\in\mathfrak{g}$, $\kappa(\varepsilon,\cdot)\in\mathfrak{g}^*$ is a regular value of J and hence $J^{-1}(\kappa(\varepsilon,\cdot))$ is an $n = \dim\mathfrak{g}$-dimensional submanifold of $\mathfrak{g}\times\mathfrak{g}$ whose tangent space at (η,ξ) is $T_{(\eta,\xi)}(J^{-1}(\kappa(\varepsilon,\cdot)))$ $= \mathrm{Ker}T_{(\eta,\xi)}J = \{(\zeta_1,\zeta_2)\in\mathfrak{g}\times\mathfrak{g}\,|\,[\eta,\xi]=\varepsilon,\ [\zeta_1,\xi]+[\eta,\zeta_2]=0\}$. It is also easy to see that $G_\varepsilon = \{g\in G\,|\,\mathrm{Ad}_g\varepsilon = \varepsilon\}$ is the isotropy group of $\kappa(\varepsilon,\cdot)$ with respect to the co-adjoint action of G on $\mathfrak{g}^*$. In order to perform the reduction procedure, assume that G_ε acts freely and properly on the manifold $J^{-1}(\kappa(\varepsilon,\cdot))$ and form the reduced manifold $M = J^{-1}(\kappa(\varepsilon,\cdot))/G_\varepsilon$. In particular examples (as will be given below) this manifold can be determined explicitly. It is easy to see that its tangent space at $\pi(\eta,\xi)$, where $\pi: J^{-1}(\kappa(\varepsilon,\cdot)) \to M$ is the canonical projection, is given by

$$T_{\pi(\eta,\xi)}M = \left(\frac{\{(\zeta_1,\zeta_2)\,|\,[\zeta_1,\xi]+[\eta,\zeta_2]=0,\ [\eta,\xi]=0\}}{\{([\zeta,\eta],[\zeta,\xi])\,|\,[\zeta,\varepsilon]=0}\right)\Big/ G_\varepsilon =$$

$$= \{(\zeta_1,\zeta_2)\,|\,[\eta,\xi]=\varepsilon,[\zeta_1,\xi]+[\eta,\zeta_2]=0,\ [\eta,\zeta_1] + [\xi,\zeta_2]\in\mathfrak{g}_\varepsilon^\perp\}/G_\varepsilon$$

where $\mathfrak{g}_\varepsilon = \{\zeta\in\mathfrak{g}\,|\,[\zeta,\varepsilon]=0\}$ is the centralizer of ε and is the Lie algebra of G_ε.

Functions in involution on M arise in the simplest way when they are induced from Ad-invariant functions on $\mathfrak{g}\times\mathfrak{g}$ independent on one of the variables, for then these functions trivially Poisson commute on $\mathfrak{g}\times\mathfrak{g}$ and by Proposition 1.1. they will do so on M. In order to be able to recognize such systems we will explicitly determine the expression of the Hamiltonian vector field X_{f_ε}

for $f(\xi,\eta) = f(\xi)$, $f_\varepsilon = f \circ \pi$. One shows first that on $\mathfrak{g} \times \mathfrak{g}$, $X_f(\xi,\eta) = (\text{grad } f(\xi), 0)$ and then by π-relation ($T\pi \circ X_f = X_{f_\varepsilon} \circ \pi$) one finds

$$X_{f_\varepsilon}(\pi(\eta,\xi)) = G_\varepsilon\text{-orbit of } ((\text{grad } f)(\xi) - [\zeta,\eta], -[\zeta,\xi])$$

where ζ has to satisfy: $[\eta, (\text{grad } f)(\xi)] - [\eta, [\zeta,\eta]] - [\xi, [\zeta,\xi]] \in \mathfrak{g}_\varepsilon^\perp$. Loosely speaking, puting tildes for G_ε-orbits, one gets a system of the form

$$\text{(2.1)} \qquad \begin{cases} \dot{\tilde{\eta}} = ((\text{grad } f)(\xi) - [\zeta,\eta])^\sim \\ \dot{\tilde{\xi}} = -[\zeta,\xi]^\sim \end{cases}$$

Such a system is the Calogero system (Kazhdan, Kostant, Sternberg [7], Moser [14]). One takes for $G = U(n)$, $\mathfrak{g} = \mathfrak{u}(n)$, $\varepsilon = -\begin{pmatrix} 0 & i & \dots & i \\ i & 0 & \dots & i \\ \vdots & \vdots & & \vdots \\ i & i & \dots & 0 \end{pmatrix}$. It is shown (see e.g. Moser [14]) that $J^{-1}(\kappa(\varepsilon,\cdot)) = \{(\eta,\xi) \in \mathfrak{g} \times \mathfrak{g} \mid \eta = \text{Ad}_g \delta, \zeta = \text{Ad}_g \lambda, \ \delta = \text{diag}(x_1,\dots,x_n),$ $\lambda_{ii} = y_i, \ \lambda_{ij} = 1/(x_i - x_j), \ g \in G_\varepsilon\}$, $G_\varepsilon = \left\{ g \in U(n) \mid ge = ae, \ |a| = 1, \ e = \begin{bmatrix} 1 \\ \vdots \\ 1 \end{bmatrix} \right\}$. Thus the reduced symplectic manifold is $M = \{(\delta,\lambda)\}$ and its symplectic structure is given by $\sum_{n=1}^{n} dy_k \wedge dx_k$. System (2.1) becomes for $f(\eta,\xi) = -(1/2)\text{ReTr}\,\xi^2$:

$$\text{(2.2)} \qquad \begin{cases} \dot{\lambda} = -[\zeta,\lambda] \\ \dot{\delta} = -\lambda - [\zeta,\delta] \end{cases}$$

$$\text{(2.3)} \qquad \zeta_{jk} = i\delta_{jk} \sum_{\substack{\ell=1 \\ \ell\neq j}}^{n} \phi'(x^j - x^\ell) - i(1 - \delta_{jk})\phi'(x^j - x^k)$$

where $\phi(x) = 1/x$. We worked here with $U(n)$ instead of $SU(n)$ by applying formally the results above for semisimple Lie groups; the degeneracy of the Killing form creates no problems here. The n rational integrals in involution are $I_k = -\frac{1}{k}\text{Tr}\,\lambda^k$. System (2.2) is the Hamiltonian system on $\mathbb{R}^{2n}$ with the Hamiltonian $H = \frac{1}{2}\sum_{k=1}^{n} y_k^2 + \sum_{k<j} 1/(x_k - x_j)^2$; remark $H = I_2$.

Example 2. Consider the lift to the tangent bundle in body coordinates of the conjugation $(g,h) \mapsto ghg^{-1}$ of G. This action is given by (Abraham, Marsden [1], §4.4) $\phi_B^T(g,(h,\xi)) = (ghg^{-1}, Ad_g\xi)$, $\phi_B^T: G \times (G \times \mathfrak{g}) \to G \times \mathfrak{g}$ and it is symplectic with respect to the symplectic structure on $G \times \mathfrak{g}$ defined by $\Omega_B = -d\Theta_B$, $\Theta_B(g,\xi)\cdot(v_g\zeta) = \kappa(T_gL_g^{-1}(v_g), \zeta)$ for $v_g \in T_gG$, $g \in G$, $\xi, \zeta \in \mathfrak{g}$. For $\zeta \in \mathfrak{g}$, the infinitesimal generator of this action is $\zeta_{G\times\mathfrak{g}}(g,\xi) = (T_eR_g\zeta - T_eL_g\zeta, \xi, [\zeta,\xi])$, where $L_g(h) = gh$, $R_g(h) = hg$. ϕ_B^T leaves Θ_B invariant and hence this action has a momentum map $J: G \times \mathfrak{g} \to \mathfrak{g}^*$, $J(g,\xi) = (i_{\zeta_{G\times\mathfrak{g}}}\Theta_B)(g,\xi) = \kappa(Ad_g\xi - \xi, \zeta)$ whose tangent map at (g,ξ) is $T_{(g,\xi)}J(v_g,\xi,\eta) = \kappa([T_gR_g^{-1}(v_g), Ad_g\xi] + Ad_g\eta - \eta, \cdot)$. For generic $\varepsilon \in \mathfrak{g}$, $J^{-1}(\kappa(\varepsilon,\cdot))$ is an n-dimensional submanifold of $G \times \mathfrak{g}$. Let G_ε, $\mathfrak{g}_\varepsilon$ be as in example 1 and form the reduced manifold

$M = J^{-1}(\kappa(\varepsilon,\cdot))/G_\varepsilon$. Since $T_{(g,\xi)}J^{-1}(\kappa(\varepsilon,\cdot)) = \operatorname{Ker} T_{(g,\xi)}J =$

$$= \{(v_g,\xi,\eta) \in T_gG \times \mathfrak{g} \times \mathfrak{g} \mid [T_gR_g^{-1}(v_g), Ad_g\xi] + Ad_g\eta - \eta = 0, \; Ad_g\xi - \xi = \varepsilon\}$$

we obtain after a short computation that the tangent space at $\pi(g,\xi)$ $(\pi: J^{-1}(\kappa(\varepsilon,\cdot)) \to M)$ of M equals $T_{\pi(g,\xi)}M =$

$$= (\{(w_g,\xi,\theta) \in T_gG \times \mathfrak{g} \times \mathfrak{g} \mid Ad_g\xi - \xi = \varepsilon, [T_gR_g^{-1}(w_g), Ad_g\xi] + Ad_g\theta - \theta = 0, \; T_gR_g^{-1}(w_g) - T_gL_g^{-1}(w_g) + [\xi,\theta] \in \mathfrak{g}_\varepsilon^\perp\})/G_\varepsilon.$$

Assume now that $f_1, f_2: \mathfrak{g} \to \mathbb{R}$ are Ad-invariant functions and extend them by "left invariance" to $G \times \mathfrak{g}$, i.e. $\tilde{f}_i(h,\xi) = f(\xi)$, $i = 1,2$, for all $h \in G$. Then clearly $\{\tilde{f}_1, \tilde{f}_2\} = 0$ on $G \times \mathfrak{g}$ and hence $\{f_\varepsilon, g_\varepsilon\}^\varepsilon = 0$ by proposition 1.1. As in example 1 we are interested in the expression of X_{f_ε} for $f: \mathfrak{g} \to \mathbb{R}$ an Ad-invariant function. Using the definition of X_f we obtain $X_f(g,\xi) = (T_eL_g(\text{grad } f)(\xi), \xi, 0) \in T_gG \times \mathfrak{g} \times \mathfrak{g}$ and then from $T\pi \circ X_f = X_{f_\varepsilon} \circ \pi$ it follows that

$X_{f_\varepsilon}(\pi(g,\xi)) = G_\varepsilon\text{-orbit of } (T_eL_g(\text{grad } f)(\xi) - T_eR_g\zeta + T_eL_g\zeta, \xi, [\xi,\zeta])$, where the condition on ζ is $Ad_g(\text{grad } f(\xi)) - 2\zeta + Ad_g\zeta + Ad_g^{-1}\zeta -$

$-(\mathrm{grad}\ f)(\xi) - [\xi, [\zeta,\xi]] \in \mathfrak{g}_\varepsilon^\perp$. Thus the equations of motion are

$$(2.4)\quad \begin{cases} \dot{\tilde{g}} = (T_eL_g(\mathrm{grad}\ f)(\xi) - T_eR_g(\xi) + T_eL_g(\xi))^{\sim} \\ \dot{\tilde{\xi}} = [\xi,\zeta]^{\sim} \end{cases}$$

An example of such a system is the Moser-Sutherland system (Kazhdan, Kostant, Sternberg [7]). Take as before $G = U(n)$, $\mathfrak{g} = u(n)$, the same ε. G_ε is as in example 1, $J^{-1}(\kappa(\varepsilon,\cdot))$ $= \{(hgh^{-1}, \mathrm{Ad}_h\xi) \mid h \in G_\varepsilon,\ g = \mathrm{diag}(e^{i\theta_1}, \ldots, e^{i\theta_n}),\ \xi_{jj} = ip_j,$ $\xi_{ij} = -i(e^{i(\theta_i-\theta_j)}-1)^{-1}$ for $i \neq j\}$ and hence $M = \{(g,\xi)\}$. This Hamiltonian system on M is completely integrable, n integrals in involution being $I_k = -\frac{1}{k}\mathrm{Tr}\,\xi^k$. For $I_2 = H = \frac{1}{2}\sum_{i=1}^{n} p_i^2$ $+ \sum_{i\neq j} \sin^{-2}\frac{1}{2}(\theta_i-\theta_j)$ the Hamiltonian equations are

$$(2.5)\quad \begin{cases} \dot g = g\xi + [g,\zeta] \\ \dot\xi = [\xi,\zeta] \end{cases}$$

where $\zeta_{jj} = i\sum_{k\neq j} e^{i(\theta_j-\theta_k)}(e^{i(\theta_j-\theta_k)}-1)^{-2}$, $\zeta_{k\ell} =$ $ie^{i(\theta_k-\theta_\ell)}(e^{i(\theta_k-\theta_\ell)}-1)^{-2}$ for $k \neq \ell$. H represents a Hamiltonian for an n-body problem on the line with $\sin^{-2}$-potential.

<u>Example 3</u>. Consider the n-body problem on the line with $\sinh^{-2}$ potential. The Hamiltonian is $H(Q,P) = \frac{1}{2}\sum_{n=1}^{n} P_k^2$ $+ \sum_{k\neq\ell} \sinh^{-2}\frac{1}{2}(Q^k-Q^\ell)$. To prove that this system is completely integrable it suffices to notice that $(q,p) \mapsto (iq, ip) = (Q,P)$ transforms H into minus the Hamiltonian for the Moser-Sutherland system and that it is an antisymplectic diffeomorphism. The equations in matrix form are (2.5) with $g = \mathrm{diag}(e^{Q^1}, \ldots, e^{Q^n})$, $\xi_{jj} =$ P_j, $\xi_{k\ell} = -i(e^{Q^k-Q^\ell}-1)^{-1}$ for $k \neq \ell$, $\zeta_{jj} = i\sum_{k\neq j} e^{Q^j-Q^k}(e^{Q^j-Q^k}-1)^{-2}$, $\zeta_{k\ell} = ie^{Q^k-Q^\ell}(e^{Q^k-Q^\ell}-1)^{-2}$ for $k \neq \ell$. The integrals in involution are $I_k = -\frac{1}{k}\mathrm{Tr}\,\xi^k$, $I_2 = H$.

We turn now to the question of when involution in one reduced manifold implies involution in another reduced manifold.

Theorem 1.2. (Marsden, Ratiu). Let G be a Lie group, $H \subset G$ a closed Lie subgroup and denote by $\mathfrak{g}$, $\mathfrak{h}$ the corresponding Lie algebras. Assume $\mathfrak{g} = \mathfrak{h} \oplus \mathfrak{k}$ for $\mathfrak{k}$ a vector subspace and let $\Pi_{\mathfrak{h}^*}: \mathfrak{g}^* \to \mathfrak{h}^*$ denote the canonical projection. Let (P,ω) be a symplectic manifold, $(R,\omega|R)$ a symplectic submanifold on which G and H respectively act symplectically with Ad^*-equivariant momentum maps $J: P \to \mathfrak{g}^*$, $j = \Pi_{\mathfrak{h}^*} \circ (J|R): R \to \mathfrak{h}^*$. Let $\mu \in \mathfrak{h}^*$ and assume that one can form the reduced manifolds R_μ and $P_{\tilde{\mu}}$ for any $\tilde{\mu} \in \mathfrak{g}^*$ such that $\Pi_{\mathfrak{h}^*}\tilde{\mu} = \mu$. Assume that J satisfies

$$\{\hat{J}(\xi''), \hat{J}(\eta'')\}|R = \{\hat{J}(\xi'')|R, \hat{J}(\eta'')|R\}$$

for all $\xi'', \eta'' \in \mathfrak{k}$ Then if $f_{\tilde{\mu}}, g_{\tilde{\mu}}$ Poisson commute in $P_{\tilde{\mu}}$ for all $\tilde{\mu}$ satisfying $\Pi_{\mathfrak{h}^*}\tilde{\mu} = \mu$, f_μ, g_μ Poisson commute in R_μ.

Proof. Since f, g are constant on $P_{\tilde{\mu}}$, $X_{f_{\tilde{\mu}}} = X_{g_{\tilde{\mu}}} = 0$ on $P_{\tilde{\mu}}$ and hence $X_f(p), X_g(p)$ are tangent to $G_{\tilde{\mu}}\cdot p$, i.e. $X_f(p) = \xi_P(p)$, $X_g(p) = \eta_P(p)$, for $\xi, \eta \in \mathfrak{g}_{\tilde{\mu}} = \{\zeta \in \mathfrak{g} \mid (ad\zeta)^*\tilde{\mu} = 0\}$. Let $\xi = \xi' + \xi''$, $\eta = \eta' + \eta''$ be the splittings of ξ and η in the direct sum $\mathfrak{h} \oplus \mathfrak{k} = \mathfrak{g}$ and notice that since H acts on R and $\xi', \eta' \in \mathfrak{h}$ we have that $\xi'_P(p) = \xi'_R(p)$, $\eta'_P(p) = \eta'_R(p)$ for $p \in R$.

To show that $\{f_\mu, g_\mu\} = 0$ on R_μ, it suffices to prove that for any $p \in j^{-1}(\mu)$, $\omega(p)(X_{f|R}(p), X_{g|R}(p)) = 0$. Since R is symplectic, $TP|R = TR \oplus E$, $E = TR^\perp$ (ω-orthogonal complement); denote by $\Pi: TP|R \to TR$ the canonical projection. Fix in all that follows $p \in j^{-1}(\mu)$. It is easy to see that $X_{f|R}(p) = \Pi_p X_f(p)$ and hence $X_{f|R}(p) = \xi'_R(p) + \Pi_p\xi''_P(p)$, $X_{g|R}(p) = \eta'_R(p) + \Pi_p\eta''_P(p)$. By the theorem of conservation of momentum j, $\omega(p)\,(\xi'_R(p), X_{g|R}(p)) = 0$; use this in the third equality below to get:

$$\begin{aligned}\omega(p)(X_{f|R}(p), X_{g|R}(p)) &= \omega(p)(\Pi_p X_f(p), X_{g|R}(p)) \\ &= \omega(p)(X_f(p), X_{g|R}(p)) \\ &= \omega(p)(\Pi_p \xi''_P(p), \eta'_R(p) + \Pi_p \eta''_P(p)) \\ &= \omega(p)(\xi''_P(p), \eta'_R(p)) + \omega(p)(\Pi_p \xi''_P(p), \Pi_p \eta''_P(p)) \\ &= \omega(p)(\xi''_P(p), \eta'_P(p)) + \omega(p)(X_{\hat{J}(\xi'')|R}(p), X_{\hat{J}(\eta'')|R}(p)) \\ &= \{\hat{J}(\xi''), \hat{J}(\eta')\}(p) + \{\hat{J}(\xi'')|R, \hat{J}(\eta'')|R\}(p) \\ &= \hat{J}[\xi'',\eta'](p) + \{\hat{J}(\xi'')|R, \hat{J}(\eta'')|R\}(p) \\ &= \tilde{\mu}[\xi'',\eta-\eta''] + \{\hat{J}(\xi'')|R, \hat{J}(\eta'')|R\}(p) \\ &= -\tilde{\mu}[\xi'',\eta''] + \{\hat{J}(\xi'')|R, \hat{J}(\eta'')|R\}(p) \\ &= -\{\hat{J}(\xi''), \hat{J}(\eta'')\}(p) + \{\hat{J}(\xi'')|R, \hat{J}(\eta'')|R\}(p)\end{aligned}$$

which by hypothesis equals zero. ■

The hypotheses in this theorem are all very natural and generalize the ones in the Kostant-Symes involution theorem which will be described next.

To derive the Kostant-Symes theorem from Theorem 1.2, we need a few remarks. Recall that the lift to T^*G of the left multiplication on G has the momentum map $J:T^*G \to \mathfrak{g}^*$, $J(\alpha_g) = (T_e^*R_g)(\alpha_g)$

Each $\mu \in \mathfrak{g}^*$ is clearly a regular value of J and $J^{-1}(\mu) = \{\alpha_g \in T^*G | \alpha_g \circ T_eR_g = \mu\}$ = graph of right-invariant one-form whose value at e is μ. Thus the isotropy subgroup G_μ of the co-adjoint action acts on $J^{-1}(\mu)$ by left translations on the base point and we get that $J^{-1}(\mu)/G_\mu \approx G/G_\mu \approx G\cdot\mu \subset \mathfrak{g}^*$. Thus the co-adjoint orbits are symplectic submanifolds of $\mathfrak{g}^*$. If one traces through the above diffeomorphisms, the symplectic form on $G\cdot\mu$ turns out to be given by

$$\omega_\mu(Ad^*_{g^{-1}}\mu)((ad\xi)^* Ad^*_{g^{-1}}\mu, (ad\eta)^* Ad^*_{g^{-1}}\mu) = -Ad^*_{g^{-1}}\mu\cdot[\xi,\eta],$$

which is the Kirillov-Kostant-Souriau symplectic structure. If

$H \in \mathcal{F}(T^*G)$ is left invariant, i.e. $H \circ T^*L_g = H$, its reduction H_μ to $G\cdot\mu$ is given by $H_\mu(Ad^*_{g^{-1}}\mu) = H(T^*_g R_{g^{-1}}(\mu))$.

Assume now $\mathfrak{g} = \mathfrak{h} \oplus \mathfrak{k}$, $\mathfrak{g}$ is the Lie algebra of G, $\mathfrak{h}$ the Lie algebra of H. Take in theorem 1.2 $P = T^*G$, $R = T^*H$. Assume in addition that $\mathfrak{k}$ is a Lie subalgebra with underlying group K. Then $K \pitchfork H$ and choose local coordinates $(q^1, \ldots, q^n, p_1, \ldots, p_n)$ in T^*G such that H is given by $q^{k+1} = \ldots = q^n = 0$. For $\xi'' \in \mathfrak{k}$, $\hat{J}(\xi'') = i_{\xi''_K}\theta$, where θ is the canonical 1-form on the cotangent bundle. Thus $\hat{J}(\xi'') = \sum_{i=k+1}^{n} X^i(q^{k+1}, \ldots, q^n)p_i$. Then clearly $\hat{J}(\xi'')|T^*H \equiv 0$ and $\{\hat{J}(\xi''), \hat{J}(\eta'')\}(q^1, \ldots q^k, 0, \ldots, 0, p_1, \ldots, p_k, 0, \ldots, 0) = 0$ and we showed $\{\hat{J}(\xi'')|T^*H, \hat{J}*\eta'')|T^*H\} = \{\hat{J}(\xi''), \hat{J}(\eta'')\}|T^*H$ for all $\xi'', \eta'' \in \mathfrak{k}$. We conclude the following:

<u>Theorem 1.3. (Kostant, Symes).</u> Let $\mathfrak{g} = \mathfrak{k} \oplus \mathfrak{h}$, $\mathfrak{k}$, $\mathfrak{h}$ Lie subalgebras. Then if $f, g: \mathfrak{g}^* \to \mathbb{R}$ are Ad^*-invariant, $\{f|H\cdot\nu\ g|H\cdot\nu\} \equiv 0$ for any $\nu \in \mathfrak{h}^*$.

§ 3. Involution Theorems on Adjoint and Co-Adjoint Orbits

This section is devoted to the Kostant-Symes theorem and its consequences. We shall give here another proof of this theorem and formulate it in terms agreeing with Hamiltonian structures which will be discussed in the next section; also explicit formulas for the Hamiltonian vector fields will be given. If G is a Lie group, $\mathfrak{g}$ its Lie algebra and $\mu \in \mathfrak{g}^*$, recall that the Kirillov-Kostant-Souriau symplectic structure on the co-adjoint orbit $G\cdot\mu$ is given by

$$(3.1) \qquad \omega_\mu(\beta)((ad\xi_1)^*\beta, (ad\xi_2)^*\beta) = -\beta\cdot[\xi_1, \xi_2],$$

where $\xi_1, \xi_2 \in \mathfrak{g}$, $\beta = Ad^*_{g^{-1}}\mu \in G\cdot\mu$ for some $g \in G$. This formula uses the fact that the tangent space at $\beta \in G\cdot\mu$ to the co-adjoint orbit $G\cdot\mu$ is given by

$$(3.2) \qquad T_\beta(G\cdot\mu) = \{(\mathrm{ad}\xi)^*\beta \mid \xi \in \mathfrak{g}\}.$$

If $H \in \mathfrak{F}(T^*G)$ is a Hamiltonian such that $H\circ T^*L_g = H$ for all $g \in G$, it induces a Hamiltonian $H|G\cdot\mu \in \mathfrak{F}(G\cdot\mu)$. Denote in all what follows by $\varepsilon_{\mathfrak{g}} : \mathfrak{g} \to \mathfrak{g}^{**}$ the canonical isomorphism between $\mathfrak{g}$ and its bidual $\mathfrak{g}^{**}$, i.e. $\varepsilon_{\mathfrak{g}}(\xi)\cdot\alpha = \alpha(\xi)$, for $\xi \in \mathfrak{g}$, $\alpha \in \mathfrak{g}^*$. Define $\xi(\bar\mu) \in \mathfrak{g}$ by $\varepsilon_{\mathfrak{g}}(\xi(\bar\mu)) = dH(\bar\mu)$. Then

$$(3.3) \qquad X_{H|G\cdot\mu}(\bar\mu) = (\mathrm{ad}\xi(\bar\mu))^*\bar\mu, \quad \bar\mu = \mathrm{Ad}^*_{g^{-1}}\mu$$

To prove this formula, let $(\mathrm{ad}\eta)^*\bar\mu \in T_{\bar\mu}(G\cdot\mu)$ be arbitrary. We have

$$\begin{aligned}(i_{(\mathrm{ad}\xi(\bar\mu))^*\bar\mu}\omega_\mu)(\bar\mu)((\mathrm{ad}\eta)^*\bar\mu) &= \omega_\mu(\bar\mu)((\mathrm{ad}\xi(\bar\mu))^*\bar\mu, (\mathrm{ad}\eta)^*\bar\mu)\\ &= \bar\mu\cdot[\eta, \xi(\bar\mu)]\\ &= (\mathrm{ad}\eta)^*\bar\mu\cdot\xi(\bar\mu)\\ &= \varepsilon_{\mathfrak{g}}(\xi(\bar\mu))\cdot(\mathrm{ad}\eta)^*\bar\mu\\ &= dH(\bar\mu)\cdot(\mathrm{ad}\eta)^*\bar\mu\end{aligned}$$

and hence (3.3) holds.

Assume now that $\mathfrak{g}$ has a non-degenerate, symmetric, bilinear, bi-invariant form $\kappa: \mathfrak{g}\times\mathfrak{g} \to \mathbb{R}$; bi-invariant means $\kappa(\mathrm{Ad}_g\xi, \mathrm{Ad}_g\eta) = \kappa(\xi,\eta)$, for all $g \in G$. Then $\mathrm{ad}\xi$ is skew with respect to κ and κ induces a diffeomorphism ϕ between the adjoint orbit $G\cdot\xi$ and the co-adjoint orbit $G\cdot\kappa(\xi,\cdot)$; $\phi(\xi) = \kappa(\xi,\cdot)$. Thus $G\cdot\xi$ becomes a symplectic manifold with symplectic form

$$(3.4) \qquad \omega_\xi(\mathrm{Ad}_g\xi)((\mathrm{ad}\eta)(\mathrm{Ad}_g\xi), (\mathrm{ad}\zeta)(\mathrm{Ad}_g\xi))$$

$$= -\kappa([\eta,\zeta], \mathrm{Ad}_g\xi) = \kappa(\zeta, [\eta, \mathrm{Ad}_g\xi]) = -\kappa(\eta, [\zeta, \mathrm{Ad}_g\xi]).$$

We used here the fact that the tangent space at $\bar\xi = \mathrm{Ad}_g\xi$ to $G\cdot\xi$ is given by

$$(3.5) \qquad T_{\bar\xi}(G\cdot\xi) = \{(\mathrm{ad}\Xi)\bar\xi \mid \Xi \in \mathfrak{g}\}.$$

If $E \in \mathfrak{F}(TG)$ is a left invariant energy function, i.e. $E\circ TL_g = E$

for all $g \in G$, it induces a Hamiltonian $E|G\cdot\xi \in \mathcal{F}(G\cdot\xi)$ which equals $\phi^*(H|G\cdot\kappa(\xi,\cdot))$, where H is the left invariant function on T^*G given by $H(\phi(\xi)) = E(\xi)$. Thus since $\phi|G\cdot\xi$ is a symplectic diffeomorphism onto $G\cdot\kappa(\xi,\cdot)$ we have $X_{E|G\cdot\xi} = \phi^*(X_{H|G\cdot\phi(\xi)})$. Hence, if $Ad_g\xi \in G\cdot\xi$,

$$\begin{aligned} X_{E|G\cdot\xi}(Ad_g\xi) &= \phi^*(X_{H|G\cdot\phi(\xi)})(Ad_g\xi) \\ &= \phi^{-1}(X_{H|G\cdot\phi(\xi)}(Ad^*_{g^{-1}}\phi(\xi))) \\ &= \phi^{-1}\{[ad\eta(Ad^*_{g^{-1}}\phi(\xi))]^*Ad^*_{g^{-1}}\phi(\xi)\}, \end{aligned}$$

where $\eta = \eta(\phi(\zeta))$ is defined by $\varepsilon_{\mathfrak{g}}(\eta(\phi(\zeta)) = dH(\phi(\zeta))$. We determine now $\eta(\phi(\zeta))$. On one hand for any $\zeta' \in \mathfrak{g}$ $dH(\phi(\zeta))\cdot\phi(\zeta') =$ $= \phi(\zeta')\cdot\eta(\phi(\zeta)) = \kappa(\eta(\phi(\zeta)), \zeta')$ and on the other hand -- using $E = H\circ\phi$ -- we have $dE(\zeta)\cdot\zeta' = \kappa((\text{grad } E)(\zeta), \zeta') = dH(\phi(\zeta))\cdot\phi(\zeta')$. Thus $\eta(\phi(\zeta)) = (\text{grad } E)(\zeta)$, and using again $Ad^*_{g^{-1}}\phi(\xi) = \phi(Ad_g\xi)$ as well as the skew symmetry of $ad\zeta$ we get:

$$\begin{aligned} X_{E|G\cdot\xi}(Ad_g\xi) &= \phi^{-1}\{[ad((\text{grad } E)\cdot Ad_g\xi)]^*\cdot\phi(Ad_g\xi)\} \\ &= \phi^{-1}\{\phi(Ad_g\xi)\circ ad((\text{grad } E)\cdot Ad_g\xi)\} \\ &= \phi^{-1}\kappa(ad((\text{grad } E)\cdot Ad_g\xi)\cdot, Ad_g\xi) \\ &= -\phi^{-1}(\cdot, [(\text{grad } E)(Ad_g\xi), Ad_g\xi]) \\ &= -[(\text{grad } E)(Ad_g\xi), Ad_g\xi] \end{aligned}$$

We proved hence

(3.6). $X_{E|G\cdot\xi}(Ad_g\xi) = -[(\text{grad } E)(Ad_g\xi), Ad_g\xi]$,

i.e. <u>any Hamiltonian system on an adjoint orbit is given by a Lax equation</u>.

Let $\mathfrak{g} = \mathfrak{k} \oplus \mathfrak{n}$ be a Lie algebra, $\mathfrak{k}$ a vector sub-space and $\mathfrak{n}$ a Lie subalgebra of $\mathfrak{g}$. Denote by $i_{\mathfrak{n}} : \mathfrak{n} \hookrightarrow \mathfrak{g}$, $i_{\mathfrak{k}} : \mathfrak{k} \hookrightarrow \mathfrak{g}$, $\Pi_{\mathfrak{n}} : \mathfrak{g} \to \mathfrak{n}$, $\Pi_{\mathfrak{k}} : \mathfrak{g} \to \mathfrak{k}$ the canonical inclusions and projections.

$\mathfrak{g}^* \cong \mathfrak{h}^* \oplus \mathfrak{n}^*$, the isomorphism being given by $\mu \mapsto (\mu|\mathfrak{h}, \mu|\mathfrak{n})$ with inverse $(\chi,\nu) \mapsto i_{\mathfrak{h}^*}(\chi) + i_{\mathfrak{n}^*}(\nu)$. Denote by $\Pi_{\mathfrak{n}^*}(\mu) = \mu|\mathfrak{n}$, $\Pi_{\mathfrak{h}^*}(\mu) = \mu|\mathfrak{h}$ the canonical projections of $\mathfrak{g}^*$ onto $\mathfrak{n}^*$ and $\mathfrak{h}^*$ respectively.

Assume that G and N are the underlying Lie groups for $\mathfrak{g}$ and $\mathfrak{n}$ respectively. Then the adjoint actions Ad^N, Ad of N and G are related by

(3.7) $$Ad^N_n(\xi) = Ad_n i_{\mathfrak{n}}(\xi), \ n \in N, \ \xi \in \mathfrak{n}$$

and thus, since $i^*_{\mathfrak{n}} = \Pi_{\mathfrak{n}^*}$, the co-adjoint actions are related by

(3.8) $$Ad^{N^*}_{n^{-1}}\nu = \Pi_{\mathfrak{n}^*}(Ad_{n^{-1}}\nu), \ n \in N, \ \nu \in \mathfrak{n}^*$$

The infinitesimal versions of (3.7) and (3.8) are

(3.9) $$ad^{\mathfrak{n}}\xi = ad\xi \circ i_{\mathfrak{n}}, \quad \xi \in \mathfrak{n}$$

(3.10) $$(ad^{\mathfrak{n}}\xi)^* = \Pi_{\mathfrak{n}^*}(ad\xi)^*, \quad \xi \in \mathfrak{n}$$

If $f: \mathfrak{g}^* \to \mathbb{R}$ is a smooth map let $\xi(\mu)$ denote the vector represenative of its differential $df(\mu)$, i.e. $df(\mu) = \varepsilon_{\mathfrak{g}}(\xi(\mu))$. Let $d^{\mathfrak{n}^*}$, $d^{\mathfrak{h}^*}$ denote the partial derivatives with respect to $\mathfrak{n}^*$ and $\mathfrak{h}^*$ respectively. It is easy to see that

(3.11) $$d^{\mathfrak{h}^*}f(\chi) = \varepsilon_{\mathfrak{h}}(\Pi_{\mathfrak{h}}\xi(\chi)), \ \chi \in \mathfrak{h}^*$$
$$d^{\mathfrak{n}^*}f(\nu) = \varepsilon_{\mathfrak{n}}(\Pi_{\mathfrak{n}}\xi(\nu)), \ \nu \in \mathfrak{n}^*$$

In the context above consider $f|\mathfrak{n}^* \in \mathcal{F}(\mathfrak{n}^*)$, $\nu \in \mathfrak{n}^*$ and the co-adjoint orbit of N in $\mathfrak{n}^*$ through ν. Then $f|N\cdot\nu$ induces a Hamiltonian vector field, equal by (3.3) to

(3.12) $$X_{f|N\cdot\nu}(\bar{\nu}) = (\Pi_{\mathfrak{n}^*} \circ ad(\Pi_{\mathfrak{n}}\xi(\bar{\nu}))^*)\cdot\bar{\nu}, \quad \bar{\nu} \in N\cdot\nu \subset \mathfrak{n}^*$$

Thus the Poisson bracket of $f|N\cdot\nu$, $g|N\cdot\nu$ is given by

(3.13) $$\{f|N\cdot\nu, g|N\cdot\nu\}(\bar{\nu}) = -\bar{\nu}[\Pi_{\mathfrak{n}}\xi(\bar{\nu}), \Pi_{\mathfrak{n}}(\bar{\nu})], \ \bar{\nu} \in N\cdot\nu$$

Case 1. in theorem below is the Kostant-Symes theorem.

Theorem 3.1. Let $\mathfrak{g} = \mathfrak{k} \oplus \mathfrak{n}$, $\mathfrak{g}, \mathfrak{n}$ Lie algebras, $\mathfrak{k}$ a vector subspace of $\mathfrak{g}$. Assume that either:

1.) $\mathfrak{k}$ is a subalgebra of $\mathfrak{g}$, or

2.) $[\mathfrak{k}, \mathfrak{n}] \subseteq \mathfrak{k}$.

Then if $f,g: \mathfrak{g}^* \to \mathbb{R}$ are smooth maps satisfying $(\mathrm{ad}\,\xi(\mu))^*\mu = 0$, $(\mathrm{ad}\eta(\mu))^*\mu = 0$ for all $\mu \in \mathfrak{g}$, where $df(\mu) = \varepsilon_{\mathfrak{g}}(\xi(\mu))$, $dg(\mu) = \varepsilon_{\mathfrak{g}}(\eta(\mu))$, then $\{f|\mathfrak{n}^*, g|\mathfrak{n}^*\} = 0$ in the Poisson bracket of $\mathfrak{n}^*$ given by the foliation in its co-adjoint orbits. The expression of the Hamiltonian vector field for such an ad^*-invariant f is:

$$X_{f|\mathfrak{n}^*}(\bar{\mu}) = \begin{cases} -\mathrm{ad}(\Pi_{\mathfrak{k}}\,\xi(\bar{\nu}))^*\cdot\bar{\nu}, & \text{in case 1.} \\ \mathrm{ad}(\Pi_{\mathfrak{n}}\,\xi(\bar{\nu}))\cdot\bar{\nu}, & \text{in case 2.} \end{cases}$$

Remarks. 1.) On purpose, the theorem was formulated such that no Lie groups appear in the statement. The reason for this will become clear in §4.

2.) $(\mathrm{ad}\xi(\mu))^*\mu = 0$ for all $\mu \in \mathfrak{g}^*$ is the infinitesimal version of Ad^*-invariance as an easy computation shows.

3.) No semisimplicity of the Lie algebras involved is required.

Proof. The hypothesis implies

$$\mathrm{ad}(\Pi_{\mathfrak{n}}\,\xi(\mu))^*\cdot\mu = -\mathrm{ad}(\Pi_{\mathfrak{k}}\,\xi(\mu))^*\cdot\mu,$$

for all $\mu \in \mathfrak{g}^*$ and similarly for $\eta(\mu)$. Using (3.13) we get

$$\begin{aligned}
\{f,g\}(\bar{\nu}) &= -\bar{\nu}\cdot[\Pi_{\mathfrak{n}}\,\xi(\bar{\nu}), \Pi_{\mathfrak{n}}\,\eta(\bar{\nu})] \\
&= -\mathrm{ad}(\Pi_{\mathfrak{n}}\xi(\bar{\nu}))^*(\bar{\nu})\cdot\Pi_{\mathfrak{n}}\,\eta(\bar{\nu}) \\
&= \mathrm{ad}(\Pi_{\mathfrak{k}}\,\xi(\bar{\nu})^*(\bar{\nu})\cdot\Pi_{\mathfrak{n}}\,\eta(\bar{\nu}) \\
&= \bar{\nu}\cdot[\Pi_{\mathfrak{k}}\,\xi(\bar{\nu}), \Pi_{\mathfrak{n}}\,\eta(\bar{\nu})]
\end{aligned}$$

If assumption 2. holds then the bracket is in $\mathfrak{k}$ and the result is zero since $\bar{\nu} \in \mathfrak{n}^*$. If assumption 1. holds we go on:

$$= -\mathrm{ad}(\Pi_{\mathfrak{n}}\,\eta(\bar{\nu}))^*(\bar{\nu})\cdot\Pi_{\mathfrak{k}}\xi(\bar{\nu})$$

$$= \mathrm{ad}(\Pi_{\mathfrak{k}}\,\eta(\bar{\nu}))^*(\bar{\nu})\cdot\Pi_{\mathfrak{k}}\,\xi(\bar{\nu})$$

$$= \bar{\nu}\cdot[\Pi_{\mathfrak{k}}\,\eta(\bar{\nu}),\ \Pi_{\mathfrak{k}}\,\xi(\bar{\nu})] = 0$$

We compute now the expressions of the Hamiltonian vector fields. If hypothesis 1. holds we show first that $(\mathrm{ad}\chi)^*\bar{\nu} \in \mathfrak{n}^*$ for any $\chi \in \mathfrak{k}$, $\bar{\nu} \in \mathfrak{n}^*$. But if $k \in \mathfrak{k}$ is arbitrary $(\mathrm{ad}\chi)^*\bar{\nu}(k) = \bar{\nu}\cdot[\chi, k] = 0$, which proves our claim. Using (3.12) and this observation we conclude $X_f(\bar{\nu}) = -(\Pi_{\mathfrak{n}^*}\circ \mathrm{ad}(\Pi_{\mathfrak{k}}\,\xi(\bar{\nu}))^*)\bar{\nu} = -\mathrm{ad}(\xi(\bar{\nu}))^*\cdot\bar{\nu}$. If assumption 2. holds, as before it is shown that $(\mathrm{ad}\eta)^*\bar{\nu} \in \mathfrak{n}^*$ for any $\eta \in \mathfrak{n}$, $\bar{\nu} \in \mathfrak{n}^*$ and again (3.12) yields $X_f(\bar{\nu}) = (\Pi_{\mathfrak{n}^*}\circ \mathrm{ad}(\Pi_{\mathfrak{n}}\,\xi(\bar{\nu})))(\bar{\nu}) = \mathrm{ad}(\Pi_{\mathfrak{n}}(\xi(\bar{\nu}))\cdot\bar{\nu}$. ■

The condition of Ad^* – invariance is sometimes too strong so we would like a weaker condition like Poisson commutation in $\mathfrak{g}^*$ only. This can easily be achieved at the expense of the generality of the splitting. The Lie algebra version of the next theorem (see theorem 3.4 below) was used in Ratiu [15] to prove the involution of the Manakov integrals for the free n-dimensional rigid body motion.

<u>Theorem 3.2</u>. Assume $\mathfrak{g} = \mathfrak{k} \oplus \mathfrak{n}$ and $[\mathfrak{k}, \mathfrak{n}] \subseteq \mathfrak{k}$. If f,g are smooth Poisson commuting functions on $\mathfrak{g}^*$ and either

1.) $\mathfrak{k}$ is a Lie subalgebra, or

2.) $\Pi_{\mathfrak{n}}[\Pi_{\mathfrak{k}}\,\xi(\bar{\nu}), \Pi_{\mathfrak{k}}\,\eta(\bar{\nu})] = 0$, for all $\bar{\nu} \in \mathfrak{n}^*$,

then $\{f|\mathfrak{n}^*, g|\mathfrak{n}^*\} = 0$ in the Poisson bracket of $\mathfrak{n}^*$.

<u>Proof</u>. Since f,g Poisson commute in $\mathfrak{g}^*$, for any $\mu \in \mathfrak{g}^*$ $\mu\cdot[\xi(\mu), \eta(\mu)] = 0$. Let $\nu \in \mathfrak{n}^*$ be arbitrary; we have hence

$$0 = \nu\cdot[\xi(\nu), \eta(\nu)] = \nu\cdot[\Pi_{\mathfrak{n}}\xi(\nu) + \Pi_{\mathfrak{k}}\eta(\nu),\ \Pi_{\mathfrak{n}}\,\xi(\nu) + \Pi_{\mathfrak{k}}\,\eta(\nu)]$$
$$= -\{f|\mathfrak{n}^*,\ g|\mathfrak{n}^*\}(\nu) + \nu\cdot[\Pi_{\mathfrak{k}}\,\xi(\nu),\ \Pi_{\mathfrak{k}}\,\eta(\nu)]$$

since by hypothesis the second and third term in the expansion of the bracket are zero. Under either of the assumptions 1., 2., the last term vanishes too. ■

We turn now to the Lie algebraic formulations of the same results. It will be assumed throughout that $\mathfrak{g}$ has a bilinear, symmetric, non-degenerate, bi-invariant 2-form κ. Then $\mathfrak{g} \cong \mathfrak{n}^{\perp} \oplus \mathfrak{k}^{\perp}$ and $\mathfrak{n}^{\perp} \cong \mathfrak{k}^{*}$, $\mathfrak{k}^{\perp} \cong \mathfrak{n}^{*}$, all isomorphisms being induced by $\phi: \mathfrak{g} \to \mathfrak{g}^{*}$, $\phi(\xi) = \kappa(\xi, \cdot)$. If N is the underlying Lie group of $\mathfrak{n}$, then the co-adjoint action of N on $\mathfrak{n}^{*}$ is isomorphically equivariant to the following action of N on $\mathfrak{n}$:

$$(3.14) \qquad (n,\xi) \mapsto \Pi_{\mathfrak{k}^{\perp}} \mathrm{Ad}_n \xi, \quad n \in N, \ \xi \in \mathfrak{k}^{\perp}.$$

Thus the orbit $N\cdot\xi$ and its tangent space at $\overline{\xi} \in N\cdot\xi$ are given by

$$(3.15) \qquad N\cdot\xi = \{\Pi_{\mathfrak{k}^{\perp}}(\mathrm{Ad}_n \xi) \mid n \in N\} \subseteq \mathfrak{k}^{\perp}, \ \xi \in \mathfrak{k}^{\perp}$$

$$T_{\overline{\xi}}(N\cdot\xi) = \{\Pi_{\mathfrak{k}^{\perp}} [\overline{\xi}, \eta] \mid \eta \in \mathfrak{n}\} \subseteq \mathfrak{k}^{\perp}.$$

Thus, from (3.4) it follows that the symplectic form on $N\cdot\xi$ is given by

$$(3.16) \qquad \omega_{\xi}(\overline{\xi})(\Pi_{\mathfrak{k}^{\perp}}[\eta,\overline{\xi}], \Pi_{\mathfrak{k}^{\perp}}[\zeta,\overline{\xi}]) = \kappa([\zeta,\eta],\overline{\xi}), \ \overline{\xi} \in N\cdot\xi.$$

Using the pull-back by ϕ, formulas (3.12) and (3.13) become

$$(3.17) \qquad X_{f|N\cdot\xi}(\overline{\xi}) = -\Pi_{\mathfrak{k}^{\perp}}[\Pi_{\mathfrak{n}}(\mathrm{grad}\, f)(\overline{\xi}), \overline{\xi}], \ \overline{\xi} \in N\cdot\xi \subseteq \mathfrak{k}^{\perp}$$

$$(3.18) \qquad \{f|N\cdot\xi, g|N\cdot\xi\}(\overline{\xi}) = -\kappa([\Pi_{\mathfrak{n}}(\mathrm{grad}\, f)(\overline{\xi}), \Pi_{\mathfrak{n}}(\mathrm{grad}\, g)(\overline{\xi})], \overline{\xi}).$$

The following is obtained either by a pull-back of theorem 3.1 or by an identical proof.

Theorem 3.3. Let $\mathfrak{g} = \mathfrak{k} \oplus \mathfrak{n}$, $\mathfrak{g}, \mathfrak{n}$ Lie algebras, $\mathfrak{k}$ a vector subspace. Assume that $\mathfrak{g}$ has a bilinear, symmetric, non-degenerate, bi-invariant 2-form κ. If either:

1) $\mathfrak{k}$ is a subalgebra of $\mathfrak{g}$, or
2) $[\mathfrak{k}, \mathfrak{n}] \subseteq \mathfrak{k}$,

then any two smooth maps $f,g: \mathfrak{g} \to \mathbb{R}$ satisfying $[\mathrm{grad}\, \varphi(\xi), \xi] = 0$ for all ξ Poisson commute in the bracket of $\mathfrak{k}^{\perp}$. For such an ad-invariant f the expression of the Hamiltonian vector field is

$$X_{f|\mathfrak{k}^\perp}(\xi) = \begin{cases} [\Pi_{\mathfrak{k}}(\text{grad } f)(\xi), \xi], & \text{in case 1.} \\ -[\Pi_{\mathfrak{n}}(\text{grad } f)(\xi), \xi], & \text{in case 2.} \end{cases}$$

The analog of theorem 3.2 is the following.

Theorem 3.4. Assume $\mathfrak{g} = \mathfrak{k} \oplus \mathfrak{n}$, $[\mathfrak{k},\mathfrak{n}] \subseteq \mathfrak{k}$ and $\mathfrak{g}$ carrying a symmetric, bilinear, non-degenerate, bi-invariant 2-form. If f,g are smooth Poisson commuting functions on $\mathfrak{g}$ and either

1.) $\mathfrak{k}$ is a Lie subalgebra, or

2.) $\Pi_{\mathfrak{n}}[\Pi_{\mathfrak{k}}(\text{grad } f)(\eta), \Pi_{\mathfrak{k}}(\text{grad } g)(\eta)] = 0$, for all $\eta \in \mathfrak{k}^\perp$, then $\{f|\mathfrak{k}^\perp, g|\mathfrak{k}^\perp\} = 0$ in the Poisson bracket of $\mathfrak{k}^\perp$.

Remark. A natural candidate for splittings in case 2. above is the Cartan decomposition with respect to a fixed Cartan subalgebra and Killing form. There $\mathfrak{k}^\perp = \mathfrak{n}$, $\mathfrak{n}^\perp = \mathfrak{k}$ and the Hamiltonian vector field given by (3.17) simplifies again to a Lax equation

$$X_{f|\mathfrak{n}}(\eta) = -[\Pi_{\mathfrak{n}}(\text{grad } f)(\eta), \eta], \quad \text{for } \eta \in \mathfrak{n}. \tag{3.19}$$

This formula was the key in applying this involution theorem to the free n-dimensional rigid body motion (see Ratiu [15]).

Example 3.5. The standard example for Theorem 3.3 is the Toda lattice. (see Adler [2], Symes [17]). There one has after a change of variables a Hamiltonian system in the form $\dot{L} = [A,L]$ where

$$L = \begin{pmatrix} b_1 & a_1 & \cdots & 0 & 0 \\ a_1 & b_1 & \cdots & 0 & 0 \\ - & - & - & - & - \\ 0 & 0 & \cdots & a_{n-1} & b_n \end{pmatrix}, \quad A = \begin{pmatrix} 0 & a_1 & \cdots & 0 & 0 \\ -a_1 & 0 & \cdots & 0 & 0 \\ - & - & - & - & - \\ 0 & 0 & & -a_{n-1} & 0 \end{pmatrix} \qquad a_i > 0$$

If one chooses N = invertible lower triangular matrices with all diagonal elements > 0, $\mathfrak{n}$ = lower triangular matrices, $\mathfrak{k}$ = so (n),

$\kappa(A,B) = \mathrm{Tr}(AB)$, then $\mathfrak{n}^{\perp}$ = strictly lower triangular matrices, $\mathfrak{k}^{\perp}$ = symmetric matrices and both $\mathfrak{k}$, $\mathfrak{n}$ are Lie subalgebras of $\mathfrak{g} = s\ell(n) = \mathfrak{k} \oplus \mathfrak{n}$. Take $\xi = \begin{pmatrix} 0 & 1 & \dots & 0 & 0 \\ 1 & 0 & \dots & 0 & 0 \\ - & - & - & - & - \\ 0 & 0 & \dots & 1 & 0 \end{pmatrix}$ and an easy computation shows that $N \cdot \xi$ are all matrices L with symplectic form $\omega = \sum_{j=1}^{n-1} db_j \wedge \left(\sum_{i=j}^{n-1} \frac{da_i}{a_i} \right)$. The functions $I_k = \frac{1}{k} \mathrm{Tr} L^k$, $k = 2, \dots, n$ are all Ad-invariant and hence Poisson commute; the Hamiltonian is $H = I_2$. Their independence follows from the Chevalley theorem on invariants; for a full treatment of non-periodic Toda lattices see Kostant [8].

§4. Hamiltonian Structures and Lenard Relations

Let M be a smooth manifold and $\mathcal{F}(M)$, $\mathcal{X}(M)$ the algebra of smooth functions, respectively the Lie algebra of smooth vector fields on M. M is said to be endowed with a Hamiltonian structure (Kupershmidt, Manin [10], Manin [11]), if the following data are given:

1) a bracket $\{,\}$ on $\mathcal{F}(M)$ making $\mathcal{F}(M)$ into a Lie algebra,

2) a Lie algebra anti-homomorphism $X : (\mathcal{F}(M)), \{,\}) \to (\mathcal{X}(M), [,])$ given by $X_f(g) = -\{f,g\}$.

Condition 2. above reads $X_{\{f,g\}} = -[X_f, X_g]$. X_f will be called the Hamiltonian vector field of the Hamiltonian f; $\{,\}$ is called a Poisson bracket.

Let $(M,\{,\},X)$ and $(N,\{,\}', Y)$ be two Hamiltonian structures on the manifolds M and N respectively. A diffeomorphism $F: M \to N$ is said to define an isomorphism of Hamiltonian structures, if $F^*\{f,g\}' = \{F^*f, F^*g\}$, $F^*Y_f = X_{F^*f}$, for any smooth maps $f,g \in \mathcal{F}(N)$.

Let $(M,\{,\}, X)$, $(M,\{,\}', X')$ be two different Hamiltonian structures on the same manifold M and Φ a family of Hamiltonians on M. Identities between the Hamiltonian vector fields in the two structures defined by Hamiltonians in Φ are called Lenard relations in Φ. Such relations were first discovered by Lenard for the Korteweg-DeVries equation and appeared as a feature in all well-studied completely integrable systems like Toda lattice, Gelfand-Dikii systems (Adler [2], Gelfand, Dikii [5]), long wave approximation equations (Manin [11], Kupershmidt [9]), free N-dimensional rigid body (Mishchenko, Fomenko [13], Ratiu [15]), a generalized Lagrange top (Ratiu [16]). Lenard relations essentially prove involution in two Hamiltonian structures at the same time. Examples will be given below.

Let $(M,\{,\}, X)$ be a Hamiltonian structure. We investigate a little to what extent this manifold is symplectic.

Lemma 4.1. For $f,g \in \mathcal{F}(M)$, $\{f,g\}(p)$ and $X_f(p)$ only depend on df_p and dg_p.

Proof. Let (V,ϕ) be a chart at p, $\phi(p) = (p^1, \ldots, p^n)$, $\phi(u) = (x^1, \ldots, x^n)$, $u \in V$. Applying the mean value theorem to $f\circ\phi^{-1}$ in the open set $\phi(V) \subseteq \mathbb{R}^n$ at $\phi(p)$, one gets

$$f(u) = f(p) + \sum_{i=1}^{n} (x^i-p^i)\alpha_i(p,u)$$

where

$$\alpha_i(p,u) = \int_0^1 \frac{\partial(f\circ\phi^{-1})}{\partial x^i} (\phi(p) + t(\phi(u) - \phi(p)))\, dt$$

are smooth functions. Both statements in the lemma follow immediately if one proves that if $df_p = 0$, $\{f,g\}(p) = 0$ for any smooth function g. Since the Poisson bracket of a function with a constant is zero, the above local representation of f implies

$$\{f,g\}(p) = \sum_{i=1}^{n} \alpha_i(p,p)\{\phi^i,g\}(p) = 0$$

since $\alpha_i(p,p) \frac{\partial f}{\partial x^i}(p) = 0$. ■

The Hamiltonian structure defines hence a contra-variant, antisymmetric 2-tensor Λ on M given by $\Lambda(df_p, dg_p) = \{f,g\}(p)$. Λ in turn induces a map $\Lambda_{\#}(p): T_p^*M \to T_pM$ with $k(p)$-dimensional kernel. Let $\mathcal{H}(p)$ denote the span of $\{X_f(p) \mid f \in \mathcal{F}(M)\}$. Clearly $\mathcal{H}(p)$ is isomorphic to $T_p^*M/\mathrm{Ker}\Lambda_{\#}(p)$ and has hence dimension $d(p) = n-k(p)$. Let $X_{f_1}(p), \ldots, X_{f_{d(p)}}$ be a basis of $\mathcal{H}(p)$. Since the vector fields X_{f_i} $i = 1, \ldots, d(p)$ are linearly independent at p they will remain so in an open neighborhood of p, i.e. $\dim \mathcal{H}(p') \geq \dim \mathcal{H}(p)$ for all p' near p. In particular, if $d = \max\{d(p) \mid p \in M\}$, the set $S = \{p \in M \mid d(p) = d\}$ is open in M. The Frobenius theorem implies then the following.

Proposition 4.2. The Hamiltonian structure $(M,\{,\}, X)$ defines a contravariant antisymmetric two tensor on M. Let k be the minimal dimension of its kernel and $n = \dim M$. The set $S = \{p \in M \mid \dim \mathcal{H}(p) = n-k\}$ is open in M, where $\mathcal{H}(p) = \mathrm{span}\{X_f(p) \mid f \in \mathcal{F}(M)\}$. Thus S is foliated by symplectic manifolds having as tangent spaces $\mathcal{H}(p)$, $p \in S$. In particular $n-k =$ even.

Assume S' is a dense set in M and is such that $\dim \mathcal{H}(p') =$ constant, for all $p' \in S'$. Then if $p \in S$ there is an open neighborhood of p intersecting S' such that $\dim \mathcal{H}(p') \geq \dim \mathcal{H}(p)$ which implies $\dim \mathcal{H}(p') = \dim \mathcal{H}(p)$ i.e. $p' \in S$. Thus in concrete examples if one finds a dense set on which $\dim \mathcal{H}(p)$ is constant, this set necessarily lies in S. In general S is not dense, but it is dense in all interesting examples.

We devote the rest of this section to examples which have proved to be useful in concrete problems.

Example 4.3. Let (M,ω) be a symplectic manifold, $\{,\}$ its Poisson bracket and $f \mapsto X_f$ the associated Hamiltonian vector field. Then $(M, \{,\}, X)$ is a Hamiltonian structure. Actually more is true. If $(M, \{,\}, X)$ is a Hamiltonian structure such that $\{,\}$ is non-degenerate, i.e. $\{f,g\} = 0$ for all $g \in \mathfrak{F}(M)$ implies f = constant (M is assumed to be connected) then there is a symplectic form ω on M whose Poisson bracket and Hamiltonian vector field coincide with $\{,\}$ and X. This is a theorem of Jost [6] whose proof we now sketch. The non-degeneracy of $\{,\}$ clearly implies the non-degeneracy of the contravariant, antisymmetric 2-tensor Λ and thus it induces a bundle isomorphism $\#: T^*M \to T^{**}M = TM$ given by $\beta(\alpha_\#) = \Lambda(\alpha,\beta)$. Let its inverse be $\flat: TM \to T^*M$, $v_\flat(w) = \Lambda(v_\flat, w_\flat)$, $\alpha(v) = \Lambda(v_\flat, \alpha)$. Define $\omega \in \Omega^2(M)$ by $\omega(v,w) = \Lambda(v_\flat, w_\flat)$ and the non-degeneracy of Λ implies the non-degeneracy of ω. Denoting $Z(f,g,h) = \{f,\{g,h\}\} + \{h,\{f,g\}\} + \{g,\{h,f\}\} = 0$ one shows first that $[u,v](h) = Z(f,g,h) + \{\{f,g\},h\}$, where $u = (df)_\#$, $v = (dg)_\#$, $w = (dh)_\#$, and then, using this formula, that $d\omega(u,v,w) = -Z(f,g,h) = 0$. Thus $d\omega = 0$ and ω is hence symplectic. Let Y_f be the Hamiltonian vector field of f defined by ω. We have $-df(v) = -(i_{Y_f}\omega)(v) = \omega(v, Y_f) = \Lambda(v_\flat, (Y_f)_\flat) = (Y_f)_\flat(v)$, i.e. $-df = (Y_f)_\flat$ and hence $Y_f(g) = dg \cdot Y_f = \Lambda((Y_f)_\flat, dg) = -\Lambda(df, dg) = -\{f,g\}$, i.e. $Y = X$. Similarly, the Poisson bracket given by $\omega(X_f, X_g) = df \cdot X_g = \Lambda((X_g)_\flat, df) = \Lambda(df, dg) = \{f,g\}$ and Jost's theorem is proved.

Example 4.4. The Kirillov-Kostant-Souriau Structure of $\mathfrak{g}^*$ is induced from the co-adjoint orbits of G on $\mathfrak{g}^*$. The Hamiltonian structure is given by formula (3.3) and the corresponding Poisson bracket, i.e.

$$\{f,g\}(\alpha) = -\alpha\cdot[\xi(\alpha), \eta(\alpha)],$$
$$X_f(\alpha) = (\mathrm{ad}\,\xi(\alpha))^* \cdot \alpha, \quad \alpha \in \mathfrak{g}^*,$$

where $df(\alpha) = \varepsilon_{\mathfrak{g}}(\xi(\alpha))$, $dg(\alpha) = \varepsilon_{\mathfrak{g}}(\eta(\alpha))$.

Let $\mathfrak{g}_\alpha = \{\xi \in \mathfrak{g} \mid (\mathrm{ad}\,\xi)^*\alpha = 0\}$. Then $\mathcal{H}(\alpha) = T_\alpha(G\cdot\alpha)$ by the very definition of this Hamiltonian structure and $\mathcal{H}(\alpha)$ is $(\dim\mathfrak{g} - \dim\mathfrak{g}_\alpha)$-dimensional. Let k be the minimal dimension of $\mathfrak{g}_\alpha$. Then the set S is foliated by $(\dim\mathfrak{g} - k)$-dimensional symplectic manifolds. A theorem of Duflo and Vergne states that $\mathfrak{g}_\alpha$ is in this case abelian.

The involution theorems as formulated in theorems 3.1 and 3.2 apply in this case. Moreover, they hold also in the context of Katz-Moody Lie algebras $\tilde{\mathfrak{g}}$, i.e. formal Laurent series $\sum_{n=-\infty}^{N} \xi_i \lambda^i$ with "coefficients" in $\mathfrak{g}$ where addition is on components and the bracket is given by

$$\left[\sum_{i=-\infty}^{N} \xi_i \lambda^i, \sum_{j=-\infty}^{M} \eta_j \lambda^j\right] = \sum_{k=-\infty}^{M+N} \left(\sum_{i+j=k} [\xi_i, \eta_j]\right)\lambda^k.$$

<u>Example 4.5. The Kirillov-Kostant-Souriau Structure of $\mathfrak{g}$</u> is induced from the adjoint orbits of G on $\mathfrak{g}$, for $\mathfrak{g}$ a semi-simple Lie algebra with Killing form κ and is given by (3.6) and its corresponding Poisson bracket, i.e.

$$\{f,g\}(\xi) = -\kappa([(\mathrm{grad}\ f)(\xi), (\mathrm{grad}\ g)(\xi)], \xi)$$
$$X_f(\xi) = -[(\mathrm{grad}\ f)(\xi), \xi].$$

If $r = \mathrm{rank}\,\mathfrak{g}$ on an open dense set of elements of ξ (the set of regular semi-simple elements in $\mathfrak{g}$), the dimension of the orbit is $(\dim\mathfrak{g} - r)$. Thus, in this example S is open and dense.

The involution Theorems 3.3 and 3.4 apply unchanged in this context even if $\mathfrak{g}$ is a Katz-Moody Lie algebra. Then one defines a 2-form κ_α on the Katz-Moody extension $\tilde{\mathfrak{g}}$ of $\mathfrak{g}$ by

$$\kappa_\alpha\left(\sum_{i=-\infty}^{M} \xi_i \lambda^i, \sum_{j=-\infty}^{N} \eta_j \lambda^i\right) = \sum_{i+j=\alpha} \kappa(\xi_i, \eta_j)$$

The choice of $\alpha = 0$ and the splitting of $\widetilde{s\ell(n)}$ in lower triangular and antisymmetric "matrices" (imagine $\widetilde{s\ell(n)}$ as an infinite periodic band matrix) gives the involution of the integrals in the periodic Toda lattices whereas the choice $\alpha = -1$ and a splitting in strictly lower ($i \leq -1$) and upper ($i \geq 0$) triangular "matrices" yields the involution of the Manakov integrals for the Dubrovin equation (and similarly for the free N-dimensional rigid body motion), but in a Hamiltonian structure different from this one of $\mathfrak{g}$, which will be given in Example 4.7 (see Adler, Van Moerbeke [3]). Lenard relations are needed to connect the two Hamiltonian structures (see Mishchenko, Fomenko [12], Ratiu [15]).

Example 4.6. The manifold is $\mathfrak{g}^*$ and $\delta \in \mathfrak{g}^*$ is fixed. Define for $f,g: \mathfrak{g}^* \to \mathbb{R}$

$$\{f,g\}_\delta(\alpha) = -\delta\cdot[\xi(\alpha), \eta(\alpha)]$$

$$X_f^\delta(\alpha) = (\mathrm{ad}\xi(\alpha))^*\cdot\delta$$

To prove that this is a Hamiltonian structure one has to do straightforward but lengthy direct computations in which the following two formulas are needed:

$$\mu(d\xi(\alpha)\cdot\nu) = \nu(d\xi(\alpha)\cdot\mu)$$

obtained from the symmetry of the second derivative of f, and

$$d\{f,g\}(\alpha) = \varepsilon_{\mathfrak{g}}(d\xi(\alpha)\cdot(\mathrm{ad}\eta(\alpha))^*\delta - d\eta(\alpha)\cdot(\mathrm{ad}\xi(\alpha))^*\delta).$$

Example 4.7. Let $\mathfrak{g}$ be a semi-simple Lie algebra with Killing form κ and $\varepsilon \in \mathfrak{g}$ a fixed element. Define

$$\{f,g\}_\varepsilon(\xi) = -\kappa([(\mathrm{grad}\ f)(\xi), (\mathrm{grad}\ g)(\xi)],\varepsilon)$$

$$X_f^\varepsilon(\xi) = -[(\mathrm{grad}\ f)(\xi), \varepsilon].$$

To see that this is a Hamiltonian structure one can proceed in two ways. The first consists of long direct computations in which the following two formulas are used repeatedly

$$\kappa(d(\text{grad } f)(\xi)\cdot\eta_1,\eta_2) = \kappa(d(\text{grad } f)(\xi)\cdot\eta_2,\eta_1)$$

$$\text{grad}\{f,g\}(\xi) = -d(\text{grad } f)(\xi)\cdot[(\text{grad } g)(\xi), \varepsilon] + $$
$$+ d(\text{grad } g)(\xi)\cdot[(\text{grad } f)(\xi), \varepsilon].$$

The second consists of the observation that this Poisson bracket and Hamiltonian vector field coincide with the ones given by the Kirillov-Kostant-Souriau structure on the invariant set $\{\xi + \varepsilon\lambda | \xi \in \mathfrak{g}\} \subset \mathcal{H}^{\perp}$, where $\tilde{\mathfrak{g}} = \mathcal{H} \oplus \mathcal{N}$, $\mathcal{H} = \{\sum_{n=0}^{N} \xi_n\lambda^n | \xi_n \in \mathfrak{g}\}$, $\mathcal{N} = \{\sum_{n=-\infty}^{-1} \xi_n\lambda^n | \xi_n \in \mathfrak{g}\}$, $\tilde{\mathfrak{g}}$ is the Katz-moody extension of $\mathfrak{g}$ and $\alpha = -1$ in the "Killing form" of $\tilde{\mathfrak{g}}$; remark $\mathcal{H}^{\perp} = \mathcal{H}$, $\mathcal{N}^{\perp} = \mathcal{N}$. The tangent space to the N-orbit through $\xi + \varepsilon\lambda$, where N is the "Lie group" underlying N obtained by exponentiation, is easily seen to be $\{[\eta,\varepsilon] | \eta \in \mathfrak{g}\}$. One then verifies that the diffeomorphism $\xi + \varepsilon\lambda \mapsto \xi$ gives an isomorphism of Hamiltonian structures.

If ε is regular semisimple, the set S is open and dense and is foliated by $(\dim\mathfrak{g} - \operatorname{rank}\mathfrak{g})$-dimensional symplectic submanifolds.

To see how Lenard relations are found and involution is proved we shall study the Dubrovin equation in $s\ell(n)$, isolate a class of Hamiltonians and connect their Hamiltonian vector fields in the structures of examples 4.5 and 4.7. (Mishchenko, Fomenko [13], Ratiu [15]).

<u>Example 4.8.</u> <u>The Dubrovin Equation</u> is

(4.1) $$[a,V]^{\cdot} = [[a,V], [b,V]],$$

where a, b are constant diagonal matrices in $s\ell(n)$ and V is a matrix with all entries on the diagonal equal to zero. Assume that all entries of a, b are distinct. Let $\mathfrak{d}$ denote the diagonal matrices in $s\ell(n)$ and $\mathfrak{a}$ the matrices with zero on the diagonal ; then $s\ell(n) = \mathfrak{d} \oplus \mathfrak{a}$. Given a, any matrix in $\mathfrak{a}$ can be written in the form $[a,V]$ and thus (4.1) is a differential equation in $\mathfrak{a}$

We shall add to it a trivial differential equation on $\mathfrak{a}$ and the resulting equation will be a Hamiltonian system in $s\ell(n)$. Define $L: s\ell(n) \to s\ell(n)$ by $(L|\mathfrak{A})(A) = \left(\frac{b_i - b_j}{a_i - a_j} a_{ij}\right)$, where $a = \mathrm{diag}(a_1, \ldots, a_n)$, $b = \mathrm{diag}(b_1, \ldots, b_n)$, $A = (a_{ij})$ and $L|\mathfrak{a}$ arbitrary but symmetric with respect to the Killing form κ. Clearly if $B = (b_{ij}) \in \mathfrak{A}$, $\kappa(L(A), B) = \kappa(A\, L(B)) = \sum_{i,j=1}^{n} \frac{b_i - b_j}{a_i - a_j} a_{ij} a_{ij}$, so that L is κ-symmetric on $s\ell(n)$. For $M \in s\ell(n)$ define the Hamiltonian $H(M) = \frac{1}{2}\kappa(M, L(M))$ whose gradient with respect to κ is $(\mathrm{grad}\, H)(M) = L(M)$ and Hamiltonian vector field $X_H(M) = [M, L(M)]$. Note that since $[\mathfrak{a}, \mathfrak{a}] = 0$, $[\mathfrak{a}, s\ell(n)] \subseteq \mathfrak{A}$, $[A, L(A)] \in \mathfrak{A}$, for all $A \in \mathfrak{A}$, if $M = D + A \in \mathfrak{a} \oplus \mathfrak{A}$, Hamilton's equations become

$$\dot{D} = [D, L(D)] = 0, \quad \dot{A} = [A, L(A)] + [D, L(A)] + [A, L(D)].$$

Put $A = [a, V]$, $D = 0$ to get Dubrovin's equation (4.1). We shall refer from now on to

$$(4.2) \qquad \dot{M} = [M, L(M)], \quad M \in s\ell(n)$$

as the <u>extended Dubrovin equation.</u>

The first step towards the complete integrability of (4.2) is the remark that it is equivalent to

$$(4.3) \qquad (M + a\lambda)^{\cdot} = [M + a\lambda, L(M) + b\lambda]$$

since $[a, L(M)] = [b, M]$. Thus $\frac{1}{k+1}\mathrm{Tr}(M + a\lambda)^{k+1} = f_{k+1}(M)$, $k = 1, \ldots, n-1$ are conserved on the flow of (4.3). But if $t \mapsto M(t)$ is the flow of (4.2), $t \mapsto M(t) + a\lambda$ is the flow of (4.3) and hence the coefficients of λ in $f_{k+1}(M)$ are conserved on the flow of (4.2). Let $u_{k+1,j}(M)$ be the coefficient of λ^j in $f_{k+1}(M)$ and remark that $u_{k+1,0}(M) = \frac{1}{k+1}\mathrm{Tr} M^{k+1}$, $u_{k+1,k+1}(M) = \frac{1}{k+1}\mathrm{Tr}\, a^{k+1}$ = constant lead to vanishing Hamiltonian vector fields. Thus the total number of conserved quantities not leading to identically zero

Hamiltonian vector fields is $n(n-1)/2$ which is $(\dim s\ell(n) - \operatorname{rank} s\ell(n))/2$. The family of Hamiltonians which we want to prove are in involution is $\Phi = \{u_{k+1,j} | k = 1, \ldots, n-1,\ j = 1, \ldots, k\}$. We will do this by finding Lenard relations between the Kirillov-Kostant-Souriau structure on $s\ell(n)$ and $(s\ell(n), \{,\}_a, X^a)$.

Let A_{kj} be the coefficient of λ^j in the development of $(M + \lambda a)^k$; it is a symmetric polynomial in M and its powers. An easy computation shows that

$$(\operatorname{grad} u_{k+1,j})(M) = A_{kj} - (\tfrac{1}{n} \operatorname{Tr} A_{kj})\mathrm{Id}.$$

The trivial relation $[M + \lambda a, (M + \lambda a)^k] = 0$, implies that all coefficients of λ in the expansion of this bracket vanish, i.e.

$$[M, A_{kj}] = [A_{k,j-1}, a], \quad \text{for } k = 1, \ldots, n-1,\ j=1, \ldots, k$$

These equalities coupled with the formulas of the gradients yield the <u>Lenard relations</u>.

(4.4) $$X_{u_{k+1,j}} = -X^a_{u_{k+1,j-1}}.$$

We show below that these identities prove the involution of the Hamiltonian $u_{k+1,j}$ in both Hamiltonian structures of $s\ell(n)$. We have

$$\begin{aligned}
\{u_{k+1,j}, u_{\ell+1,i}\}(M) &= \kappa((\operatorname{grad} u_{k+1,j})(M), X_{u_{\ell+1,i}}(M)) \\
&= -\kappa((\operatorname{grad} u_{k+1,j})(M), X^a_{\ell+1,i-1}(M)) \\
&= -\{u_{k+1,j}, u_{\ell+1,i-1}\}_a(M) \\
&= -\kappa([A_{k,j}, a], A_{\ell,i-1}) \\
&= -\kappa([M, A_{k,j+1}], A_{\ell,i-1}) \\
&= \{u_{k+1,j+1}, u_{\ell+1,i-1}\}(M)
\end{aligned}$$

and hence

(4.5) $$\{u_{k+1,j}, u_{\ell+1,i}\} = -\{u_{k+1,j}, u_{\ell+1,i-1}\}_a$$

which says that involution in the two Hamiltonian structures is equivalent, and

(4.6) $\{u_{k+1,j}, u_{\ell+1,i}\} = \{u_{k+1,j+1}, u_{\ell+1,i-1}\}$.

Applying (4.6) repeatedly we come to a stop whenever j increases to reach $k+1$, or i decreases to reach zero. In the first case $\{u_{k+1,j}, u_{\ell+1,i}\} = \{u_{k+1,k+1}, u_{\ell+1,i-(k+1)+j}\} = 0$ since $X_{u_{k+1,k+1}} = 0$ and in the second case the Poisson bracket is again zero, since $X_{u_{\ell+1,0}} = 0$.

The independence of these integrals is proved in Theorem 4.2 of Mishchenko, Fomenko [13].

<u>Example 4.9</u>. <u>The Euler-Poisson Structure on $\mathfrak{g} \times \mathfrak{g}$</u> is defined by

$$\begin{aligned}\{f,g\}(\xi,\eta) &= \kappa([\mathrm{grad}_1 f(\xi,\eta),\xi], \mathrm{grad}_2 g(\xi,\eta)) \\ &\quad - \kappa([\mathrm{grad}_1 g(\xi,\eta), \xi], \mathrm{grad}_2 f(\xi,\eta)) \\ &\quad + \kappa(\eta, [\mathrm{grad}_2 g(\xi,\eta), \mathrm{grad}_1 f(\xi,\eta)])\end{aligned}$$

$$X_f(\xi,\eta) = (-[\mathrm{grad}_2 f(\xi,\eta), \xi], -[\mathrm{grad}_1 f(\xi,\eta), \xi] + [\eta, \mathrm{grad}_2 f(\xi,\eta)])$$

Here $(\mathrm{grad}_1 f, \mathrm{grad}_2 f)$ denotes the gradient of f with respect to $\kappa \times \kappa$ on $\mathfrak{g} \times \mathfrak{g}$. The proof that these data define indeed a Hamiltonian structure consists of quite long but straightforward computations, or the observation due to B. Kupershmidt that they are the Kirillov-Kostant-Souriau structure on the semidirect product of $\mathfrak{g}$ with itself. The name of this structure is due to the fact that it models Euler-Poisson equations generalizing the ones for the rigid body motion under gravity. More precise, if $L: \mathfrak{g} \to \mathfrak{g}$ is a κ-symmetric isomorphism, define the energy function $E(\xi,\eta) = \frac{1}{2}\kappa(\eta, L(\eta)) + \kappa(\chi,\xi)$. It is easy to see that $(\mathrm{grad}_1 E)(\xi,\eta) = \chi$, $(\mathrm{grad}_2 E)(\xi,\eta) = L(\eta)$ and thus Hamilton's equation become the <u>Euler-Poisson equations</u>:

$$(4.7) \qquad \dot{\xi} = [\xi, L(\eta)], \ \dot{\eta} = [\eta, L(\eta)] + [\xi, \chi].$$

It is easily shown that the dimension of $\mathcal{H}(\xi,\eta)$ at each pair ξ,η, where ξ,η are regular semi-simple and $\alpha(\xi) \neq 0$ for each root α of $\mathfrak{g}$ is $2(\dim\mathfrak{g} - \text{rank}\,\mathfrak{g})$. Thus in this Hamiltonian structure the set S is open and dense and the generic symplectic leaf is $2(\dim\mathfrak{g} - \text{rank}\,\mathfrak{g})$-dimensional.

This Hamiltonian structure and the next turn out to be the ones modeling the generalization of the Lagrange spinning top in $so(N) \times so(N)$ as is shown in Ratiu [16].

Example 4.10. On $\mathfrak{g} \times \mathfrak{g}$ define for $\varepsilon \in \mathfrak{g}$ fixed

$$\begin{aligned}\{f,g\}_\varepsilon(\xi,\eta) = {} & \kappa([\text{grad}_2 f(\xi,\eta),\varepsilon], \text{grad}_1 g(\xi,\eta)) \\ & -\kappa([\text{grad}_2 g(\xi,\eta),\varepsilon], \text{grad}_1 f(\xi,\eta)) \\ & + \kappa(\eta, [\text{grad}_1 g(\xi,\eta), \text{grad}_2 f(\xi,\eta)]) \\ X_f^\varepsilon(\xi,\eta) = {} & ([\eta, \text{grad}_1 f(\xi,\eta)] + [\varepsilon, \text{grad}_2 f(\xi,\eta)], \\ & [\varepsilon, \text{grad}_1 f(\xi,\eta)]).\end{aligned}$$

Long direct computations show that this is a Hamiltonian structure. Choosing ε to be regular semisimple, it can be proved as in the previous example that S is open and dense and is foliated by $2(\dim\mathfrak{g} - \text{rank}\,\mathfrak{g})$- dimensional manifolds. In Ratiu [16] it is shown that this Hamiltonian structure is induced from an invariant set of a certain Katz-Moody Lie algebra with the Kirillov-Kostant-Souriau structure.

Example 4.11. The N-Dimensional Lagrange Top has the equations of motion (see Ratiu [16])

$$(4.17) \qquad \dot{\Gamma} = [\Gamma, \Omega], \ \dot{M} = [M,\Omega] + [\Gamma, \chi]$$

for $\Gamma, M, \Omega, \chi \in so(N)$, $M = \Omega J + J\Omega$, $J = \text{diag}(J_1, \ldots, J_N)$ $J_i + J_j = I_{ij} > 0$ (the $N(N-1)/2$ "moments of inertia"), χ, J fixed matrices, χ representing the "center of mass". The conditions on J and χ are : $a = J_1 = J_2, b = J_3 = \ldots = J_N$, $\chi_{12} \neq 0$, $\chi_{ij} = 0$ for all $i,j \neq 1,2$. Equations (4.7) can be put in the following form

$$(\Gamma + M\lambda + C\lambda^2)^{\cdot} = [\Gamma + M\lambda + C\lambda^2, \Omega + \chi\lambda]$$

for $C = (a+b)\chi \in so(N)$; this was first observed for $N = 3$ by Van Moerbeke and Ratiu. Thus $u_{k+1,j}$ = coefficient of λ^j in $\frac{1}{k+1}\operatorname{Tr}(\Gamma + M\lambda + C\lambda^2)^{k+1}$, k = odd, $j \neq 0,1,k+1$ are conserved quantities of (4.7). For $j = 0, 1, k+1$ the corresponding Hamiltonian vector fields areidentically zero. The number of these integrals is half the dimension of the generic symplectic leaf for N odd and bigger by N if N is even. Proceeding exactly as in example 4.8 one can show that $X_{u_{k+1,j}} = -X^C_{u_{k+1,j-2}}$, k = odd, which Lenard relations prove then involution in both Poisson brackets $\{,\}$ and $\{,\}_C$; for details see Ratiu [16].

§5. Involution Theorems on Translated Invariants

Another class of involution theorems in Hamiltonian structures can be obtained by translation of the argument of real valued functions which are natural invariants of the structure. The idea behind these theorems is due to Kostant [8], Mishchenko, Fomenko [13] and Symes [17]. As will become clear below, the method of proof is that of Theorem 3.1 with some modifications depending on each case. All theorems below can be stated in $\mathfrak{g}^*$, with no restrictions on $\mathfrak{g}$, or in $\mathfrak{g}$ assuming semisimplicity. We prefer here the version on $\mathfrak{g}$ having in mind the applications, but all statements and proofs can be "copied" to hold in the dual of a not necessarily semisimple Lie algebra.

Theorem 5.1. Let $\mathfrak{g}$ be a Lie algebra with a non-degenerate, bilinear, symmetric, bi-invariant 2-form κ. Assume $\mathfrak{g} = \mathfrak{k} \oplus \mathfrak{n}$, $\mathfrak{n}$ a subalgebra, $\mathfrak{k}$ a vector subspace of $\mathfrak{g}$. Let

$\varepsilon \in \mathfrak{g}$ be fixed. Assume that either:

1.) $\mathfrak{k}$ is a Lie subalgebra, $\varepsilon \perp [\mathfrak{k}, \mathfrak{k}]$ and A.) $\varepsilon \perp [\mathfrak{n}, \mathfrak{n}]$,
or B.) $\varepsilon \perp [\mathfrak{k}, \mathfrak{n}]$, or

2) $[\mathfrak{k}, \mathfrak{n}] \subseteq \mathfrak{k}$, $\varepsilon \perp [\mathfrak{k}, \mathfrak{n}]$.

Let $f, g: \mathfrak{g} \to \mathbb{R}$ be ad-invariant maps, i.e. $[\operatorname{grad} f(\xi), \xi] = 0$, $[\operatorname{grad} g(\xi), \xi] = 0$, for all $\xi \in \mathfrak{g}$ and denote $f_a(\xi) = f(\xi + a\varepsilon)$ $g_b(\xi) = g(\xi + b\varepsilon)$. Then in hypotheses 1.B.) or 2., $\{f_a|\mathfrak{k}^\perp, g_b|\mathfrak{k}^\perp\} = 0$, for any parameters a, b, whereas in hypothesis 1.A.) $\{f_1|\mathfrak{k}^\perp, g_1|\mathfrak{k}^\perp\} = 0$, the Poisson bracket being given by the Kirillov-Kostant-Souriau structure on $\mathfrak{k}^\perp$. The expression of the Hamiltonian vector field for an ad-invariant f is

$$X_{f_a|\mathfrak{k}^\perp}(\xi) = \begin{cases} [\Pi_{\mathfrak{k}} \operatorname{grad} f(\xi+\varepsilon), \xi+\varepsilon] + \Pi_{\mathfrak{k}^\perp}[\Pi_{\mathfrak{n}} \operatorname{grad} f(\xi+\varepsilon), \varepsilon], & \text{for } a=1 \text{ in case 1.A.} \\ [\Pi_{\mathfrak{k}} \operatorname{grad} f(\xi+a\varepsilon), \xi] + a[\operatorname{grad} f(\xi+a\varepsilon), \varepsilon], & \text{in case 1.B,} \\ -[\Pi_{\mathfrak{n}} \operatorname{grad} f(\xi+a\varepsilon), \xi], & \text{in case 2.} \end{cases}$$

<u>Proof</u>. We start with 2. since some of the computations in this case will be used later. Let $\xi \in \mathfrak{k}^\perp$, $a \neq b$. Use (3.18) to get

$$\begin{aligned} \{f_a|\mathfrak{k}^\perp, g_b|\mathfrak{k}^\perp\}(\xi) &= -\frac{1}{b-a}\kappa([\Pi_{\mathfrak{n}} \operatorname{grad} f(\xi+a\varepsilon), \Pi_{\mathfrak{n}} \operatorname{grad} g(\xi+b\varepsilon)], \\ &\qquad b(\xi+a\varepsilon) - a(\xi+b\varepsilon)) \\ &= \frac{b}{b-a}\kappa([\Pi_{\mathfrak{n}} \operatorname{grad} f(\xi+a\varepsilon), \xi+a\varepsilon], \\ &\qquad \Pi_{\mathfrak{n}} \operatorname{grad} g(\xi+b\varepsilon)) \\ &+ \frac{a}{b-a}\kappa(\Pi_{\mathfrak{n}} \operatorname{grad} f(\xi+a\varepsilon), [\Pi_{\mathfrak{n}} \operatorname{grad} g(\xi+b\varepsilon), \xi+b\varepsilon]) \\ &= \frac{-b}{b-a}\kappa([\Pi_{\mathfrak{k}} \operatorname{grad} f(\xi+a\varepsilon), \xi+a\varepsilon], \\ &\qquad \Pi_{\mathfrak{n}} \operatorname{grad} g(\xi+b\varepsilon)) - \\ &- \frac{a}{b-a}\kappa(\Pi_{\mathfrak{n}} \operatorname{grad} f(\xi+a\varepsilon), [\Pi_{\mathfrak{k}} \operatorname{grad} g(\xi+b\varepsilon), \\ &\qquad \xi+b\varepsilon]) \\ &= \frac{b}{b-a}\kappa([\Pi_{\mathfrak{k}} \operatorname{grad} f(\xi+a\varepsilon), \Pi_{\mathfrak{n}} \operatorname{grad} g(\xi+b\varepsilon)], \\ &\qquad \xi+a\varepsilon) \end{aligned}$$

$$- \frac{a}{b-a} \kappa([\Pi_{\mathfrak{n}} \text{ grad } f(\xi+a\varepsilon), \; \Pi_{\mathfrak{k}} \text{ grad } g(\xi+b\varepsilon)], \; \xi + b\varepsilon).$$

Both terms vanish since $[\mathfrak{k}, \mathfrak{n}] \subseteq \mathfrak{k}$, $\xi \in \mathfrak{k}^{\perp}$, $\varepsilon \perp [\mathfrak{k}, \mathfrak{n}]$. The case $a = b$ is obtained by a passage to limit $a \to b$ in $\{f_a | \mathfrak{k}^{\perp}, g_b | \mathfrak{k}^{\perp}\} \equiv 0$. To obtain the Hamiltonian vector field, (3.17) is used. Let $\xi \in \mathfrak{k}^{\perp}$.

$$X_{f_a | \mathfrak{k}^{\perp}}(\xi) = -\Pi_{\mathfrak{k}^{\perp}}[\Pi_{\mathfrak{n}} \text{ grad } f(\xi+a\varepsilon), \xi] = $$
$$= -[\Pi_{\mathfrak{n}} \text{ grad } f(\xi+a\varepsilon), \xi],$$

since $[\mathfrak{n}, \mathfrak{k}^{\perp}] \subseteq \mathfrak{k}^{\perp}$ under assumption 2.

Assume hypothesis 1.A. holds and let $\xi \in \mathfrak{k}^{\perp}$. By (3.18) we have

$$\{f_1 | \mathfrak{k}^{\perp}, g_1 | \mathfrak{k}^{\perp}\}(\xi) = -\kappa([\Pi_{\mathfrak{n}} \text{ grad } f(\xi+\varepsilon), \Pi_{\mathfrak{n}} \text{ grad } g(\xi+\varepsilon)], \; \xi)$$
$$= \kappa([\Pi_{\mathfrak{n}} \text{ grad } f(\xi+\varepsilon), \; \xi+\varepsilon], \; \Pi_{\mathfrak{n}} \text{ grad } g(\xi+\varepsilon))$$
$$= -\kappa([\Pi_{\mathfrak{k}} \text{ grad } f(\xi+\varepsilon), \; \xi+\varepsilon], \; \Pi_{\mathfrak{n}} \text{ grad } g(\xi+\varepsilon))$$
$$= \kappa([\Pi_{\mathfrak{n}} \text{ grad } g(\xi+\varepsilon), \xi+\varepsilon], \; \Pi_{\mathfrak{k}} \text{ grad } f(\xi+\varepsilon))$$
$$= -\kappa([\Pi_{\mathfrak{k}} \text{ grad } g(\xi+\varepsilon), \xi+\varepsilon], \Pi_{\mathfrak{k}} \text{ grad } f(\xi+\varepsilon))$$
$$= \kappa([\Pi_{\mathfrak{k}} \text{ grad } g(\xi+\varepsilon), \; \Pi_{\mathfrak{k}} \text{ grad } f(\xi+\varepsilon)], \; \xi+\varepsilon)$$

which vanishes since $\xi \in \mathfrak{k}^{\perp}$, $\varepsilon \perp [\mathfrak{k}, \mathfrak{k}]$ and $\mathfrak{k}$ is a Lie algebra. The Hamiltonian vector field is given by (3.17).

$$X_{f_1 | \mathfrak{k}^{\perp}}(\xi) = -\Pi_{\mathfrak{k}^{\perp}}[\Pi_{\mathfrak{n}} \text{ grad } f(\xi+\varepsilon), \xi]$$
$$= -\Pi_{\mathfrak{k}^{\perp}}[\Pi_{\mathfrak{n}} \text{ grad } f(\xi+\varepsilon), \xi+\varepsilon] + \Pi_{\mathfrak{k}^{\perp}}[\Pi_{\mathfrak{n}} \text{ grad } f(\xi+\varepsilon), \varepsilon]$$
$$= -\Pi_{\mathfrak{k}^{\perp}}[\Pi_{\mathfrak{k}} \text{ grad } f(\xi+\varepsilon), \xi+\varepsilon] + \Pi_{\mathfrak{k}^{\perp}}[\Pi_{\mathfrak{n}} \text{ grad } f(\xi+\varepsilon), \varepsilon]$$
$$= [\Pi_{\mathfrak{k}} \text{ grad } f(\xi+\varepsilon), \; \xi+\varepsilon] + \Pi_{\mathfrak{k}^{\perp}}[\Pi_{\mathfrak{n}} \text{ grad } f(\xi+\varepsilon), \varepsilon],$$

since $[\mathfrak{k}, \xi+\varepsilon] \in \mathfrak{k}^{\perp}$.

Finally, assume 1.B. holds. For $\xi \in \mathfrak{k}^{\perp}$, $a \neq b$, pick up the formula for the Poisson bracket where it was left in Case 2. Since

$\varepsilon \perp [\mathfrak{k}, \mathfrak{n}]$ the second term in the expansion of each summand vanishes and thus one can write:

$$\{f_a|\mathfrak{k}^\perp, g_b|\mathfrak{k}^\perp\}(\xi) = \frac{b}{b-a}\kappa([\Pi_{\mathfrak{k}} \text{ grad } f(\xi+a\varepsilon), \Pi_{\mathfrak{n}} \text{ grad } g(\xi+b\varepsilon), \xi+b\varepsilon)$$

$$- \frac{a}{b-a}\kappa([\Pi_{\mathfrak{n}} \text{ grad } f(\xi+a\varepsilon), \Pi_{\mathfrak{k}} \text{ grad } g(\xi+b\varepsilon)], \xi+a\varepsilon)$$

$$= \frac{b}{b-a}\kappa([\Pi_{\mathfrak{k}} \text{ grad } f(\xi+a\varepsilon), [\Pi_{\mathfrak{n}} \text{ grad } g(\xi+b\varepsilon), \xi+b\varepsilon])$$

$$+ \frac{a}{b-a}\kappa([\Pi_{\mathfrak{k}} \text{ grad } g(\xi+b\varepsilon), [\Pi_{\mathfrak{n}} \text{ grad } f(\xi+a\varepsilon) \xi+a\varepsilon])$$

$$= - \frac{b}{b-a}\kappa([\Pi_{\mathfrak{k}} \text{ grad } f(\xi+a\varepsilon), \Pi_{\mathfrak{k}} \text{ grad } g(\xi+b\varepsilon)], \xi+b\varepsilon)$$

$$- \frac{a}{b-a}\kappa([\Pi_{\mathfrak{k}} \text{ grad } g(\xi+b\varepsilon), \Pi_{\mathfrak{k}} \text{ grad } f(\xi+a\varepsilon)], \xi+a\varepsilon)$$

which vanishes since $\mathfrak{k}$ is a Lie algebra, $\xi \in \mathfrak{k}^\perp$, $\varepsilon \perp [\mathfrak{k}, \mathfrak{k}]$.

$$X_{f_a|\mathfrak{k}^\perp}(\xi) = -\Pi_{\mathfrak{k}^\perp}[\Pi_{\mathfrak{n}} \text{ grad } f(\xi+a\varepsilon), \xi]$$

$$= -\Pi_{\mathfrak{k}^\perp}[\Pi_{\mathfrak{n}} \text{ grad } f(\xi+a\varepsilon), \xi+a\varepsilon] + \Pi_{\mathfrak{k}^\perp}[\Pi_{\mathfrak{n}} \text{ grad } f(\xi+a\varepsilon), a\varepsilon]$$

$$= \Pi_{\mathfrak{k}^\perp}[\Pi_{\mathfrak{k}} \text{ grad } f(\xi+a\varepsilon), \xi+a\varepsilon] + a\Pi_{\mathfrak{k}^\perp}[\Pi_{\mathfrak{n}} \text{ grad } f(\xi+a\varepsilon), \varepsilon]$$

$$= [\Pi_{\mathfrak{k}} \text{ grad } f(\xi+a\varepsilon), \xi+a\varepsilon] + a[\Pi_{\mathfrak{n}} \text{ grad } f(\xi+a\varepsilon), \varepsilon]$$

$$= [\Pi_{\mathfrak{k}} \text{ grad } f(\xi+a\varepsilon), \xi] + a[\text{grad } f(\xi+a\varepsilon), \varepsilon],$$

the fourth equality holding since $[\mathfrak{k}, \xi+a\varepsilon] \subseteq \mathfrak{k}^\perp$, $[\mathfrak{n}, \varepsilon] \subseteq \mathfrak{k}^\perp$. ■

Remark. For $\varepsilon = 0$ one recovers Theorem 3.3.

Let $\mathfrak{k} = 0$ in the above theorem. One gets:

Corollary 5.2. (Mishchenko, Fomenko [13]. If f, g are ad-invariant functions on $\mathfrak{g}$, then $\{f_a, g_b\} = 0$ in the Kirillov-Kostant-Souriau structure of $\mathfrak{g}$.

Let $\theta: \mathfrak{g} \to \mathfrak{g}$ be a Cartan involution on $\mathfrak{g}$ and $\mathfrak{a}$ a subalgebra. $\mathfrak{a}$ is called a Lie summand if $\theta(\mathfrak{a}^\perp)$ is also a subalgebra. Recall that a Cartan involution is defined in terms of a Cartan decomposition $\mathfrak{g} = \mathfrak{q} \oplus \mathfrak{p}$, where $\mathfrak{q}$ is a maximal subalgebra of $\mathfrak{g}$ on which the Killing form κ is negative definite. Then by definition $\theta(\xi+\eta) = \xi-\eta$, for $\xi \in \mathfrak{q}$, $\eta \in \mathfrak{p}$ and $\langle\zeta_1,\zeta_2\rangle = -\kappa(\theta(\zeta_1), \zeta_2)$ becomes a positive definite inner product on $\mathfrak{g}$. It is then immediate that $\theta(\mathfrak{a}^\perp)$ is the orthogonal of $\mathfrak{a}$ in the inner product $\langle,\rangle$. In particular $\mathfrak{g} = \mathfrak{a} \oplus \theta(\mathfrak{a}^\perp)$. Let now ε be κ-orthogonal to $[\mathfrak{a},\mathfrak{a}]$ and $[\theta(\mathfrak{a}^\perp), \theta(\mathfrak{a}^\perp)]$ and apply Theorem 5.1.1.A. with $\mathfrak{n} = \mathfrak{a}$, $\mathfrak{k} = \theta(\mathfrak{a}^\perp)$ to conclude that any ad-invariant functions on $\mathfrak{g}$ Poisson commute in $\mathfrak{k}^\perp = \theta(\mathfrak{a})$. We obtained thus:

Corollary 5.3 (Kostant [8]). If $\mathfrak{a}$ is a Lie summand of the semisimple Lie algebra $\mathfrak{g}$ and $\varepsilon \in \mathfrak{g}$ is a fixed element κ-orthogonal to $[\mathfrak{a},\mathfrak{a}]$ and $[\theta(\mathfrak{a}^\perp), \theta(\mathfrak{a}^\perp)]$, then any translates by ε of ad-invariant functions on $\mathfrak{g}$, (i.e. $\xi \mapsto f(\xi+\varepsilon)$) Poisson commute on $\theta(\mathfrak{a})$ with the Kirillov-Kostant-Souriau structure induced from $\mathfrak{a}^* \cong \theta(\mathfrak{a})$.

Example 5.4. We prove a different way the involution of the integrals of the Dubrovin equation of example 4.8 following [13]. Let $u_{k+1,j}$ be the coefficient of λ_{k+1}^j in $f_{k+1}(M) = \frac{1}{k+1} \cdot \mathrm{Tr}(M+\lambda_{k+1}a)^{k+1}$. By Corollary 5.2 $\{f_{k+1}, f_{\ell+1}\} = 0$ on $s\ell(n)$ for any λ_{k+1}, $\lambda_{\ell+1}$, i.e. f_{k+1} is conserved on the flow defined by $f_{\ell+1}$ for any λ_{k+1}, $\lambda_{\ell+1}$. Thus $u_{k+1,j}$ are constant on the flow defined by $f_{\ell+1}$ for all $\lambda_{\ell+1}$ and hence by symmetry, $\{u_{k+1,j}, u_{\ell+1,i}\} = 0$ on $s\ell(n)$.

Example 5.5. We prove the involution of the integrals of the Toda system (see example 3.5) following Kostant [8] and using Corollary 5.3. Recall that the Toda Hamiltonian is given by

$H = \frac{1}{2}\sum_{i=1}^{n} b_i^2 + \sum_{i=1}^{n-1} a_i^2$. Denote $c_i = a_i^2$ and then the Toda-Hamiltonian becomes $\frac{1}{2}\,\mathrm{Tr}M^2$, where

$$M = \begin{bmatrix} b_1 & c_1 & \cdots & & 0 \\ 1 & b_2 & \cdots & & 0 \\ \cdot & \cdot & \cdots & & \cdot \\ 0 & 0 & \cdots & & c_{n-1} \\ 0 & 0 & \cdots & 1 & b_n \end{bmatrix}$$

Choose $\mathfrak{g} = s\ell(n)$, $\kappa(A,B) = \mathrm{Tr}(AB)$, $\mathfrak{a}$ = lower triangular matrices,

$$\varepsilon = \begin{bmatrix} 0 & & & 0 \\ 1 & 0 & & \\ & \ddots & \ddots & 0 \\ 0 & & 1 & 0 \end{bmatrix}$$

, θ = transpose of matrices. Then $\mathfrak{a}^\perp$ = stricity lower triangular matrices, $\theta(\mathfrak{a}^\perp)$ = strictly upper triangular matrices and the orbit through ε^t in $\theta(\mathfrak{a}^\perp)$ has typical element M. The functions $\mathrm{Tr}M^k$, $k = 2, \ldots, n$ Poisson commute on this orbit by Corollary 5.3.

The following theorem is used in Ratiu [16] to give a proof of the involution of the constants of the motion in the N-dimensional Lagrange top problem (see next example).

<u>Theorem 5.6.</u> Let f,g be ad-invariant functions on the semisimple Lie algebra $\mathfrak{g}$. Denote $f_a(\xi,\eta) = f(\xi+a\eta+a^2\varepsilon)$, $g_b(\xi,\eta) = g(\xi+b\eta+b^2\varepsilon)$, for $\varepsilon \in \mathfrak{g}$ fixed and a,b arbitrary parameters. Then $\{f_a,g_b\} = 0$ in the Euler-Poisson structure of $\mathfrak{g}\times\mathfrak{g}$ for any a,b.

<u>Proof.</u> $\mathrm{grad}_1 f_a(\xi,\eta) = \mathrm{grad}\ f(\xi+a\eta+a^2\varepsilon)$

$\mathrm{grad}_2 f_a(\xi,\eta) = a\ \mathrm{grad}\ f(\xi+a\eta+a^2\varepsilon)$

For $a \neq b$, we have

$$\begin{aligned}\{f_a,g_b\}(\xi,\eta) = {} & b\kappa([\mathrm{grad}\ f(\xi+a\eta+a^2\varepsilon),\xi],\ \mathrm{grad}\ g(\xi+b\eta+b^2\varepsilon)) \\ & - a\kappa([\mathrm{grad}\ g(\xi+b\eta+b^2\varepsilon),\xi],\ \mathrm{grad}\ f(\xi+a\eta+a^2\varepsilon)) \\ & + ab\kappa(\eta,\ [\mathrm{grad}\ g(\xi+b\eta+b^2\varepsilon),\ \mathrm{grad}\ f(\xi+a\eta+a^2\varepsilon)])\end{aligned}$$

$$= -\kappa((a+b)+ab\eta,\ [\operatorname{grad} f(\xi+a\eta+a^2\varepsilon),\ \operatorname{grad} g(\xi+b\eta+b^2\varepsilon)])$$

$$= -\kappa\Bigl(-\frac{b^2}{a-b}(\xi+a\eta+a^2\varepsilon) + \frac{a^2}{a-b}(\xi+b\eta+b^2\varepsilon)\ ,$$

$$[\operatorname{grad} f(\xi+a\eta+a^2\varepsilon),\ \operatorname{grad} g(\xi+b\eta+b^2\varepsilon)]$$

$$= \frac{b^2}{a-b}\kappa([\xi+a\eta+a^2\varepsilon,\ \operatorname{grad} f(\xi+a\eta+a^2\varepsilon)],\ \operatorname{grad} g(\xi+b\eta+b^2\varepsilon))$$

$$+ \frac{a^2}{a-b}\kappa([\xi+b\eta+b^2\varepsilon,\ \operatorname{grad} g(\xi+b\eta+b^2\varepsilon)],\ \operatorname{grad} f(\xi+a\eta+a^2\varepsilon))$$

Both terms vanish by ad-invariance of f and g.■

Example 5.7. Consider the N-dimensional Lagrange top of Example 4.11 and recall that $u_{k+1,j}(\Gamma,M)$ is the coefficient of λ_{k+1}^j in $f_{k+1}(\Gamma,M) = \frac{1}{k+1}\operatorname{Tr}(\Gamma+M\lambda_{k+1}+C\lambda_{k+1}^2)^{k+1}$, k = odd. By Theorem 5.6, $\{f_{k+1}, f_{\ell+1}\} = 0$ on $so(N) \times so(N)$ for any $\lambda_{k+1}, \lambda_{\ell+1}$, i.e. f_{k+1} is conserved on the flow of $f_{\ell+1}$ for any $\lambda_{k+1}, \lambda_{\ell+1}$, etc. Conclude as in Example 5.4 that on $so(N) \times so(N)$ $\{u_{k+1,j}, u_{\ell+1,i}\} = 0$.

BIBLIOGRAPHY

[1] R. Abraham, J. Marsden: Foundations of Mechanics, Benjamin/Cummings, 1978.

[2] M. Adler: On a trace functional for formal pseudo-differential operators and the symplectic structure of the Korteweg-deVries equations, Inventions Math, 1979.

[3] M. Adler, P. van Moerkbeke: Algebraic curves and the classical Katz-Moody algebras, preprint, 1979.

[4] M. Adler, J. Moser: preprint on the geodesic problem on the ellipsoid and the Neumann problem, 1979.

[5] I.M. Gelfand, L.A. Dikii: The resolvent and Hamiltonian systems, Funct. Anal. and its Applications 11, 11-27, 10, 1977.

[6] R. Jost: Poisson brackets (an unpredagogical lecture), Rev. Mod. Phys. 36, 572-579, 1964.

[7] D. Kazhdan, B. Kostant, S. Sternberg: Hamiltonian group actions and dynamical systems of Calogero type, Comm. Pure Appl. Math, 31 (1978) 481-568.

[8] B. Kostant: The solution to a generalized Toda lattice and representation theory, preprint, MIT, 1979.

[9] B. Kupershmidt: Deformations of Hamiltonian structures, prinprint, MIT, 1979.

[10] B. Kupershmidt, Yu. Manin: Equations of long waves with a free surface, Funkts. Analiz. Prilozhen., 11, No. 3, 31-42, 1977.

[11] Yu, Manin: Algebraic aspects of non-linear differential equations, Journal of Soviet Math., Vol. 11, No. 1, 1-122, 1979.

[12] J. Marsden, A. Weinstein: Reduction of symplectic manifolds with symmetry, Rep. Math. Phys. 5, 121-130, 1974 .

[13] A.S. Mishchenko, A.T. Fomenko: Euler equations on finite dimensional Lie groups, Math. USSR, Izvestija, Vol. 12, No. 2, 371-389, 1978.

[14] J. Moser: Various aspects of integrable Hamiltonian systems, C.I.M.E., Bressanone, 1978.

[15] T. Ratiu: The motion of the free n-dimensional rigid body, preprint, Berkeley, 1979.

[16] T. Ratiu: Euler-Poisson equations on Lie algebras and the generalized Lagrange top, preprint, Berkeley, 1979.

[17] W. Symes: On systems of Toda, type, MRC Technical Summary Report, #1957, University of Wisconsin-Madison, 1979.

[18] W. Symes: Relations among generalized Korteweg-deVries systems, J. Math. Phys. 20 (4), April 1979.

MIX
Papier aus verantwortungsvollen Quellen
Paper from responsible sources
FSC® C105338

If you have any concerns about our products,
you can contact us on
ProductSafety@springernature.com

In case Publisher is established outside the EU,
the EU authorized representative is:
Springer Nature Customer Service Center GmbH
Europaplatz 3, 69115 Heidelberg, Germany

Printed by Libri Plureos GmbH
in Hamburg, Germany